Jiří Popovský · Lois Ann Pfiester

Dinophyceae

Süßwasserflora von Mitteleuropa

Begründet von/Founded by A. Pascher

Herausgegeben von/Edited by
H. Ettl · J. Gerloff · H. Heynig ·
D. Mollenhauer

Band/Volume 6:
Popovský/Pfiester
Dinophyceae
(Dinoflagellida)

Süßwasserflora von Mitteleuropa

Bd./Vol. 1/1	Chrysophyte and Haptophyte Algae, 2nd ed., Part 1 (Chrysophyceae s. str., Dictyochophyceae, Phaeothamniophyceae and Haptophyceae)
Bd./Vol. 1/2	Chrysophyte and Haptophyte Algae, 2nd ed., Part 2 (Synurophyceae) (2007)
Bd./Vol. 2/1	Bacillariophyceae (Naviculaceae) (1999)
Bd./Vol. 2/2	Bacillariophyceae (Bacillariaceae, Epithemiaceae, Surirellaceae) (1997)
Bd./Vol. 2/3	Bacillariophyceae (Centrales; Fragilariaceae, Eunotiaceae) (2004)
Bd./Vol. 2/4	Bacillariophyceae (Achnanthaceae, Literaturverzeichnis) (2004)
Bd./Vol. 2/5	Bacillariophyceae (English and French translation of the keys) (2000)
Bd./Vol. 3	Xanthophyceae 1. Teil (1978)
Bd./Vol. 4	Xanthophyceae 2. Teil (1980)
Bd./Vol. 5	Cryptophyceae und Raphidophyeae
Bd./Vol. 6	Dinophyceae (Dinoflagellida) (1990)
Bd./Vol. 7	Phaeophyceae und Rhodophyceae
Bd./Vol. 8	Euglenophyceae
Bd./Vol. 9	Chlorophyta I (Phytomonadina) (1983)
Bd./Vol. 10	Chlorophyta II (Tetrasporales, Chlorococcales, Gloeodendrales) (1988)
Bd./Vol. 11	Chlorophyta III (Chlorophyceae p. p.: Chlorellales, Protosiphonales)
Bd./Vol. 12	Chlorophyta IV (Chlorophyceae p. p.: Stichococcales, Microsporales; Codiolophyceae: Ulotrichales, Monostromatales)
Bd./Vol. 13	Chlorophyta V (Chlorophyceae p. p.: Chaetophorales, Trentepohliales, Chlorosphaerales)
Bd./Vol. 14	Chlorophyta VI (Oedogoniophyceae: Oedogoniales) (1985)
Bd./Vol. 15	Chlorophyta VII (Bryopsidophyceae: Cladophorales, Sphaeropleales)
Bd./Vol. 16	Chlorophyta VIII (Conjugatophyceae I: Zygnemales) (1984)
Bd./Vol. 17	Chlorophyta IX (Conjugatophyceae II: Zygnematales: Mesotaeniaceae; Desmidiales)
Bd./Vol. 18	Charales (Charophyceae) (1997)
Bd./Vol. 19/1	Cyanoprokaryota I (Chroococcales) (1998)
Bd./Vol. 19/2	Cyanoprokaryota II (Oscillatoriales) (2005)
Bd./Vol. 19/3	Cyanoprokaryota III (Nostocales, Stigonematales)
Bd./Vol. 20	Schizomycetes (1982 vergriffen/out of print)
Bd./Vol. 21/1	Fungi (Lichens)
Bd./Vol. 21/2	Fungi
Bd./Vol. 22	Bryophyta
Bd./Vol. 23	Pterido- und Anthophyta 1. Teil: Lycopodiaceae bis Orchidaceae (1980)
Bd./Vol. 24	Pterido- und Anthophyta 2. Teil: Saururaceae bis Asteraceae (1981)

Dinophyceae (Dinoflagellida)

Jiří Popovský
Lois Ann Pfiester

256 Abbildungen/figures

Spektrum
AKADEMISCHER VERLAG

Authors' adresses
Doz. Dr. Jiří Popovský, V Podhoří 270/3, 171 00 Praha 7 Troja, Czech Republic
Prof. Dr. Lois Ann Pfiester †, formerly University of Oklahoma, Norman, USA

Notice for the reader
The Authors, the Editors, and the Publisher of this work have made every effort to ensure that the information herein is accurate and in accord with the standards accepted at the time of publication. The Publishers have made an extensive effort to trace original copyright holders for permission to use borrowed material. If any has been overlooked, it will be corrected at the first reprint.

Bibliographic information published by die Deutsche Bibliothek
Die Deutsche Bibliothek lists this publication in the Deutsche Nationalbibliografie; Detailed bibliographic data is available in the Internet at http://dnb.d-nb.de.

First published 1990, Gustav Fischer Verlag, Jena Stuttgart
unaltered reprint 2008

Spektrum Akademischer Verlag is an imprint of Springer

Springer is a part of Springer Science+Business Media

08 09 10 11 5 4 3 2

Editor: Johanna Schlüter
Book production: Detlef Mädje
Cover Design: SpieszDesign, Neu-Ulm
Composed by: Druckhaus Jena GmbH
Printed: Books on Demand GmbH, Norderstedt
Graphics and Photographs: by the authors if not stated otherwise in the figure captions

Printed in Germany

ISBN 987–3–8274–2098–5

Vorwort der Herausgeber

Die von A. Pascher herausgegebene „Süßwasserflora Deutschlands, Österreichs und der Schweiz“, deren 2. Auflage bereits den Titel „Süßwasserflora Mitteleuropas“ erhielt, galt und gilt noch immer als Standardwerk für die Bestimmung der Süßwasserpflanzen. Die Süßwasserflora hat sich seit dem Erscheinen ihrer ersten Bände auf der ganzen Welt einen berechtigten Ruf erworben. Nicht nur Anfänger, sondern auch Spezialisten, die sich mit Süßwasserpflanzen und besonders mit Algen befassen, greifen immer wieder zu dieser Bücherreihe. Dies darf als Beweis für die Richtigkeit der Pascherschen Grundkonzeption dienen, die zu verändern auch heute noch keine Veranlassung besteht. Dies bedeutet jedoch nicht, daß die Neubearbeitung der Süßwasserflora nicht wesentliche Erweiterungen und Veränderungen erfahren mußte, liegen doch die Erstbearbeitungen z. T. mehr als 70 Jahre zurück. Die heutigen Anschauungen über die Verwandschaftsverhältnisse haben sich gegenüber früheren Vorstellungen ebensosehr gewandelt, wie die Zahl der Arten, die Kenntnis ihrer Entwicklungsgeschichte oder ihrer ökologischen Ansprüche zugenommen haben. All dies war in einer Neuauflage zu berücksichtigen. Es ist daher verständlich, daß sowohl Zahl und Umfang der Bände als auch die Anzahl der Abbildungen vermehrt werden mußten. Trotzdem soll die Flora die monographischen Bearbeitungen der einzelnen taxonomischen Gruppen nicht ersetzen. In diesem Sinne mußte auch manches kompilatorisch zusammengetragen werden, obwohl eine kritische Durcharbeitung stets angestrebt wurde. Dennoch werden immer gewisse Unsicherheiten bleiben – wie stets in der Taxonomie –, die sich schließlich in der Bewertung mancher Sippen widerspiegeln. Wenn auch der Titel des Werkes beibehalten wurde, so sind doch wegen der weiten Verbreitung mancher Gruppen (besonders unter den Algen), wie z. T. auch schon in den ersten beiden Auflagen, ganz Europa und vielfach auch die übrigen Kontinente berücksichtigt worden.

Das verstärkte Interesse der Gesellschaft an Fragen des Umweltschutzes drückt sich auch in einer Zunahme hydrobiologischer Forschungen aus, für die eine zuverlässige Bestimmungsflora die Voraussetzung ist. Wir glauben, daß hier die Neuauflage der Süßwasserflora eine Lücke schließen kann, die sowohl dem theoretisch als auch dem praktisch arbeitenden Hydrobiologen meist schmerzlich bewußt wird.

Diese Lücke haben bisher auch andere ähnliche Werke nicht beseitigen können. Auch aus diesem Grund wurde die Gliederung des Werkes auf 24 Bände erhöht. Sie geht zwar von der grundlegenden Einteilung Paschers aus, berücksichtigt jedoch auch die neuesten Klassifikationen. Die Einteilung der Bände soll zwar auch den vermutlichen Verwandtschaftsbeziehungen Rechnung tragen, vor allem jedoch der praktischen Orientierung dienen und solche taxonomischen Gruppen zusammenfassen, die durch bestimmte, eindeutig charakterisierende Merkmale zu unterscheiden sind (Bandübersicht siehe Vorsatzblatt!).

Bei einem Sammelwerk, dessen Erscheinen sich zwangsläufig über einen längeren Zeitraum erstreckt, werden sich jedoch Verschiebungen in der Umgrenzung einzelner

Sippen ebensowenig vermeiden lassen wie Änderungen der wissenschaftlichen Namen – sei es aus taxonomischen, sei es aus nomenklatorischen Gründen. In den Jahren seit der Veröffentlichung des 1. Bandes dieser Neuauflage (1978) haben vor allem zahlreiche Untersuchungen zur Ultrastruktur und Entwicklungsgeschichte sowie die Analysen assimilatorischer und akzessorischer Pigmente zu gewandelten Vorstellungen über die verwandschaftlichen Beziehungen sowohl der verschiedenen Algengruppen untereinander als auch innerhalb dieser selbst geführt, denen Rechnung zu tragen war. Hierdurch, aber auch infolge der unterschiedlichen Bewertung derartiger Ergebnisse durch die Bearbeiter der einzelnen Bände, wird es zu Veränderungen des ursprünglichen Konzepts und damit zu einer gewissen Uneinheitlichkeit in der Einteilung des Gesamtwerkes kommen. In Abhängigkeit vom jeweiligen Stand der biologischen Kenntnisse war die Systematik von jeher einem Wandel unterworfen und wird es auch in Zukunft sein. Trotzdem wird die Süßwasserflora konservativeren Vorstellungen über die systematische Gliederung verhaftet bleiben. So wurde beispielsweise von der Ausgliederung der Cyanophyceae aus den Algen abgesehen, obwohl sich die Herausgeber der tiefgreifenden Unterschiede dieser Gruppe gegenüber den Eukaryoten bewußt sind. Am besten scheinen heute von den höheren taxonomischen Rangstufen die Abteilungen und Klassen definiert zu sein. Aber auch hier sind Änderungen in deren Umgrenzung zu erwarten oder haben, wie bei den früher als mehr oder weniger einheitlich angesehenen Chlorophyceae, bereits Eingang bei der Bearbeitung dieser Gruppe gefunden (vgl. Allgemeiner Teil und Systematische Übersicht in Bd. 9).

Um die notwendigen Veränderungen im Bestimmungsschlüssel der Algenklassen mit der früheren Bandeinteilung in Einklang zu bringen, wurden Hinweise auf die entsprechenden Bände hinzugefügt.

Da fast alle früheren Mitarbeiter der Süßwasserflora verstorben sind, mußten für die entsprechenden Gruppen neue Bearbeiter gewonnen werden. Allen, die bereit waren, an der Neuauflage mitzuwirken, haben die Herausgeber sehr herzlich zu danken – vor allem, da sie wissen, wie undankbar eine derartige Aufgabe ist und wieviel Zeit sie beansprucht. Beiden Verlagen, die bereit waren, mancherlei Wünsche zu akzeptieren, auch wenn sie zusätzliche finanzielle Belastungen bedeuteten, und die stets bemüht waren, auftretende Schwierigkeiten zu überbrücken, gebührt besonderer Dank. Im Laufe des bisherigen Erscheinens dieser Bücherreihe hat sich naturgemäß auch die Zusammensetzung des Herausgebergremiums geändert. Herrn Prof. B. Schussnig, der sich für diese Neuauflage voll eingesetzt hatte, war es leider nicht mehr vergönnt, den Neubeginn der Süßwasserflora zu erleben. An seine Stelle trat Herr Dr. H. Heynig 1976 in die Redaktion ein. Erhöhter Arbeitsaufwand zur Bewältigung dieses so umfangreichen Werkes hat die Mitarbeit eines weiteren Herausgebers erforderlich gemacht. Mit Band 9 ist Herr Dr. D. Mollenhauer 1983 der Redaktion beigetreten.

Die Herausgeber haben den Wunsch, daß sich die neue Edition der Süßwasserflora ebenso bewähren möge wie die beiden früheren Auflagen. Sie hoffen darüber hinaus, damit dem Andenken an A. Pascher zu dienen. Mit seinen Worten möchten wir uns zum Schluß an die Benutzer dieses Werkes wenden: „Irrtümer lassen sich beim besten Willen nicht vermeiden, weder für den speziellen Bearbeiter, noch für den Herausgeber, der ein schwer übersehbares großes Gebiet unmöglich gleichmäßig übersehen kann. Für jede sachliche und wohlgemeinte Anregung und Berichtigung werden Herausgeber und Bearbeiter immer dankbar sein."

1984 H. Ettl · J. Gerloff · H. Heynig · D. Mollenhauer

Bestimmungsschlüssel für die Klassen der Algen

(Neufassung 1984)

Der folgende kurze Bestimmungsschlüssel soll es auch dem weniger Erfahrenen ermöglichen, eine gefundene Alge zunächst in die richtige Klasse einzuordnen, ehe die nähere Bestimmung vorgenommen wird. Es war selbstverständlich nicht möglich, in dem kurzen Bestimmungsschlüssel bei der Beschreibung der Klassen auch alle stark abweichenden Gattungen zu berücksichtigen. Ebenso wurden submikroskopische Strukturen nur dann zur Differenzierung der einzelnen Klassen herangezogen, wenn die üblichen Merkmale eine eindeutige Charakterisierung nicht ermöglichen. Die Herausgeber hoffen, daß damit auch dem Nichtspezialisten der praktische Gebrauch der Süßwasserflora erleichtert wird.

1a Zellen ohne typischen Zellkern und ohne Chloroplasten (Prokaryonten). Chromatinelemente unregelmäßig im zentralen Teil der Zelle verteilt. Pigmente blaugrün, olivgrün oder purpurfarbig, im peripheren Chromatoplasma. Zellen oft mit Pseudovakuolen. Reservestoffe: glykogenähnliche Polyglukane oder proteinartiges Cyanophycin. Einzellige, kolonie- oder fadenbildende Organismen . . . . **Cyanophyceae** (Band 19)

1b Zellen mit typischem Zellkern sowie durch Membranen abgegrenzten Chloroplasten (Eukaryonten) **2**

2a Zellen mit überwiegend grasgrünen Chloroplasten. Chlorophyll a und b vorhanden, jedoch kann das Chlorophyll durch akzessorische Pigmente maskiert sein (z. B. *Euglena, Haematococcus, Trentepholia)* **3**

2b Zellen mit anders gefärbten Chloroplasten. Chlorophyll b stets fehlend. Farbe der Assimilationspigmente bei einigen der hierher gehörenden Klassen (Raphidophyceae, Eustigmatophyceae, Xanthophyceae) dennoch gelbgrün bis grün . **11**

3a Einzellige, stark spezialisierte Flagellaten, meist freischwimmend, z. T. in festen Gehäusen, oft metabolisch. Assimilatbildung (Paramylum) außerhalb der Chloroplasten. Paramylum körnchen-, stab- oder ringförmig, durch Jod nicht färbbar. Zahlreiche farblose Gattungen . . . . **Euglenophyceae** (Band 8)

3b Einzellige oder mehrzellige Algen. Assimilatbildung (Stärke) innerhalb der Chloroplasten. Stärkekörnchen werden durch Jod blau gefärbt **4**

4a Sexuelle Fortpflanzung durch Konjugation unbegeißelter Gameten. Algen einzellig, koloniebildend oder fadenförmig, oft mit kompliziert gebauten, großen Chloroplasten (Megaplasten).
Begeißelte Stadien fehlen . . . . **Conjugatophyceae** (Band 16, 17)

4b Sexuelle Fortpflanzung niemals durch Konjugation, sondern stets durch begeißelte Gameten **5**

5a Makrophytische, quirlartig verzweigte Algen, deren Vegetationskörper aus Knoten- und Internodialzellen aufgebaut ist. Mit kompliziert gebauten, eigenartigen Geschlechtsorganen. Sexuelle Fortpflanzung durch Oogamie, asexuelle Fortpflanzungszellen (z. B. Zoosporen) fehlen . **Charophyceae** (Band 18)

5b Algen nicht aus Knoten- und Internodialzellen aufgebaut, sehr vielgestaltig; einzellig, kolonie- oder zönobienbildend, fädig oder vorwiegend unbeweglich. Asexuelle Fortpflanzungszellen werden gebildet **6**

6a Bewegliche, geißeltragende oder unbewegliche, zystäre Sippen, einzeln lebend oder koloniebildend. Begeißelte Zellen von charakteristischem asymmetrischem Bau, apikal mit vestibulärem Krater, aus dem 2 oder 4, seltener 1 oder 8 Geißeln entspringen. Zellhülle in Form eines Schuppenpanzers oder einer Theka ausgebildet . . . . **Prasinophyceae** (Band 9)

6b Begeißelte Zellen ohne vestibulären Krater, primär radiär- oder bilateralsymmetrisch, nackt oder mit Zellwand **7**

7a Bewegliche, geißeltragende oder unbewegliche, capsale oder coccale Sippen, einzeln oder koloniebildend. Begeißelte Zellen mit Zellwand aus Glykoproteinen; nackte Protoplasten nur vorübergehend und kurzfristig (Protoplastenteilung, Gametenverschmelzung) auftretend. Bei allen Sippen (auch bei Flagellaten) werden Sporangien und Gametangien gebildet
Chlamydophyceae (Band 9, 10)

7b Vorwiegend unbewegliche, ein- bis mehrzellige coccale oder trichale Sippen. Begeißelte Fortpflanzungszellen und Flagellaten nackt; letztere vermehren sich ausschließlich durch Zweiteilung **8**

8a Algen in Form einfacher oder verzweigter Fäden. Zellteilung erfolgt polar unter Bildung charakteristischer Kappen. Zoosporen und Spermatozoiden stephanokont begeißelt.
Sexuelle Fortpflanzung durch Oogamie . . . . **Oedogoniophyceae** (Band 14)

8b Bewegliche Stadien mit 2 oder 4 Geißeln, nicht stephanokont. Zellteilung der trichalen Formen ohne Kappenbildung **9**

9a Grünalgen mit trichalem oder thallösem Bau und mit heteromorphem Generationswechsel. Als Sporophyt entwickelt sich das einzellige *Codiolum*-Stadium
. . . . **Codiolophyceae** (Band 12)

9b Grünalgen ohne heteromorphen Generationswechsel **10**

10a Grünalgen in Form einfacher oder verzweigter Fäden, die durch Querwände in vielkernige Abschnitte (Siphonoblasten) geteilt sind . . . .
. . . . **Bryopsidophyceae** (Band 15)

10b Ein- oder mehrzellige, monadoide, coccale oder trichale Algen, deren begeißelte Stadien stets nackt sind und nur vom Plasmalemma begrenzt werden
Chlorophyceae (Band 9, 10, 11, 12, 13)

11a Chloroplasten „maigrün" oder gelbgrün **12**

11b Chloroplasten gelblich oder braun, graurot oder blau **14**

12a Chloroplasten häufig durch Xanthophylle und Karotine gelbgrün gefärbt, durch Zusatz verdünnter Salzsäure in einen bläulich-grünen Farbton umschlagend. Reservestoffe: Chrysolaminarin und fette Öle. Endogen gebildete Zysten verkieselt und zweiteilig. Einzellige, koloniebildende, fädige oder siphonale Algen **13**

12b Chloroplasten sehr zahlreich, gewöhnlich radial ausgerichtet, „maigrün". Zellen häufig mit Trichozysten. Durch 2 seitenständige Geißeln bewegliche, hoch entwickelte und spezialisierte Flagellaten . . . . . **Raphidophyceae** (Band 5)

13a Stigma der Zoosporen im Chloroplasten eingebettet. Pyrenoid in den Chloroplasten eingesenkt, von zahlreichen Lamellen durchzogen. Zellwand zweiteilig, oft ± deutlich (H-Stücke!). Zellkern immer mit einem oder mehreren Chloroplasten assoziiert . . . . **Xanthophyceae** (Band 3, 4)

13b Stigma der Zoosporen außerhalb des Chloroplasten. Pyrenoid knospenartig vorgestülpt, mit einem schwachen Stiel der Innenseite des Chloroplasten aufsitzend, homogen. Zellwand aus einem Stück bestehend. Zellkern nicht mit den Chloroplasten assoziiert . . . . **Eustigmatophyceae** (Band 3)

14a Meist als Flagellaten vorkommend, seltener in zystärer Ausbildung. Zellkern in Form eines Dinokaryons, mit dicht gedrängten, in der Interphase überdauernden Chromosomen. Mit 2 in verschiedenen Ebenen (Längs- und Querfurche) angeordneten Geißeln. Zellwand bei manchen Gattungen aus skulpturierten polygonalen Platten gebildet. Häufig mit kompliziertem Lebenszyklus. Reservestoffe: Stärke und Öl. Zahlreiche farblose Gattungen **Dinophyceae** (Band 6)

14b Sowohl begeißelte als auch unbewegliche Algen. Zellkern als normales Eukaryon ausgebildet **15**

15a Chloroplasten gelblich oder braun **16**

15b Chloroplasten graurot, rötlich oder blau **19**

16a Algen fadenförmig (einfach oder verzweigt) oder kleine parenchymatöse Thalli bildend, mit individualisierten Sporangien. Reservestoff: Chrysolaminarin . . . . **Phaeophyceae** (Band 7)

16b Algen einzellig oder koloniebildend, nur sehr selten fadenförmig und dann ohne individualisierte Sporangien **17**

17a Zellen mit einer zweiteiligen, stark verkieselten und skulpturierten Zellwand (Frustel). Sexuelle Fortpflanzung durch Auxosporen. Reservestoffe: Chrysolaminarin und fette Öle . . . . **Bacillariophyceae** (Band 2)

17b Zellen keine zweiteilige und stark verkieselte Zellwand bildend. Sexuelle Fortpflanzung durch Hologamie **18**

18a Meist einzellige, seltener kolonie- oder fadenförmige Algen. Bewegliche Zellen mit 2 heterokonten und heteromorphen Geißeln, bisweilen auch nur mit einer Geißel. Manche Sippen Kieselschuppen tragend oder in Gehäusen lebend. Endogen gebildete, verkieselte Zysten mit einem Stöpsel. Mit zahlreichen farblosen Gattungen . . . . **Chrysophyceae** (Band 1)

18b Vorwiegend einzellige, seltener mehrzellige Flagellaten oder Algen. Neben 2 oft ± gleichlangen Geißeln findet sich ein fadenförmiges Haptonema verschiedener Länge. Zelloberfläche gewöhnlich mit Schuppen bedeckt, die ausschließlich aus organischem Material bestehen oder verkalkt sind. Nur wenige Arten im Süßwasser; vorwiegend marin . . . . **Haptophyceae** (Band 1)

19a Algen mit violetten oder graugrünen, bisweilen blaugrünen oder leuchtend roten Pigmenten. Keine begeißelten Zellen bildend. Einzellige, fadenförmige oder kompliziert zusammengesetzte Thalli. Reservestoff: Florideenstärke, die durch Jod rötlich gefärbt wird . . . . **Rhodophyceae** (Band 7)

19b Algen mit grünen, blauen, bräunlichen oder auch roten Pigmenten; meist einzellige Flagellaten mit 2 verschieden langen Geißeln, die subapikal in einem Schlund oder einer Furche entspringen. Reservestoffe: Cryptophyceen-Stärke und Öle. Jodtest auf Stärke oft positiv.
Manche Sippen mit Trichozysten. . . . . **Cryptophyceae** (Band 5)

Preface

This volume is a compilation of organisms which are of interest to phycologists and protozoologists. Since it is very difficult to classify of the unicellular organisms as plants or animals in the classical sense two approaches can be found in the literature to the classification of dinoflagellates – the botanical and zoological.

We have employed the phycological (botanical) approach in our textbook from a practical aspect. However, we view the dinoflagellates as protists which exhibit both plant and animal features. This also holds true for the marine dinoflagellates which are richer in number of genera and species than their freshwater counterparts.

Classically phycologists divided dinoflagellates into monads, rhizopods, coccals, capsals and trichals based on morphology. During our studies of freshwater dinoflagellates we have observed and recorded stages transcendent among these forms.

We have also lumped many poorly described species. In so doing we have tried to include all available information (including figures). We have listed in the appendix all the freshwater species known. Thus we face a dilemma-wether to compile dinoflagellates according to classical systematics the danger being that our knowledge is still too limited or-to use the classical approach inserting new information where available with subsequent suggestions. We have chosen the latter for we believe it will take many years to gain the knowledge needed to correctly systematize the freshwater dinoflagellates.

We hope this textbook will be usefull and helpfull – not only for limnologists but also for specialists in protistology. We thank Prof. Dr. H. Ettl, CSc. and Dr. Dieter Mollenhauer, for their valuable help, suggestions and encouragement during the compilation of this manuscript. We are very grateful to Mrs. Marcela Dobiášová (Prague, Czechoslovakia) for her invaluable technical support mainly in the form of drawings and figure arrangements.

Dedicated to Pavla and Dee, our spouses.

J. Popovský · L. A. Pfiester

Contents

Dinophyceae (Dinoflagellida)

I. General part

1. Introduction

The Dinophyceae are an unique group of protists in which at least during one part of their life cycle their chromosomes are always condensed and little histone is present. For this reason the dinoflagellate nucleus is referred to as a dinokaryon or mesokaryon. Their zoospores and/or gametes have a morphological structure that is species specific. At the present time the class Dinophyceae is divided into 4 subclasses: the Dinophycidae, Ebriophycidae, Ellobiophycidae, and Syndiniophycidae with 16 orders and 56 families. Many of the taxa in this class are also classified by zoologists as members of the protozoan order Dinoflagellida. For further information, the authors refer the reader to Loeblich (1982).

Motile representatives from the family Peridiniaceae and the zoospores and gametes of taxa from the orders Peridiniales and Dinococcales (subclass Dinophycidae) are flattened dorsiventrally, are spherical or egg-shaped with two heteromorphic flagella. One of the flagella lies within a transverse furrow (the girdle, or cingulum) which encircles the organism, the other lies along a longitudinal furrow (the sulcus) and trails beyond it. The beating motion of these two flagella cause the cell both to turn and to be propelled. Members of the subclass Adinophycidae are characterized by the absence of furrows and by two hetermorphic flagella inserted on the cell apex. If a theca (obvious cell wall) is present it is in the form of two identical parts (Bourrelly 1970). Monads from the family Protaspidaceae are characterized by the presence of a sulcus in which the trailing flagellum is located, the second flagellum is directed forward as in the other members of the Adinophycidae (Skuja 1939). Colored members of the class Dinophyceae possess chloroplasts of varying number, shape, color and position in the cell. Their color, ranging from yellow-brown, olive, red-brown to dark brown, evidently depends on the relationship of the photosynthetic pigment (chlorophyll *a)* and the accessory pigments (chlorophyll *c,* β-carotene, peridinin, dinoxanthin and diadinoxanthin (Bold and Wynne 1978, Ettl 1980). Color intensity is also related to the intensity of sunlight striking the cell and on the relation between its autotrophic and/or heterotrophic mode of nutrition. The green colored chloroplast are probably cyanells. Several members of the class Dinophyceae possess pyrenoids lacking starch sheaths. Food is stored in the form of starch which originates from the chloroplasts and oil. The oil referred to here is not the red pigment spot found in some dinoflagellates (Pfiester and Popovský 1979).

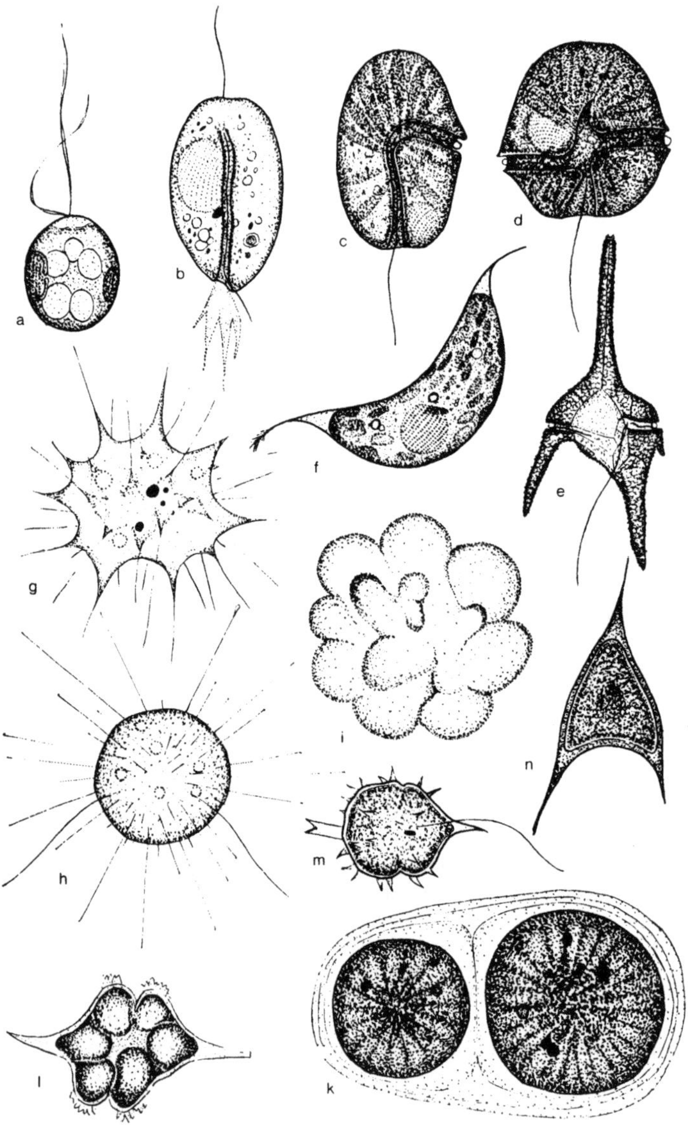

Fig. 1. Types of dinophycean plant body; diagrammatic, *a. Desmomastix globosa, b. Protaspis metarhiza, c. Hemidinium nasutum, d. Gymnodinium uberrimum, e. Ceratium hirundinella, f. Cystodinium cornifax, g.* a trophozoid-amoeboid of *Stylodinium*

The dinokaryon or nucleus is also unusual in that it is relatively large compared to the size of the cell. Its shape may be spherical, horseshoe-like, peanut-like, our obtuse conical. Its position in the protoplasm differs in relationship to the stage of the life cycle and the presence or absence of a food vacuole. If the dinokaryon is visible in the light microscope then the condensed chromosomes appear bead-like in structure (Ettl 1980, Kudo 1971). Contractile vacuoles are known to be present in the family Protaspidaceae and in several reproductive stages (trophozoids) in the family Dinococcales (Skuja 1939, Pfiester and Popovský 1979). Pusules are special organelles present mainly in marine species. Their function is not yet fully known. In marine species two saclike pusules, often with finger-like protuberances (vesicles) are connected by a channel which also has finger-like protuberances (vesicles) which are then connected by a channel to a flagellar canal (Grassé 1952, Starmach 1974). In freshwater taxa the pusules are not as complicated in structure. They consist of a system composed of one large and several small vacuoles. If a stigma or eyespot is present it is usually located in the sulcal region outside of a chloroplast. The cell covering is termed the amphiesma (Schütt 1895) or theca. Its presence has not been observed in dinococcalian trophozoids and sporocytes. A dinoflagellate cell may be unarmored (naked) or armored (plated). In naked taxa the theca usually lacks plates, or sculpturing and is passively metabolic. In thecate taxa on which plates are present, they are often finely sculptured. Spines, teeth, horn-like projections and hyaline wings may also be present. Cysts have not been observed in all taxa. Cysts have a very thick wall which may be irregular in shape. They are ovoid or two to four horned. Several cysts may be enveloped in a very thick, multilayered mucilage and always lack furrows (Starmach 1971). The mucilage stage, well known as "dinocapsal *Gloeodinium montanum*" may serve as survival cysts during unfavorable conditions of for reproduction of monads from the genera *Hemidinium* and *Glenodiniopsis*. Some monads lose both their flagella following phagocytosis. This stage has been incorrectly termed a "digestive cyst" or "food cyst".

Dinophycean morphology was originally characterized by the evolutionary structure of its vegetative stages which go from monadoid to trichale forms; (Klebs 1912, Pascher 1927, Schiller 1933/37, Kiselev 1951, Fott 1971, Christensen 1962). However, it is impossible with our present knowledge to detect a true separation among the monadoid, rhizopoid and coccal structures. There is only one form, that is the monadoid – rhizopoid structure with all transcendent types in the class Dinophycae. Presently named coccal forms are rhizopodial (sometimes without pseudopodia) stages or cysts of other dinophyceans. Also the trichal structure is probably a free aggregation of cysts or temporarily nonmotile vegetative cells which persists for a short time period (Schussnig 1960, Pfiester and Popovský 1978). Cysts with mucilagenous envelopes may be found in almost all freshwater genera of dinoflagellates. Such mucilagenous forms (capsal) as *Gloeodinium montanum,* and *G. cracoviense* are morphologically identical to cysts of *Glenodiniopsis steinii* and *Hemidinium nasutum* (Popovský 1971a, 1971b, Pfiester, unpublished observations). Thus the presence of mucilage is not a generic character.

---

h. a trophozoid-heliozoid of *Cystodinedria inermis, i.* a sporocyte of *Cystodinedria inermis, k.* a mucilage reproductive stage of *Glenodiniopsis steinii (= "Gloeodinium montanum"), l.* cysts of *Woloszynskia coronata, m. W. tenuissima, n. Ceratium hirundinella.*

Dinophyceans were originally classified solely as Protozoa. At the present time some still classify them as Dinoflagellida Bütschli, class Mastigophora Diesing. In this class there are more colorless, heterotrophic algal species present than among all the other algae. These protists sometimes produce strong toxins and may act as parasites or phagotrophs. Dinoflagellates are probably more phagocytic than is commonly known (Kudo 1971, Grell 1973). Two marine species *Kryptoperidinium foliaceum* and *P. balticum* are known to possess a second eukaryotic nucleus as well as the usual dinokaryon present in dinoflagellages (Dodge 1971, Tomas et al. 1973) *Peridiniopsis balticum* has a chrysophycean chloroplast (Tomas and Cox 1973). Thus some believe that this represents an invasion of an autotrophic organism into a dinoflagellate cell which otherwise would be apochloric (Raikov 1978) or on the other hand the dinoflagellate cell could store certain of its prey's organelles or storage products for food or maintain them functioning as a reserve metabolic mechanism.

Following Klebs (1912) work Pascher (1927) placed the Dinophyceae in the phylum Pyrrhophyta together with the Cryptophyceae and Desmokontae on the basis of certain similar morphological features, photosynthetic and accessory pigments and the presence of starch. Chadefaud (1960) and Christensen (1962) placed the class Dinophyceae and Cryptophyceae into the phylum Chromophyta. Fott (1971) excluded the Cryptophyceae from the Chromophyta thus emphasizing their nonrelatedness to the Dinophyceae. The position of the Dinophyceae as an isolated group in the algae is unusual. Thus, at the present time it seems appropriate to classify the Dinophyceae among the Protozoa (Mycetozoa) since the presence of coccal and trichal forms is questionable.

2. Cell structure

2.1. Cell covering

Naked dinoflagellate cells have definitive body shapes or forms (some monads) during certain phases of their life cycles or are metamorphic in shape such as *Dinamoebidium*, dinococcalian members and their rhizopodial stages (Popovský and Pfiester 1982). The cell covering of all dinophyceans is called the amphiesma (Schütt 1895, Loeblich 1968) or theca. In naked members (unarmored or athecated) the theca is composed of several membrane layers as seen with an electron microscope and may be passively metabolic. Some species of the genera *Gymnodinium, Katodinium* and *Amphidinium* have a delicate cell cover which is sensitive to chemical and physical changes in the environment. These genera cannot be well preserved. In other naked taxa the cell cover is rather rigid and often sculptured. This protects the cell against chemical and physical changes in the environment. They retain their cell shape when preserved.

Flattened thecal vesicules may be found between the outer membrane and the plasma membrane. Armored forms have structural cellulose or other polysachrides in these vesicules. The inner membrane is interpreted as the plasma membrane. Microtubules occur on the inner side of the plasma membrane. A 5th membrane that may represent a cytoplasmic vesicular component can occur below the microtubules. Those armored species which shed the epi- and hypothecae at cell division often have a pellicular layer. There is disagreement on whether the outermost or inner membrane is the functional plasmalemma (Loeblich 1968).

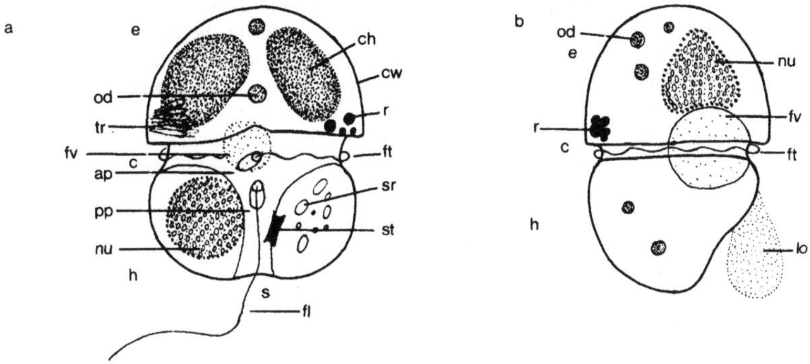

Fig. 2. Cell body, diagrammatic; *a.* ventral, *b.* lateral section; e epicone, h hypocone, c cingulum or girdle, s sulcus, ft transverse flagellum, fl longitudinal flagellum, ap anterior flagellar pore, pp posterior flagellar pore, ch chloroplast, r red remnants, red oil droplets, sr starch, fv food vacuole, nu nucleus, tr trichocyste, cw cell wall, lo lobopode. Organelles and structures that may be present but are not illustrated here include: contractile vacuole and all types of pseudopodia and tentacles.

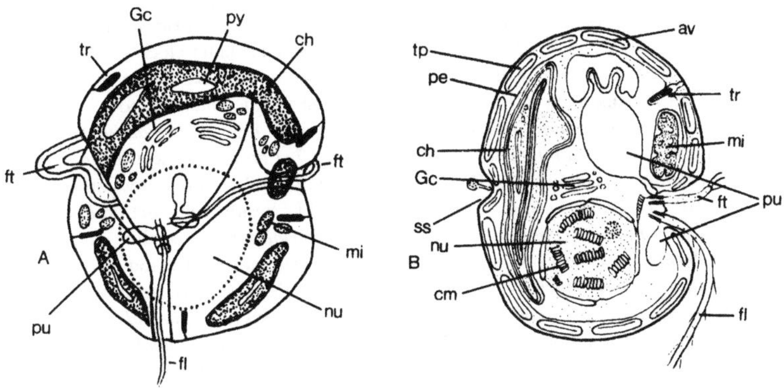

Fig. 3. A. Ultrastructure of marine *Woloszynskia micra* (ventral view); ch chloroplast, py pyrenoid, nu nucleus, tr trichocyst, Gc Golgi complex, mi mitochondria, pu pusule, ft transverse flagellum, fl longitudinal flagellum (after Leadbeater and Dodge).
B. Ultrastructure of a gymnodinioid dinoflagellate (diagrammatic lateral section); av amphiesmal vesicle, ch chloroplast, cm chromosome, Gc Golgi complex, ft transverse flagellum, fl longitudinal flagellum, ss striated strand, mi mitochondrion, nu nucleus, pe pellicular layer, pu pusule, tp thecal plate, tr trichocysts (after Taylor).

In the dinococcalians the cell wall is composed of cellulose and is relatively thick. In *Cystodinedria, Stylodinium* and *Cystodinium* the cell wall of the phagocytic phase is metabolic enabling the cell to move by changing its shape and to enlarge its cell volume during predation. These cells also possess haustoria at some stages in their life history (figs. 54, 55). In *Stylodinium* and *Cystodinium* the stalk (haustorium) is similar to that found in suctorians. It may be short or three times the cell diameter (see chapter 2.6.) *Cystodinedria inermis* rarely exhibits a rhizopodium (haustorium) by means of which it phagocytizes *Oedogonium* cells (Popovský and Pfiester 1981). One of the horns in *Cystodinium* probably has an opening through which very short pseudopodia anchor it as shown on figures of *C. cornifax* (fig. 243). Stalks figured on drawings of *Dinococcus* (figs. 236–238) and *Tetradinium* (figs. 239–241) probably function as haustoria. However, at this stage of investigation into the now recognized complex life of members of the order Dinococcales one cannot be certain that these latter two genera are not zygotic stages within the life histories of the other previously mentioned dinococcalian genera. *Stylodinium tarnum* at times produced a single zoospore (Baumeister 1943) which has a plate pattern formula identical to that of *Peridiniopsis dinobryonis* (fig. 221). In our opinion it is a monad, which forms a haustorium through its apical pore which it uses in

---

Figs. 4–31. Diagrammatic plates arrangements in the following genera and species. Projections of the epitheca *(a)* and hypotheca *(b)*. Each projection in the genera *Peridinium and Peridiniopsis* represents a section of species which exhibits the same plate formula. Symbols used: S sulcus, G girdle, Xa anterior intercalary plates , X′ apical plates, X″ precingular plates, X‴ postcingular plates, X⁗ antapical plates, X_p accessory posterior intercalary plate, sa upper ventral plate in the junction, sp lower plate, sd right plate, ss left plate, c_x cingulum plates, po pore plate; i. e. *Peridinium willei* plate formula is: 4′, 3a, 7″; 5‴, 2⁗. Plates are numbered in a counter clockwise fashion in apical view and clockwise in the antapical view. In the formula, however, they are ordered from the cingulum to the (a) apical pole, and (b) the antapical pole.
Fig. 4. *Peridinium,* section *willei* (Cleistoperidinium), Fig. 5. *Peridinium,* section *striolatum,* Fig. 6. *Peridinium,* section *cinctum,* Fig. 7. *Peridinium,* section *palatinum,* Fig. 8. *Peridinium,* section *bipes* (Poroperidinium), Fig. 9. *Peridinium,* section *gutwinskii,* Fig. 10. *Peridinium,* section *umbonatum,* Fig. 11. *Peridinium,* section *lomnickii,* Fig. 12. *Peridinium,* section *allorgei,* Fig. 13. *Peridiniopsis,* section *polonicum,* Fig. 14. *Peridiniopsis* section *lindemannii,* Fig. 15 *Peridiniopsis,* section *elpatiewskyi,* Fig. 16. *Peridiniopsis,* section *cunningtonii,* Fig. 17. *Peridiniopsis,* section *penardii,* Fig. 18. *Glenodiniopsis steinii,* Fig. 19. *Peridiniopsis,* section *quadridens,* Fig. 20. *Peridiniopsis,* section *borgei,* Fig. 21. *Peridiniopsis,* section *thompsonii.* Fig. 22. *Sphaerodinium cinctum,* Fig. 23. *Diplopsalis acuta,* Fig. 24. *Dinosphaera palustris.* Fig. 25. *Peridiniopsis,* section *dinobryonis,* Fig. 26. *Gonyaulax apiculata,* Fig. 27. *Thompsodinium intermedium.* Fig. 28. shows the numbered plates in *Hemidinium nasutum, a* ventral view, *b* dorsal view, *c* apical projection, *d* antapical projection. Fig. 29. shows the numbered plates in *Ceratium hirundinella, a* ventral view, *b* dorsal view, *c* apical projection, *d* antapical projection. Fig. 30. the numbered plates in *Peridinium bipes,* a view from the *a ventral, b* dorsal, *c* lateral position. Fig. 31. the arrangement of the sulcus and girdle juction in *a Peridinium willei, b P. cinctum, c P. palatinum.*

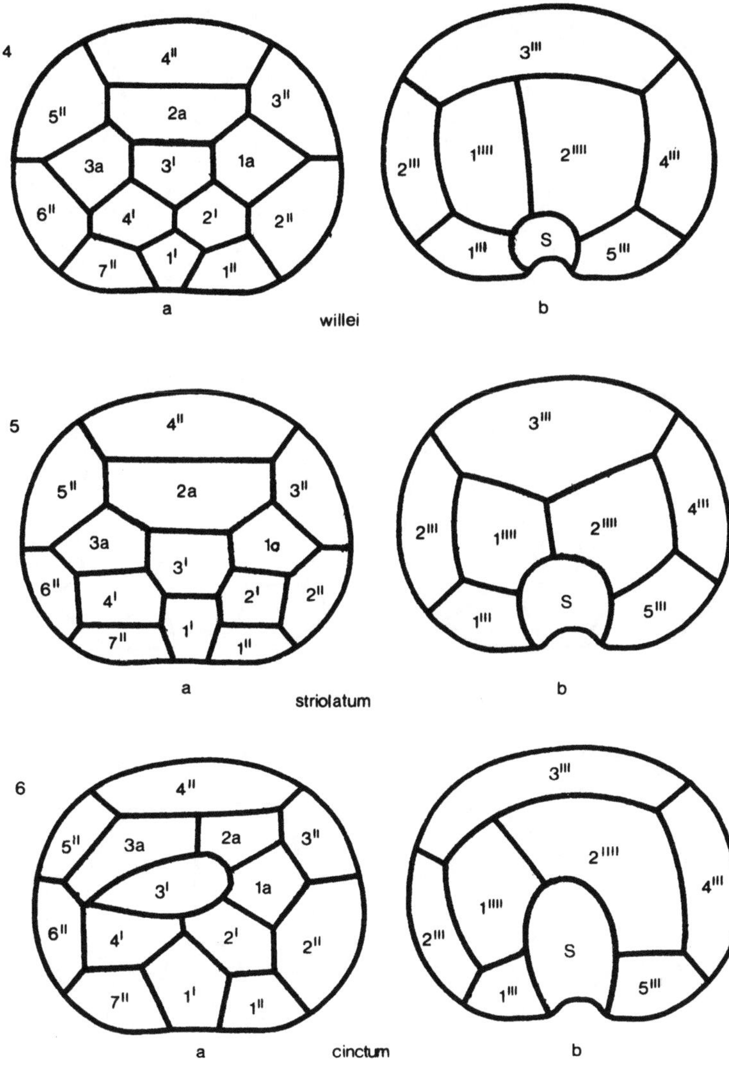

Fig. 4–6

7

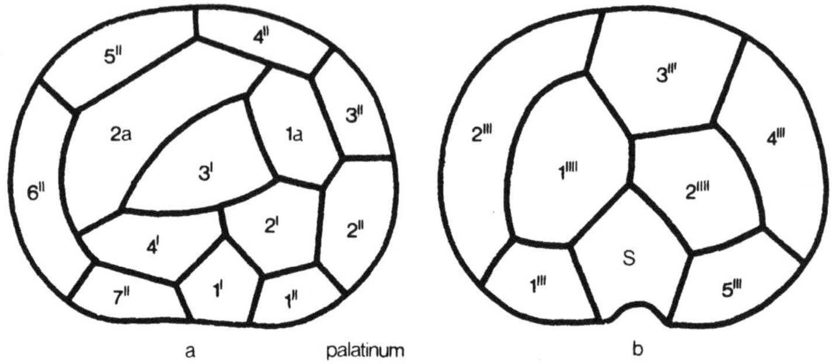

8

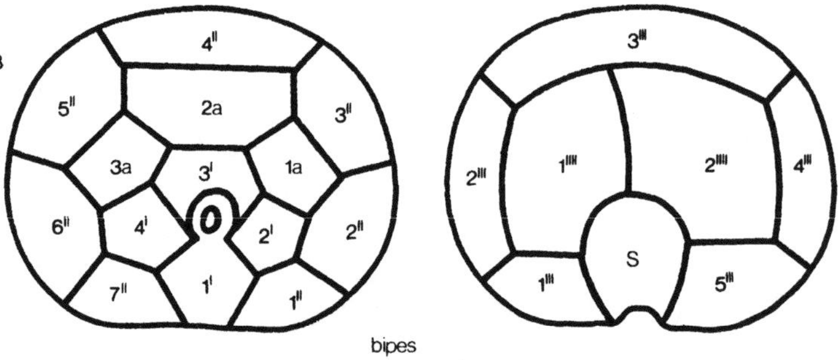

9

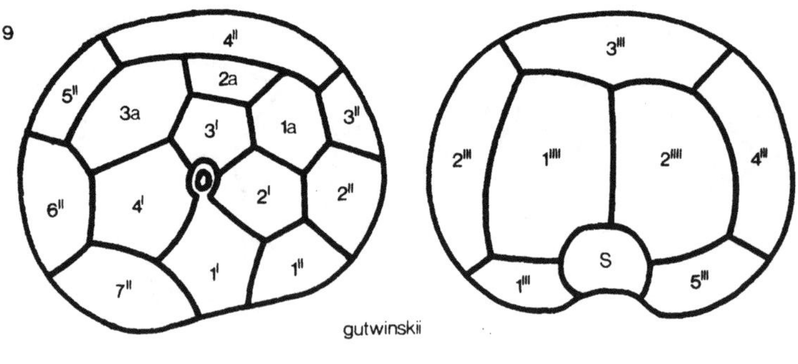

Fig. 7–9

10

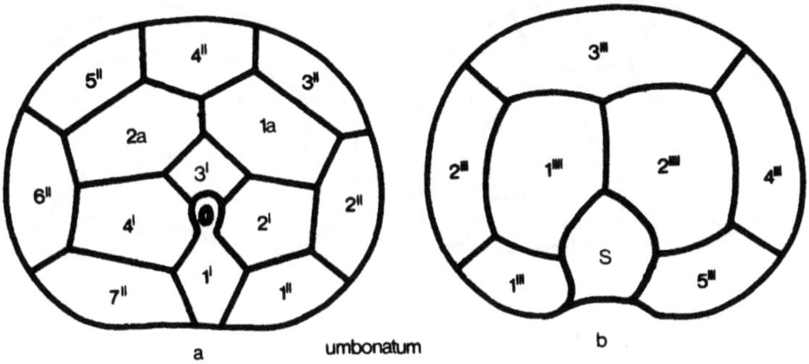

11

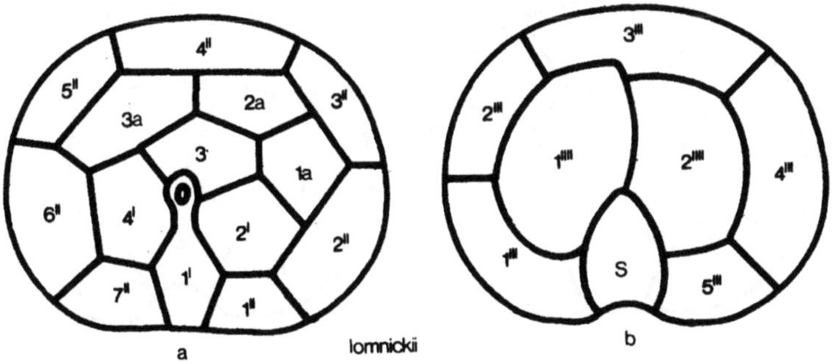

12

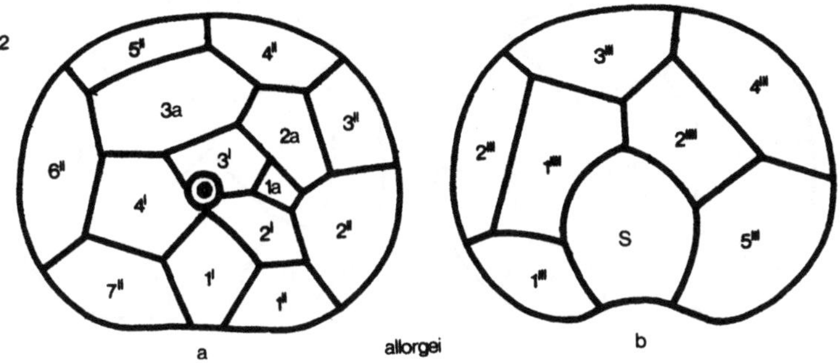

Fig. 10–12

13

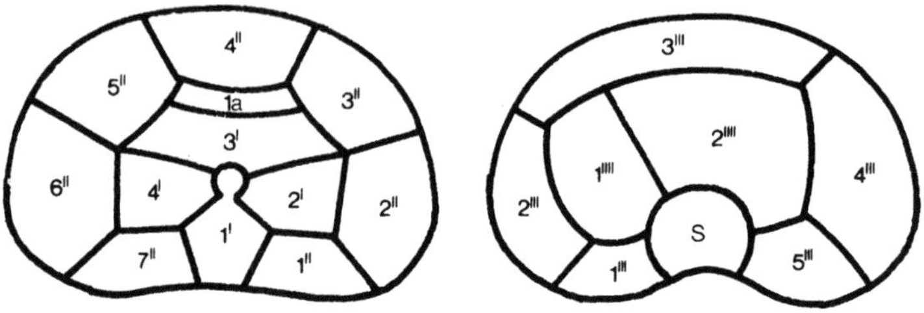

a polonicum b

14

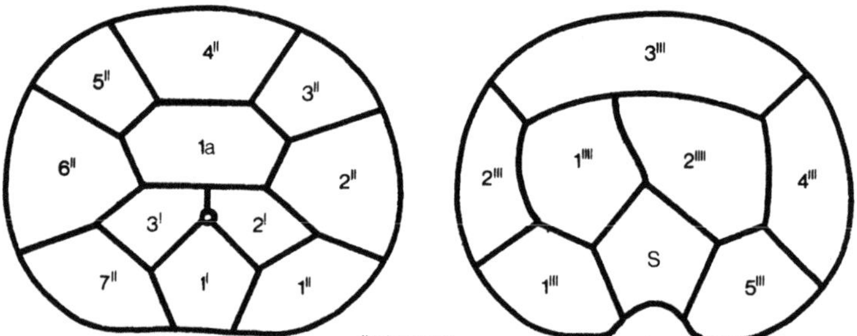

a lindemannii b

15

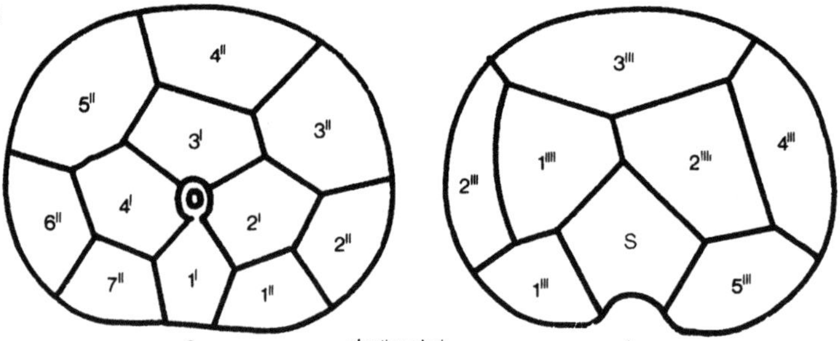

a elpatiewskyi b

Fig. 13–15

16

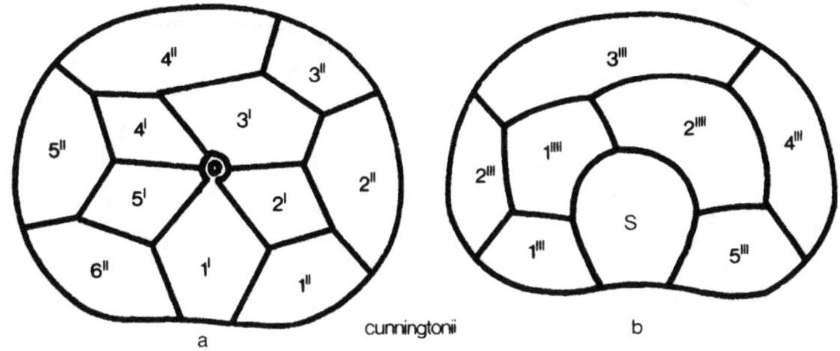

17

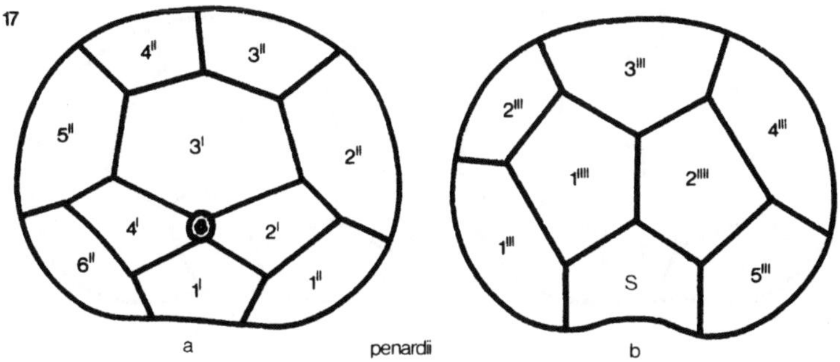

18

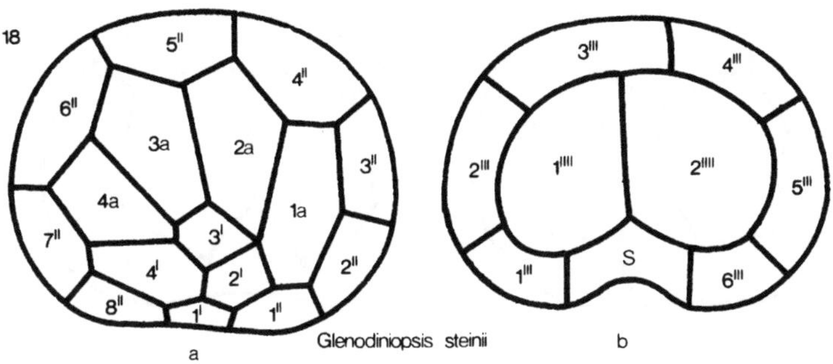

Fig. 16–18

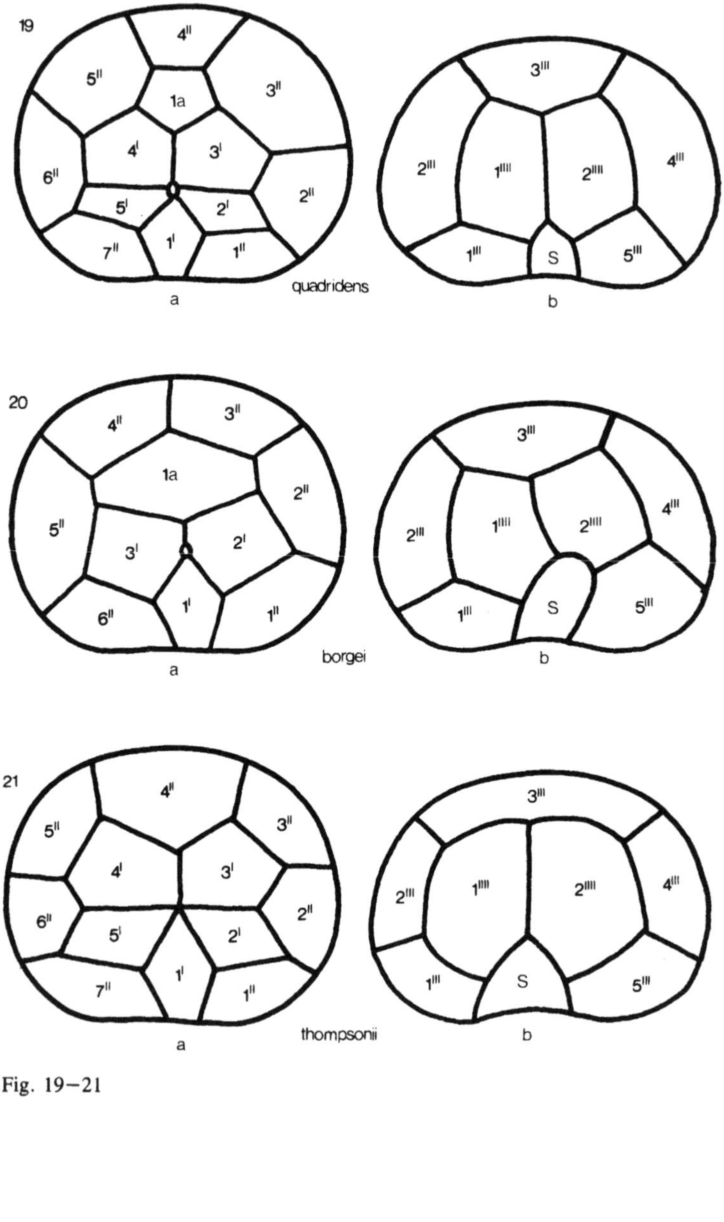
19
4''
5''
1a
3''
4'
3'
6''
2''
5'
2'
7''
1'
1''
quadridens
a
3'''
2'''
1''''
2''''
4'''
1'''
S
5'''
b
20
4''
3''
1a
2''
5''
3'
2'
6''
1'
1''
borgei
a
3'''
2'''
1''''
2''''
4'''
1'''
S
5'''
b
21
4''
5''
3''
4'
3'
6''
2''
5'
2'
7''
1'
1''
thompsonii
a
3'''
2'''
1''''
2''''
4'''
1'''
S
5'''
b

Fig. 19–21

22

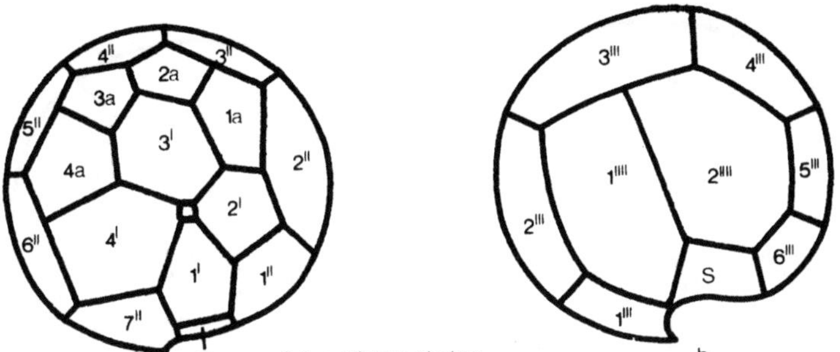

a S Sphaerodinium cinctum b

23

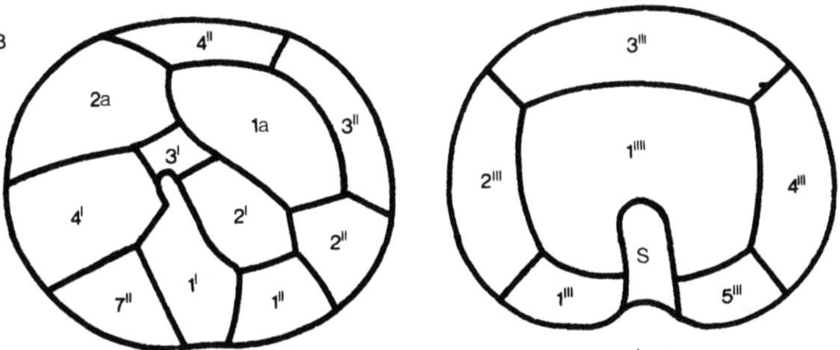

a Diplopsalis acuta b

24

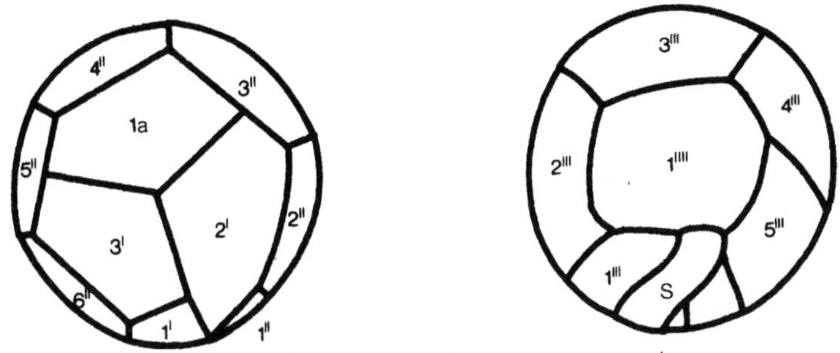

a Dinosphaera palustris b

Fig. 22–24

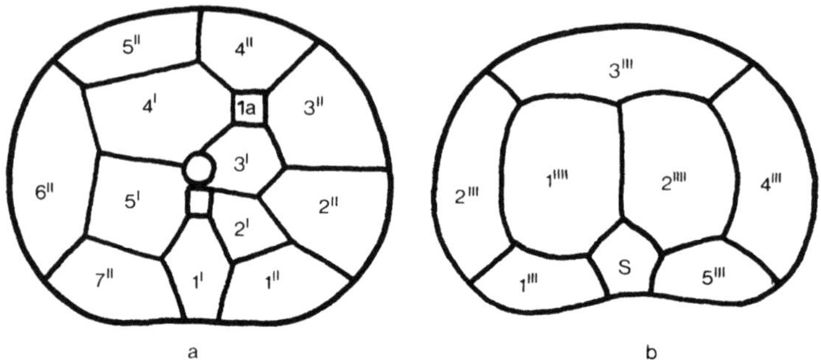

Peridiniopsis dinobryonis

Fig. 25

26

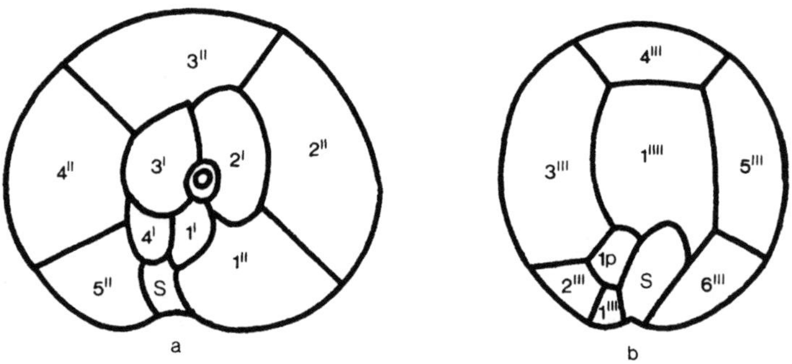

27

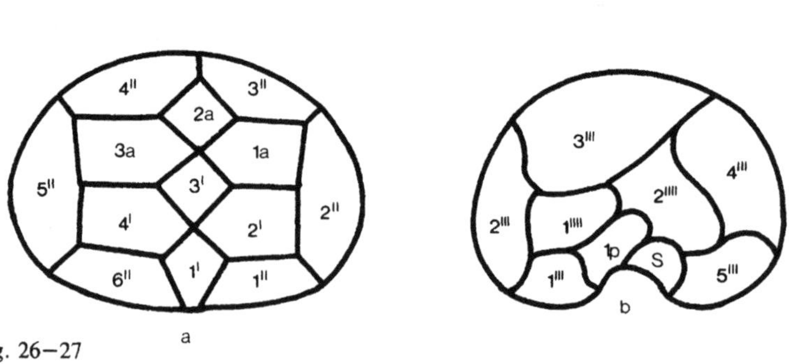

Fig. 26–27

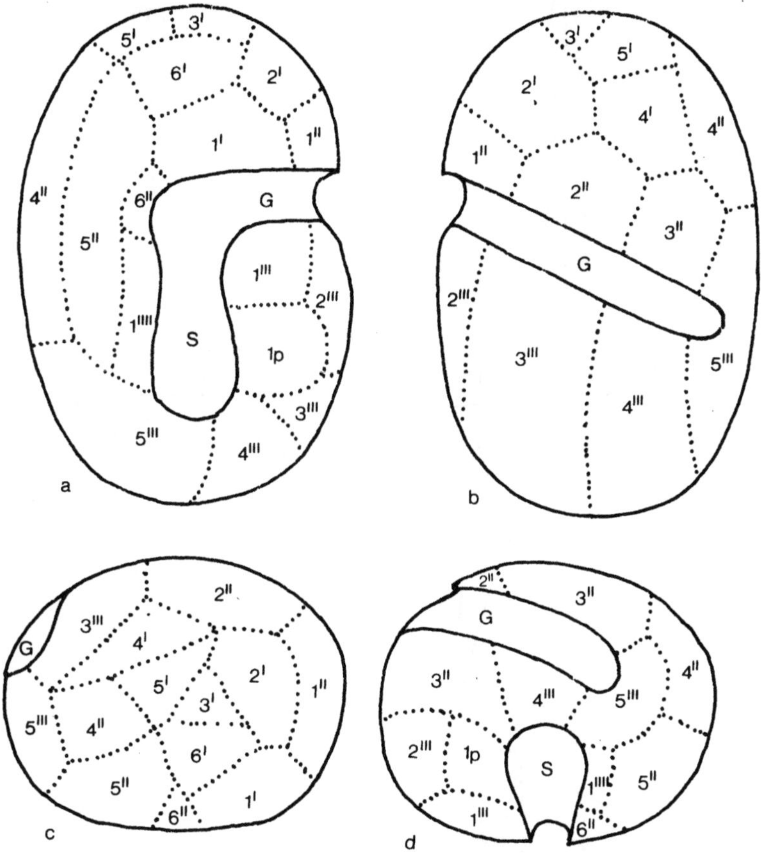
3'
5'
6'
2'
1'
1''
4''
6''
G
5''
1'''
2'''
1''''
S
1p
3'''
5'''
4'''
a
3'
5'
2'
4'
4''
1''
2''
3''
G
2'''
3'''
5'''
4'''
b
2''
3'''
4'
G
5'
2'
1''
3'
5'''
4''
6'
5''
1'
6''
c
2''
3''
G
3''
4''
4'''
5'''
2'''
1p
S
1''''
5''
1'''
6''
d

Fig. 28

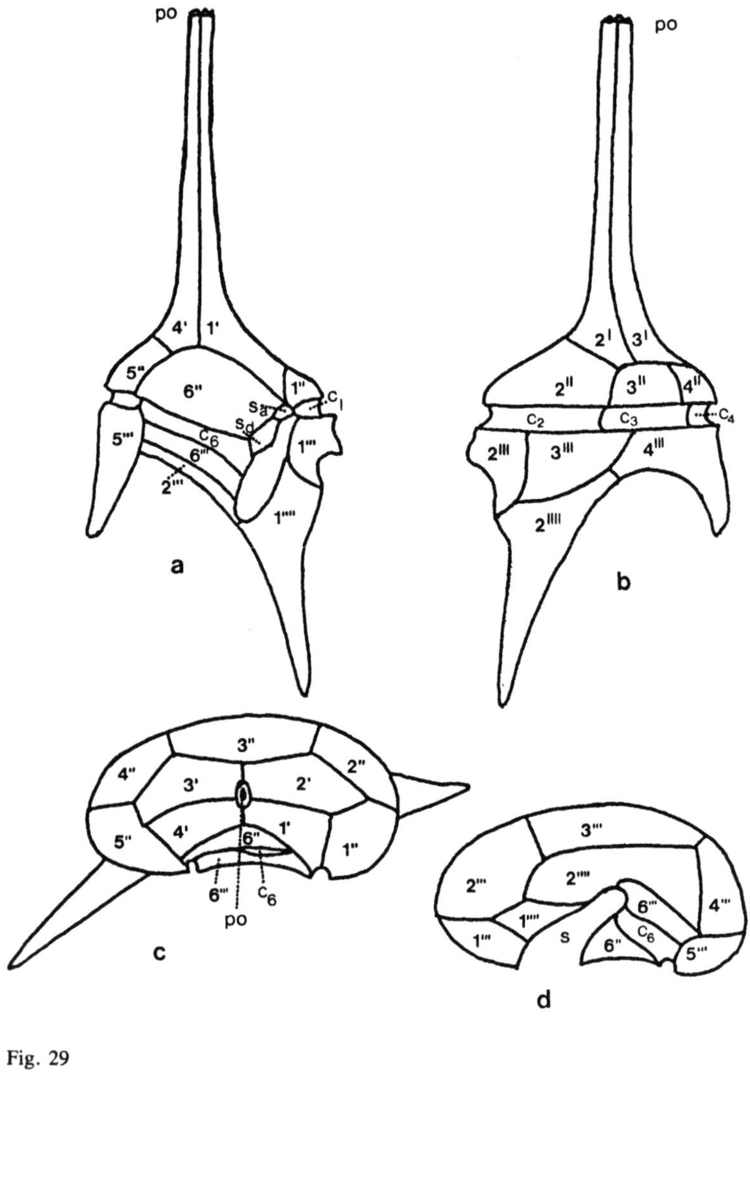
po
4'
1'
5"
6"
1"
sa
c1
sd
c6
5'''
6'''
1'''
2''''
1''''
a
po
2'
3'
2"
3"
4"
c2
c3
c4
2'''
3'''
4'''
2''''
b
3"
4"
3'
2'
2"
5"
4'
6"
1'
1"
6'''
c6
po
c
3'''
2'''
2''''
6'''
4'''
1''''
c6
1'''
s
6"
5'''
d

Fig. 29

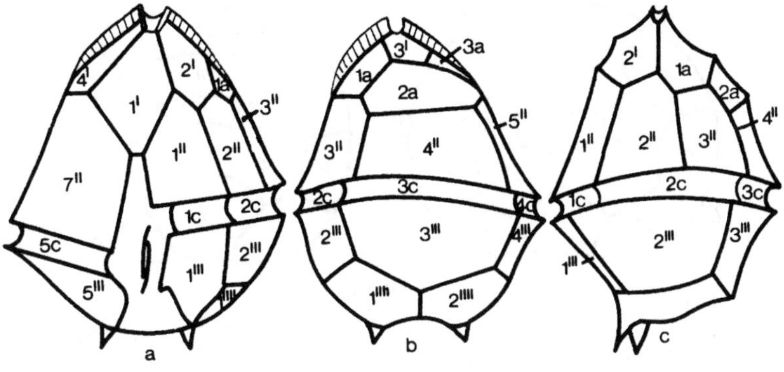
4'
1'
2'
1a
3''
7''
1''
2''
1c
2c
5c
5'''
1'''
2'''
a
3a
3'
1a
2a
5''
3''
4''
3c
2c
4c
2'''
3'''
4'''
2''''
b
2'
1a
2a
4''
1''
2''
3''
2c
1c
3c
2'''
3'''
1'''
c

Fig. 30

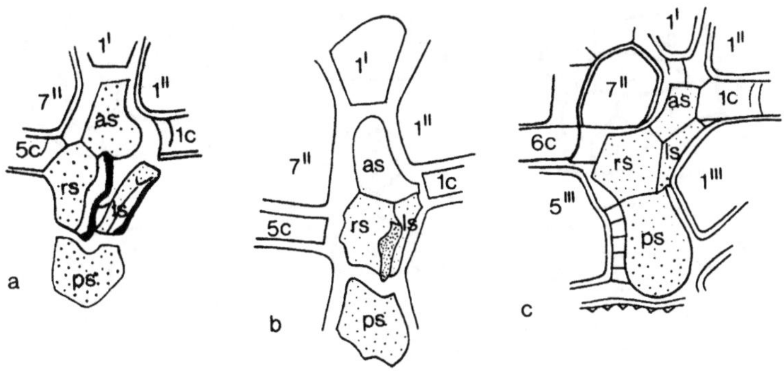
1'
7''
1''
as
5c
1c
rs
ls
ps
a
1'
1''
7''
as
1c
5c
rs
ls
ps
b
1'
1''
7''
as
1c
6c
rs
ls
1'''
5'''
ps
c

Fig. 31

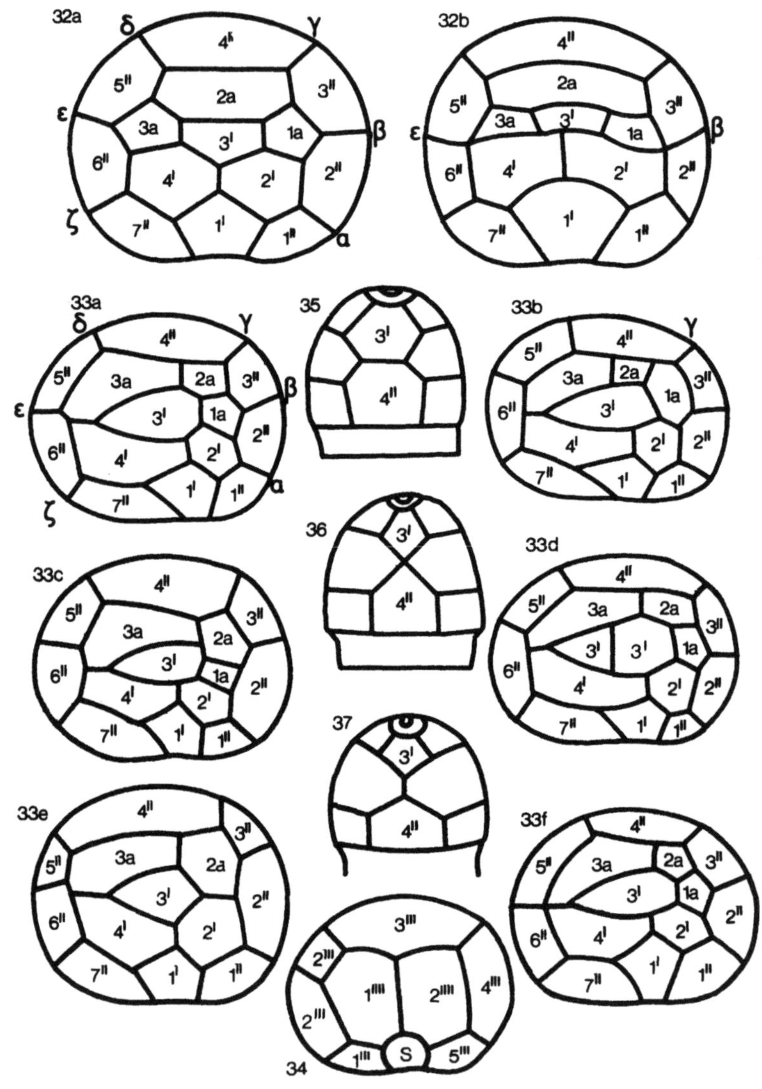

Fig. 32.–37. Projections on the epitheca of *Peridinium willei* and *P. cinctum* show the variations which may occur along the sutures, and in the numbers and position of the plates. Greek script of sutures for plates 1'' and 7'' differ except for plate 1' (the rhomboidal plate). This is also true for the hypotheca. Fig. 32. *a Peridinium willei, b P.*

predation. *Stylodinium polymorphum* also produces a single "zoospore" with plates and an apical pore. Its plate formula, however, is unknown and thus cannot be compared to existing formulae for other taxa (Baumeister 1957).

Spiromonas perforans also has a metabolic cell covering by means of which it phagocytizes *Chilomonas paramaecium*. This species possesses a haustorium (rostrum) during its trophozoid habit in its life history (fig. 60).

In armored species the celullose wall is very thick and may contain some chitin. As stated previously the cell covering is composed of a number of membranes as viewed through the transmission electron microscope: (1) an outer membrane which is common to all species, (2) a vesicular layer which is also common to all taxa. This layer is composed of flattened vesicles which stain with silver and appear to have a hexangular structure; the cellulosic plates are contained in these vesicles, (3) another membrane which is probably the plasma membrane and (4) a pellicular layer of acid polysaccharides (Loeblich 1969) common only to armored species. This layer is homogenous and structureless. This complex membrane system composes the amphiesma or theca. In some genera such as *Woloszynskia, Glenodiniopsis* and *Hemidinium* the theca appears to be divided into thin hexagonal plates, supposedly of uniform thickness (Bold and Wynne 1978). *Woloszynskia reticulata* (fig. 142) is unusual in the genus by having a thin, smooth, bell-shaped epitheca with a carina (ridge formed by small plates) over the apex from the ventral to the dorsal margin of the girdle and a hemispherical hypotheca composed of smooth, concave plates (fig. 134b, c) (Pfiester et al. 1980) with thick ridges. These plates differ from those found in such genera as *Peridinium, Peridiniopsis, Gonyaulax, Ceratium, Sphaerodinium, Diplopsalis,* and *Dinosphaera* in lacking suture bands joining polyangular plates. The sutures represent zones of thecal growth (figs. 38e, 153, 170). A such they enable the cell wall to accommodate protoplast enlargement. This is extremely important during sexual reproduction when the planozygote doubles in size. Thus in the latter stages of planozygote development the sutures are extremely widened and striated (Pfiester and Skvarla 1979, 1980).

Resting vegetative cysts possess a two layered cell wall and hypnozygotes have a three layered cell wall. The exact chemical composition of these walls is as yet unknown but it has been suggested that hypnozygotic walls may possess chitin (Pfiester 1975) and/or sporopollenin (Spector et al. 1981). That opening which forms in the zygote through which the meiocyte "escapes" is termed the archeopyle. It is formed when one or more plates disassociate and create an opening or pore. If these plates remain attached to the theca along one suture they form what is called an operculum or lid. The plates which separate from the theca forming the archeopyle are species specific. No information is as yet available concerning the cell wall structure of the Adinophycidae, Protaspidaceae and rhizopodial stages (trophozoides and sporocytes) of the Dinococcales.

---

willei β ε bicollineatum, Fig. 33. *a Peridinium cinctum, b P. cinctum* γ travectum, *c P. cinctum* travectum, *d P. cinctum* 23′ complexum (the apical plate 3′ is double) *e P. cinctum* la simplex (the intercalary plate 1a is absent), *f. P. cinctum* ε collineatum, Fig. 34. shows the doubled plate on the hypotheca-22′′′′ complexum in *Peridinium cinctum*, Fig. 35–37, three variations are shown for *Peridinium umbonatum* from the dorsal view and depict the arrangement of sutures of plates 3′ and 4′′ referred to as – 35 conjunctum, 36 contactum, 37 remotum.

The number, shape and arrangement of thecal plates is the main taxonomic character used in the identification of armored dinoflagellates (Desmokontae excepted – see chapter 3.1). We have used the epithecal and/or hypothecal plate formulae in our key. These plate formulae are genera and species specific. There are numerous freshwater *Peridinium* and *Peridiniopsis* species, each of which has a distinct plate pattern. Some possess an apical pore. These genera are subdivided into sections on the basis of plate pattern formulae and the presence of apical pore. Figures 4–29 show the generalized epi- and hypothecal plate arrangements for the following genera: *Peridinium, Peridiniopsis, Sphaerodinium, Diplopsalis, Dinosphaera, Gonyaulax, Thompsodinium, Hemidinium,* and *Ceratium.* Figures 30 and 31 show plates from the ventral, dorsal, and lateral views.

Plates are numbered in a counter clockwise fashion in apical view and clockwise in antapical view. In the formula, however, they are ordered from the cingulum to (a) the apical pole, and (b) the antapical pole; i.e. *Peridinium willei* plate formula is: 4′, 3 a, 7′′; 5′′′, 2′′′′ – four apical plates, three anterior intercalary plates, seven precingular plates, five postcingular plates, and two antapical plates.
Several systems for naming and numbering plates have been used in the older literature. In Bütschli's system *Peridinium willei* would be given as: 7pr + 1r + 2vap + 3map + 1dap; 5pst + 2at – seven precingular, one rhomboidal ventral ("Rautenplatte"), two ventral, three medium, one dorsal plates; five postcingular, and two antapical plates. Accordingly Lefèvre's system the same formula would read: p_7 a a_3 p'_5 a'_2 – seven precingular, one rhomboid ventral, three apical, three medium plates, five postcingular, and two antapical plates. Kiselev would list it as 7pr + 4a + 4sr; 5zb + 2a′ – seven precingular, four apical, three medium plates; five postcingular, and two antapical plates.
Most plates are ornamented with spines and ridges. Pores, poroid, and trichocysts pores may also be present and arranged in rows. Plate reticulation is common in many species (fig. 38). The margins of furrows are sometimes ornamented with hyaline ridges and wings. There is a great deal of variability found on small structures on the epi -and hypothecae. Common examples have been illustrated by Lefèvre (1927) and are referred to in many textbooks. Because these variations, common though they may be, are usually present in low numbers in a population or when present are seldom documented, they have not been credited with taxonomic value. Only a few such variations are depicted in figures 32–37. Recently more attention is being given to the plate formula in the girdle and sulcus. Plates forming the junction of the sulcus and girdle have two flagellar pores or openings. The number of plates forming the girdle or cingulum is usually the same as the number of postequatorial plates on the hypotheca. The arrangement of plates forming the junction between the girdle and sulcus is given in figures 29–31.
In contrast to the external theca of most Dinophyceae, a complex internal skeleton (a dominant taxomomical feature for the Foraminifera) is present in the family Actiniscaceae (Zimmermann 1930, Schiller 1933/37). Composition of this internal skeleton consists of opaline silica in freshwater species of the genus *Actiniscus* Ehrenberg and *Pseudoactiniscus* Bursa. Adachi (1965) described an internal plate-like structure located beneath the theca in *Peridiniopsis polonicum.* This "internal skeleton" is in contact with the ventral side of the cell and has the following formula: 3′,t,2c,4c,2′′. Its origin and position is quite different from that of the internal skeleton in the Actiniscaceae.

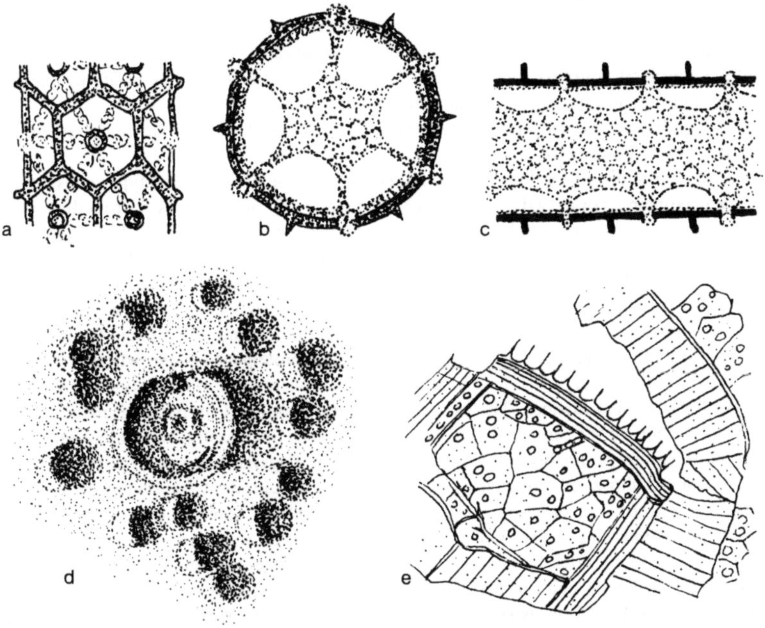

Fig. 38. Arrangement of pores and cytoplasm in the apical horns of *Ceratium hirundinella; a* plate reticulation with pores (dark lines) and cytoplasmic connection (light lines), *b* cross section near the apex, *c* longitudinal section (after Entz), *d* single trichocyst pore surrounded by spines in marine *Ensiculifera loeblichii*, 500 ×, *e* single plate of *Peridinium cinctum* showing reticulations and trichocysts pores and region of plates expansion – 4650 × (redrawn from EM microphotographs in Bold and Wynne).

2.2. Flagella and motion

Monads have two flagella. They may be apical as in *Desmomastix globosa* (fig. 1a) or subapical in *Protaspis metarhiza* (fig. 1b) and *Spiromonas perforans* (fig. 60). Most commonly the flagella are inserted in pores on the ventral side of the organism where the sulcus and cingulum join as in *Ceratium hirundinella*. The longitudinal flagellum is either used to pull the organism trough the water (as in the Desmomastigales) or to push it (Peridiniales). The other flagellum encircles the monad lying within the cingulum (fig. 2). Its movement causes the cell to rotate or swirl as it moves through the water.
The longitudinal flagellum is coiled in the sulcus when the cell is at rest and stretches out past the antapex trailing the cell when it is in motion. Its length which varies with specific taxa may reach a length two and a half times the length of the cell. Hypnozygotes may be recognized also by the presence of two trailing flagella. The transverse flagellum arises from a pore located just above the pore for the longitudinal flagellum (fig. 2). It encircles

the cell extending to the left a minimum distance of 2/3 the cell's circumferance. It remains in the girdle as the cell moves through the water. It has been described in the past as (1) a row cilia, and (2) an undulating membrane. We now know it is shaped like a flattened, spiralled ribbon (figs. 3, 39a). Cells may lose flagella during predation, reproduction, or digestion and reform them.

Dinoflagellates possess basal bodies which are identical with those found in other flagellates or plant cells.

The ultrastructure of the longitudinal flagellum is of the conventional type, constisting of an axonema composed of a ring of nine pairs of microtubules surrounding two central ones. This microtubular core is surrounded by a membrane. Externally the flagellum appears to consist of two regions – a distal narrow region about 1/2 of the length and wider basal portion. The entire flagellum is covered by fine mastigonemata probably arranged bilaterally. The narrow tip and mastigonemata may be lacking in certain other dinoflagellates. The transverse flagellum is more complex. Besides the conventional axonema it also contains a bonded proteinaceous thread, the striated strand, which appears to be held in tension. The axonema is wound around the striated strand in a helical manner. The two structures are kept apart by packing material, yet held together by the stretched flagellar sheath. Long fine mastigonemata are attached to the flagellum, probably in a unilateral array. The flagellar bases and their associated root systems show some diversity of organization although the basal bodies themselves have a uniform construction. Just before the flagellum enters the cell two central tubules terminate at a disc which completely blocks the axonema center. Just within the cell one or two diaphragms partially occlude the space surrounded by the peripheral doublets. These pairs of tubules are then joined by a third making triplet associations. At this point fibrous structures that end in dense blobs project from the triplets. A single tubule is present from this point to the bottom of the basal body. The characteristically angled triplets are joined together by radial and peripheral septa (Dodge 1971).

The free swimming monad moves by means of the flagellum. The longitudinal flagellum is either extended in the direction of movement *(Desmomastix,* Protaspidaceae, and several Peridiniales) or is extended backwards (most common in Peridiniales). The second flagellum is coiled and away from the longitudinal axis *(Desmomastix),* or positioned in the sulcus (Protaspidaceae), or in monads and motile hypnozygotes in the cingulum (Peridiniales). The transverse flagellum causes the cell to move in a smooth or jerky pattern. Some rhizopodial stages of dinococcalians move by using two thin rhizopodia as flagella. Some colorless monads lose their flagella and furrows after predation, become rounded (a stage incorrectly referred to as a food cyst) only to develop the furrows and flagella again once the food has been digested. The method

---

Fig. 39. *a submicroscopical structure of a transverse flagellum, b* transverse section, *c* a transverse section of a longitudinal flagellum, *d* a fusiform pyrenoid situated between the chloroplast lamellae, *e* a more complex interlamellar pyrenoid, *f* a stalked pyrenoid; the chloroplast is packed with pyrenoid matrix, *g* a talked pyrenoid suspended by protrusions from a peripheral chloroplast, *h* pyrenoid attached to chloroplast with numerous tubular invaginations of the pyrenoid envelope *(a–c* after Leadbeater and Dodge, *d–h* after Dodge and Crawford).

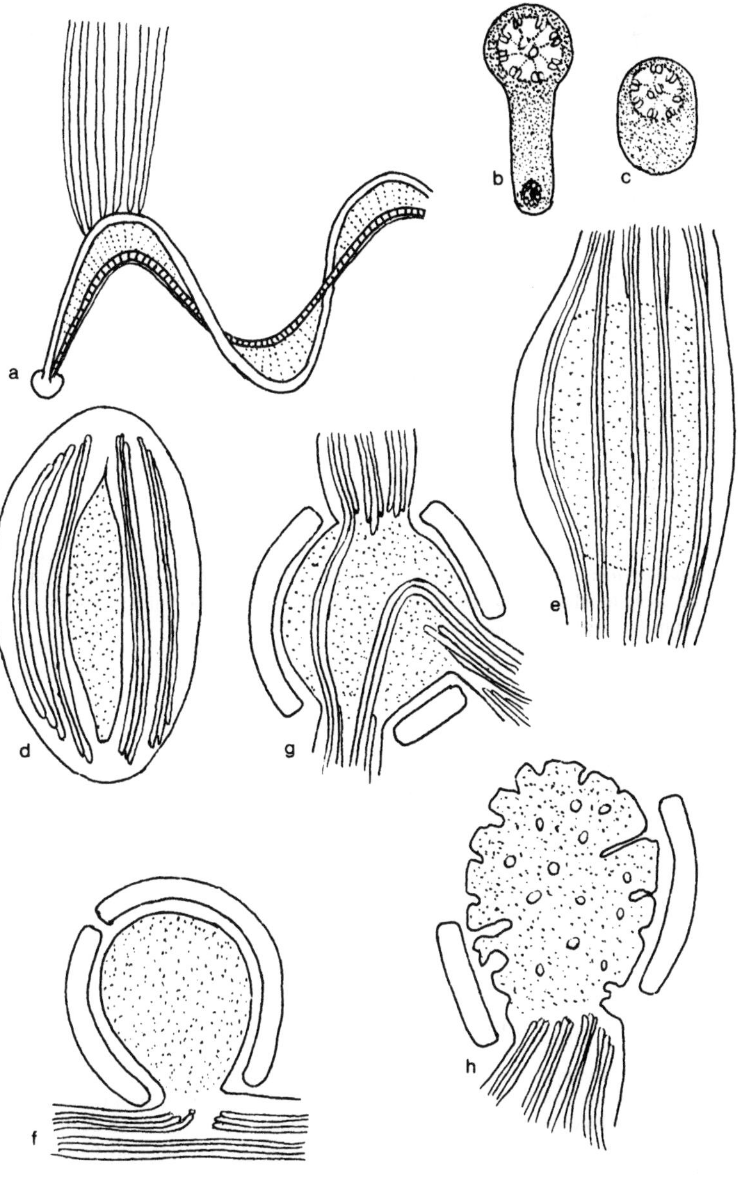
a
b
c
d
e
f
g
h

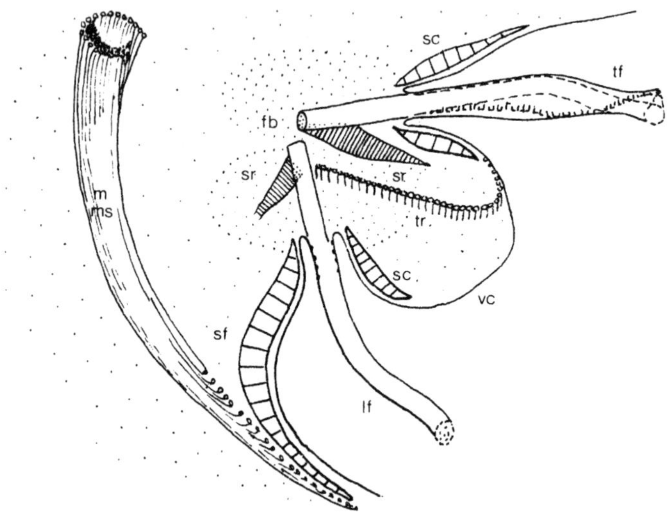

Fig. 40. A diagrammatic representation on the arrangement of the basal bodies in *Ceratium hirundinella;* ms microtubular strand, sf striated fibrils, sc striated collar, sr striped root, tr tubular root, fb flagellar base, tf transverse flagellum, lf longitudinal flagellum, vc ventral chamber (after Dodge and Crawford).

by which some stages within the life history of *Cystodinedria inermis* move has not yet been determined.

When in their typical "coccal form" *Cystodinedria inermis* and *Cystodinium bataviense* change their position on a filament by rolling. The coccal form of *Stylodinium globosum* is able to change its position by means of a modified hypertrophied tentacle (stalk) (figs. 53–55). The sessile trophozoides in the life history of these dinococcalians move upon a filament by means of rhizopodia and axopodia (Popovský and Pfiester 1982, Pfiester and Lynch 1980). *Hemidinium nasutum* and *Glenodiniopsis steinii,* are also referred to as *Gloeodinium montanum* (nonmotile stage). These nonmotile cells move by mucilage formation (fig. 146e).

2.3. Protoplasm and dinokaryon

Dinophycean protoplasm is mainly colorless and divided into ecto- and endoplasm; the ectoplasm is more consistent, bladder-like to cirrus-like, the endoplasm is granular. In some dinoflagellates both plasma layers are transcendent without visible borders. Rarely the plasma is rose-colored. Coloration is usually restricted to the ectoplasm. Chloroplasts, pyrenoids, storage products and remnants are located in the ectoplasm as are also the flagellar bases, pusules, mitochondria, trichocysts, vacuoles, and stigma. The dinokaryon, Golgi complex, food vacuoles and pusule system are located in the endoplasm (figs. 2,3) (Starmach 1974). Their Golgi complex (dictyosomes) form a hemisphere above the nucleus. Their endoplasmic reticulum is not extensive. The protoplasm has been noted to rotate as fast as ten rotation per minute before cell division. The ectoplasm is always moving slowly as can be noted by change in chloroplasts position. The protoplasm which surrounds flagellar bases is also relatively mobile and may sometimes extend through pores as pseudopodia (figs. 44, 86e,f, 110b) (Zacharias 1910, Hofeneder 1930, Fott 1938, Popovský 1980). In *Ceratium hirundinella* a thin layer of ectoplasm has been observed on the cell wall (Entz 1927), and in *Peridinium aciculiferum* a hyaline cap has been observed above the apical pore which has probably been formed from ectoplasm.

One and sometimes more than one *(Paradinium pouchetii)* dinokaryons or mesokaryons are observable in most stages of dinoflagellate life histories. In dinococcalian trophozoids and sporocytes the nucleus cannot be seen with the light microscope. However, all the trophozoids give a positive nuclear stain (Popovský and Pfiester 1982). The dinokaryon is large in relation to nuclear size observed in other unicells. The chromatin is visible at the

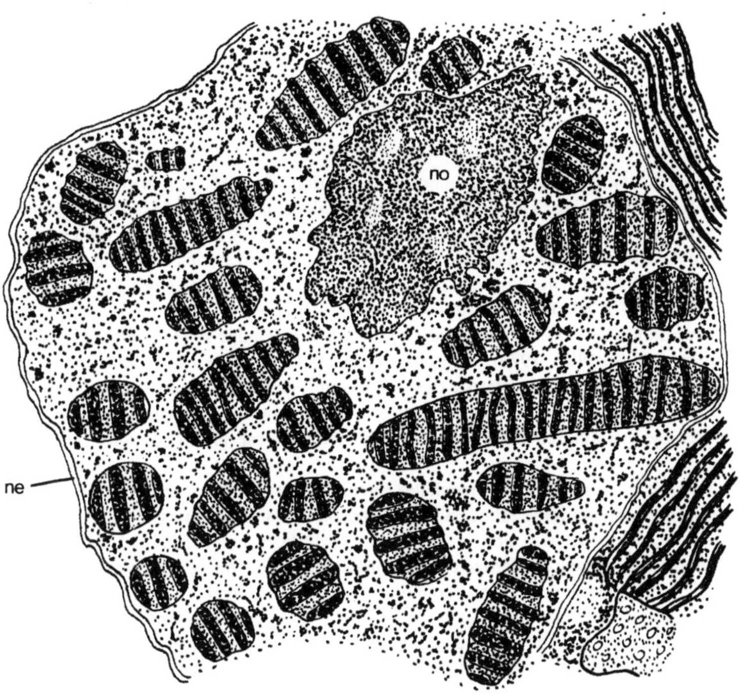

Fig. 41. A section through the dinokaryon showing the chromosomes, nucleolus (no), and the nuclear envelope (ne) (redrawn after Dodge).

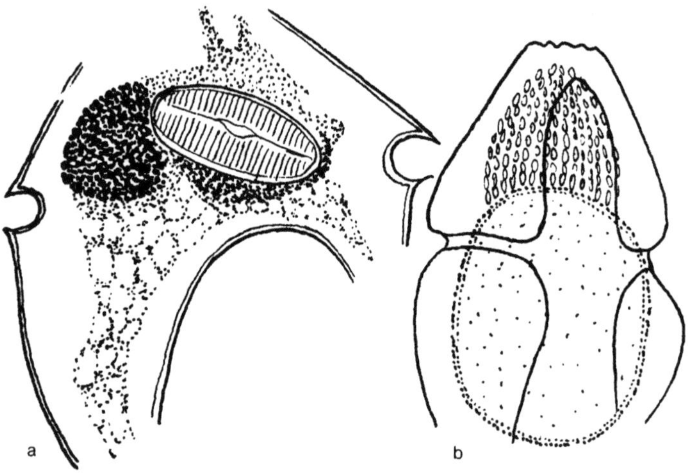

Fig. 42. A triangular shaped dinokaryon attached to the food vacuole, *a Ceratium hirundinella*, *b Gymnodinium helveticum* fo. *achroum* (*a* after Entz, *b* after Popovský).

light microscope level appearing as beads on a string (figs. 41, 42, 44). On to several nucleoli are also visible. The dinokaryon is round, developing an ovoid or dumbell shape before dividing. It is usually located in the posterior part of the cell. When a food vacuole contiguous with the girdle-sulcus juncture is present the dinokaryon is pyramidal in shape and located adjacent to the food vacuole (Entz 1927, Popovský 1980) (figs. 42, 44, 86).

An eukaryon (in addition to a dinokaryon) is also present in two marine species: *Kryptoperidinium foliaceum* and *Peridiniopsis balticum* (Dodge 1971, Tomas et al. 1973, Loeblich and Loeblich 1973). The eukaryotic nucleus divides amitotically with the dinokaryon. As previously stated, this eukaryotic nucleus probably resulted from an endosymbiotic chrysophyte as pigments present in these two taxa are chrysophycean. It is also possible that dinoflagellate predators may retain and use some organelles from their prey.

The majority of free-living dinoflagellates (including marine members) differ from other eukaryotes in having large numbers of chromosomes, large amounts of DNA per cell, uniform chromosome size and in having no or little histone. They, along with the euglenophyceans, have condensed chromosomes throughtout the cell cycle. Several types of nuclear organization can be found in dinoflagellates: (1) nuclei with condensed, rod-shaped chromosomes, numbering from 25 to 275. No centrioles are associated with dinomitosis; (2) large vegetative nucleus with uncondensed chromosomes. Mobile spores with smaller nuclei and condensed chromosomes (*Noctiluca* and an amphiaster involved in dinomitosis; (3) chromosomes U shaped and variable with the life cycle phase (4) chromosomes numbering 4 to 6 and V shaped. Centrioles involved in dinomitosis *(Syndinium)* (Loeblich et al. 1974) and (5) nuclei without condensed chromosomes, relatively small, or more than one nucleus present (trophozoids in Dinococcales).

Chromosome characteristics suggest three possible models for nuclear DNA organization: (1) a normal pattern of alternation of haploid and diploid generations in which haploid chromosomes are not repeated; (2) polyploidy with three or more sets of homologous chromosomes, and (3) polyteny, a multistranded chromosome condition which may occur in a haploid or diploid state (Roberts et al. 1974). Dinoflagellates were at first assumed to be diploid, but to date all except *Noctiluca* have cytologically been determined to be haploid. Further their sexual reproduction has been documented, meiosis is known to follow zygote excystment with some species known to have mating types (heterothallic) and others not (homothallic) (Beam and Himes 1980).

Because of the following nuclear characteristics the dinoflagellates are unique among the protist: (1) absence of significant amounts of histone; (2) 2–6 nm diameter of the chromatin; (3) membrane-attached chromosomes; (4) absence of a mitotic spindle (5) continuous DNA; and (6) an ultrastructural pattern of chromatin fibrils arranged in an arch similar to that found in bacterial nucleoids (Dodge 1966).

2.4. Chloroplasts

The number, shape, color and position of chloroplast does not appear to be an important taxonomic/diagnostic feature. Chloroplast may be plate-like, ribbon-like, discoidal (different size) and vary in number and position within the cell.

Chloroplasts are parietal (rarely radial) in most dinoflagellates (Starmach 1971). A radial arrangement is sometimes simulated in predacious dinococcalians following ingestion of green filamentous algae. They will also give a positive starch test with iodine at that time (Popovský and Pfiester 1982). Chloroplasts often change position within the cell since the cytoplasm streams rapidly. Chloroplasts color is varied among members of the same species. Color within a species may range from light, yellowish-brown to olive and dark brown. Some dinoflagellates are known to have bluish chloroplasts (cyanells ?) and others to have green ones. Their photosynthetic pigment is chlorophyll *a*; accesory pigments are chlorophyll *c*, β-carotene, γ-carotene, peridinin, neoperidinin, dinoxanthin, pyrrhoxanthin, fucoxanthin, diadinoxanthin and diatoxanthin (Ettl 1980). Species are known in which part or all of a population lack visible chloroplasts at the light microscope level such as *Amphidinium elenkinii, (A. tatrae, A. larvale), Gymnodinium lantzschii (G. albulum), G. triceratium, Peridiniopsis penardiforme, Ceratium hirundinella,* and some stages of *Cystodinedria inermis* and *Stylodinium globosum.* Depending upon available food sources dinoflagellates may be able to develope or reduce the chloroplasts visible at the light microscopical level. In some cases this change may occur rapidly within minutes or hours.

The ultrastructure of the chloroplasts (figs. 39d,h,43) shows lamellae which run somewhat parallel to each other across the chloroplast. Each lamella is composed of two to four thylakoids with three the most common number. Sometimes the lamellae appear to be almost radially arranged. In other species lamellae can be seen to be attached to the chloroplast envelope. This latter condition has not been reported in other algae. The chloroplast membrane is unique in that it actually consists of three membranes not connected to the endoplasmic reticulum or nuclear membrane (Dodge 1971).

Pyrenoids may be present in photosynthetic freshwater dinoflagellates but are absent in colorless, non-photosynthetic members. Dinoflagellate pyrenoids are unique among the algae in that they lack a starch sheath and possess proteins different from those within the

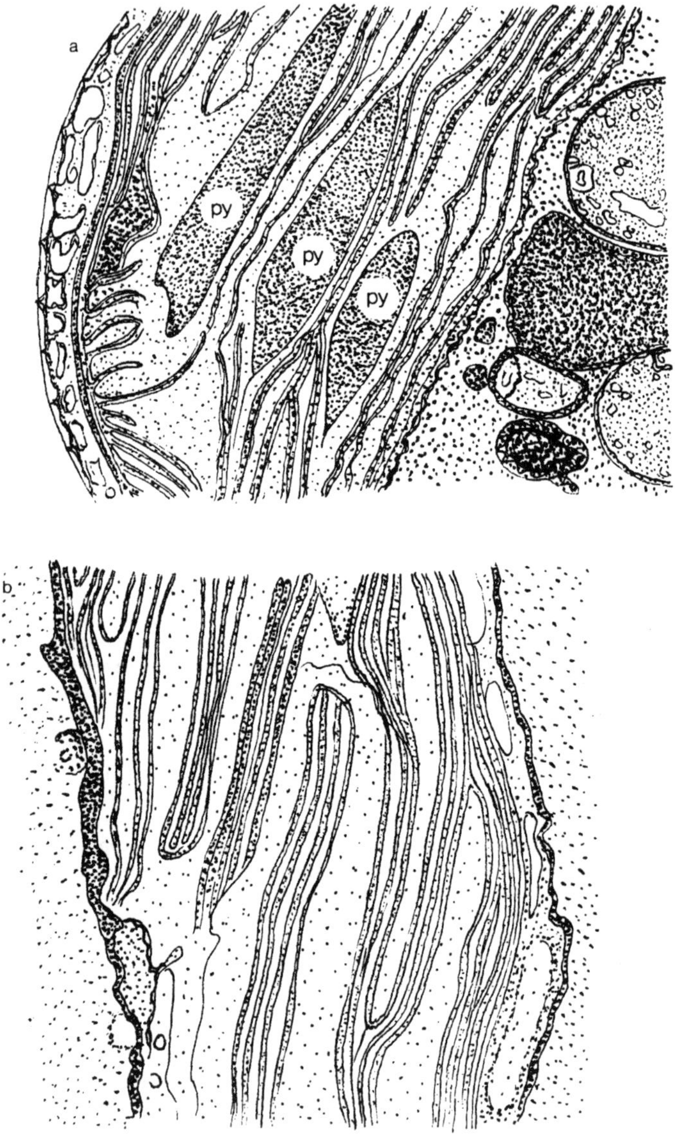

Fig. 43. *a* a section through part of one chloroplast and adjacent region showing part of the chloroplast lamellae and two fusiform pyrenoids; (py) within the chloroplast, *b* a transverse section through a chloroplast showing the three thylakoids which compose each lamella (redrawn from Leadbeater and Dodge).

chloroplast. Five pyrenoid types (fig. 39) are known to occur in marine and freshwater dinoflagellates: (1) simple interlamellar, (2) compound interlamellar, (3) single-stalked, (4) multiple-stalked, and (5) stalked with invaginations. Most of these pyrenoid types are also found in other alga groups. The function of the dinoflagellate pyrenoid is unknown. Perhaps, it is simply to store proteins (Dodge and Crawford 1971).

Dinoflagellates are also unique among the algal groups in that more than one type of **eyespot** or **stigma** is known to occur. A stigma is not present in all dinoflagellate species; when present, ist does not necessarily occur in all stages of the life cycle. Some populations of *Gymnodinium paradoxum* possess a stigma and other populations will lack it. When present a stigma is usually located close to the sulcus. It is most often described as rounded, oval or horseshoe shaped; less often shaped as a twofolded hyperboloid. It has usually been described as bright to dark red in color.

Two types of stigmas (based on their position within the cell at the electron microscopical level) are found in freshwater taxa: (1) a very simple stigma located outside the chloroplast composed of carotenoid globules which are mainly composed of lipids and are not surrounded by a membrane (i.e. *Woloszynskia coronata*). This typ of stigma is also found in the Euglenophyceae and Eustigmatophyceae; (2) the second type of stigma is also composed of carotenoid-lipid globules, but as in the green algae it is located to one side of the chloroplast as in *Peridinium cinctum* (fo. *westii*) and *Woloszynskia tenuissima* (originally described without a stigma) (Dodge 1971).

Determination of whether or not a species possesses a stigma may be very difficult because the presence of red remnant globules in the sulcal area may simulate one. It is known that lipid (?) globules, called plastoglobules accumulate inside the chloroplast in the shape of pseudostigma (Geitler 1969).

2.5. Vacuoles and pusules

Vacuoles are more visible in older cells than in younger ones. They are often visible in members of the Dinococcales. A vacuole system has been described for *Peridiniopsis borgei* which consistens of a large centrally located vacuole surrounded by radially arranged smaller vacuoles (Entz 1926). **Food vacuoles** are present before phagocytosis takes place and have been documented in *Gymnodinium helveticum* (Popovský 1980), *Ceratium hirundinella* (Entz 1927). *Cystodinedria inermis*, *Stylodinium globosum* (Pfiester and Popovský 1927) (figs. 42, 54, 55, 86f.), and *Cystodinium bataviense* (Pfiester and Lynch 1980). At times they may constitute 90 % of the total cell volume. In *Gymnodinium helveticum* fo. *achroum* the nucleus is located adjacent to the food vacuole from which a lobopode protrudes. This is also true in *Ceratium hirundinella*. In *Stylodinium globosum* the food vacuole forms just before the prey's cell wall is digested enzymatically. *Gymnodinium helveticum* fo. *achroum* has been reported capable of digesting diatom frustules (fig. 86). **Contractile vacuoles** have only been reported in a few freshwater dinoflagellates. One to two contractile vacuoles commonly occur in members of the family Protaspidaceae (Skuja 1939). Some trophozoids that occur in the life histories of *Stylodinium globosum* and *Cystodinedria inermis* have as many as 8 contractile vacuoles scattered thoughout the cell (Pfiester and Popovský 1979). They disappear very abruptly with changing stages in the life cycle. One contractile vacuole has been reported in *Spiromonas perforans* and in the epicone of *Katodinium woloszynskae*.

Pusules have only been observed in dinoflagellates. They are observable at the light microscope in marine taxa and at the electron microsope level have been found in some freshwater taxa (figs. 3, 40). The pusules consist of a system of canals, vesicules, and a collecting chamber, which are connected with flagellar canals. The pusule system within a dinoflagellate is either extremely simple (10 vesicles) or complex (up to 100). In *Gymnodinium fuscum* the collecting chamber is connected to flagellar canals by a spinal tube (the pusule canal) all of which is lined with fibrous material (Dodge and Crawford 1969). In *Woloszynskia coronata* the entire collecting chamber is long and convoluted. Vesicles are present with small finger-like projections of unknown structure and function. Its entire pusule system is surrounded by thin sheets of smooth endoplasmatic reticulum (Dodge and Crawford 1971). The fluid volume within the pusules fluctuates slowly at irregular intervals. Their function is unknown. They may serve in osmoregulation since the largest pusules systems are found in marine species.

2.6. Miscellaneous organelles and structures

Hofeneder (1930) reported on the formation of **pseudopodia** in armored monads in detail as well as on algal predation by *Ceratium hirundinella*. In the latter, pseudopodia were formed in the area where the girdle and sulcus joined and used to engest *Microcystis aeruginosa* (fig. 44). The function of the apical pore found in species of *Peridinium, Peridiniopsis, Diplopsalis, Gonyaulax* and *Ceratium* is not known. However, we now know (unpublished observations) that the zoospore of *Stylodinium tarnum* (= *Peridiniopsis dinobryonis*) has the same plate formula as *Peridiniopsis dinobryonis* which has an apical pore. Already published observations (Pfiester and Popovský 1979) have shown that the stalk found in *Peridiniopsis dinobryonis* (= *Stylodinium tarnum*) had probably at one time served as a haustorium and is connected to the apical pore region (fig. 222). Thus, we assume that the apical pore which is found opened in the above named genera may indeed be a pore through which a haustorium or pseudopodium may emerge.
Trichocysts have been reported in several freshwater dinoflagellate genera such as *Gymnodinium, Woloszynskia, Peridinium, Ceratium, Spiromonas,* and *Cystodinium.* They are usually colorless and elongate. One end usually terminated in a point ellipsoidal rod. They are radially arranged in the periphery of the ectoplasm (fig. 173g). Their function is unknown but they are thought to swell creating mucilagenous strings around the cell. Various speculations on their function include predation, osmoregulation, attachment organelles, avoidance and reducing sinking rates were given (Steidinger and Cox 1980).

2.7. Nutrition

Photosynthetic dinoflagellates produce starch. However, it is not stored in the chloroplast but rather is placed outside it in granula which are layered in *Hemidinium nasutum.* Oil may also be present and rarely glycogen. The red droplets present in cells are presumed to be oil colored by carotenes. In the Dinococcales these red bodies appear to be the undigested remnants of cells phagocytized by the dinoflagellate. These red bodies are left behind in the parent cell wall when division occurs in these phagocytic dinococcalians. Rarely, they are retained in the daughter cells. (Popovský and Pfiester 1982).

All modes of nutrition have been documented in dinoflagellates: autotrophy, heterotrophy, osmotrophy, saprotrophy, predation and parasitism. Few dinoflagellates are totally autotrophic. Studies of ^{14}C uptake during the light period by *Ceratium hirundinella* and *Peridinium willei* in a reservoir have shown that it is only 1 to 10 % of that other algal taxa present. The highest primary production value for *C. hirundinella* was 1.5 $\mu g \cdot l^{-1} C$ and for *P. willei* 1.7 $\mu g \cdot l^{-1} C$ (Desortová 1976). Dinoflagellates composed 47 % of the phytoplankton in the high mountain lake, Finsterthaler See: *Gymnodinium uberrimum* was the dominant phytoplankter. *Gymnodinium lacustre* and a colored *Gymnodinium* species were also present. Most of the dinoflagellates were found below the ice covering at a depth of 5 to 10 m or during the warm growing season at a depth of 20 to 25 m where only 0.1 % of the photosynthetic radiation was available. Apparently the dinoflagellates present were living heterotrophically. The *Gymnodinium* were found containing frustules of *Synedra nana* (Pechlaner 1971). Thus, at least some of the dinoflagellates present were obtaining their energy by predation.

Predation has been observed in a number of freshwater dinoflagellates such as *Gymnodinium helveticum* fo. *achroum* (Irish 1979, Popovský 1980) (fig. 86), *Ceratium hirundinella* (Entz 1972, Hofeneder 1930) (fig. 44), *Cystodinedra inermis, Stylodinium globosum* (Pfiester and Popovský 1979) (figs. 54, 55), *Cystodinium bataviense (Dinococcus oedogonii)* (Pfiester and Lynch 1980) (fig. 53), and *Spiromonas perforans* (Brugerolle and Mignot 1979) (fig. 60).

A bloom of the marine dinoflagellate, *Ceratium tripos,* has been reported to have been presented in the subsaturated light zone below the thermocline in New York Harbour for two months, existing either on organic matter in waste waters from New Jersey and New York or on phytodetritus (Malone 1978).

Heterotrophy appears to be a significant energy source for dinoflagellates (Loeblich 1967). Dinoflagellates may function heterotrophically at certain stages in their life histories and autotrophically in others. During the heterotrophic phase of nutrition chloroplasts may not be visible and the cell may exhibit many animal-like characteristics (see chapter 2.5, 2.7, and 6).

Freshwater monads (with or without chloroplasts and thecae) from the subclass Dinophycidae phagocytize solid particles through the opening where the sulcus and girdle join. In this area the cytoplasm is mobile and extended pseudopodia have been observed in a number of taxa such as *Ceratium hirundinella* (Hofeneder 1930), *Katodinium schilleri* (Harris 1939/40), *K. vorticella* (Baumeister 1943), *Gymnodinium palustre, G. helveticum* fo. *achroum* (Popovský 1980), *Katodinium bohemicum* (Fott 1938), and members of the genus *Protaspis* (Skuja 1939). *Gymnodinium helveticum* fo. *achroum* has been observed to engest cells of *Melosira italica* and *Cyclotella comta* by means of a lobopode. As the lobopode enlarged the dinoflagellates flagella disappeared and a food vacuole became visible with a pyramidal-like dinokaryon above it pointed toward the cell apex. Reportedly the *Gymnodinium* digested the silicious diatom shell inside the food vacuole (Popovský 1980). A food vacuole and dinokaryon has been observed in *Ceratium hirundinella* in exactly the same shape and position (figs. 42, 86). The disappearance of both flagella at the same time the cell is preying by means of rhizopodes had been reported in *Gyrodinium hyalinum* (Taylor, per. communication). However, both flagella have been observed in motion in *Katodinium vorticella* during predation (Starmach 1974). The nutritional mode is unknown in other colorless monads. It has been suggested that osmotrophy of dissolved organic matter may plan an important role in their nutrition. Saprotrophy has been assumed to occur in a number of colorless taxa such as members of

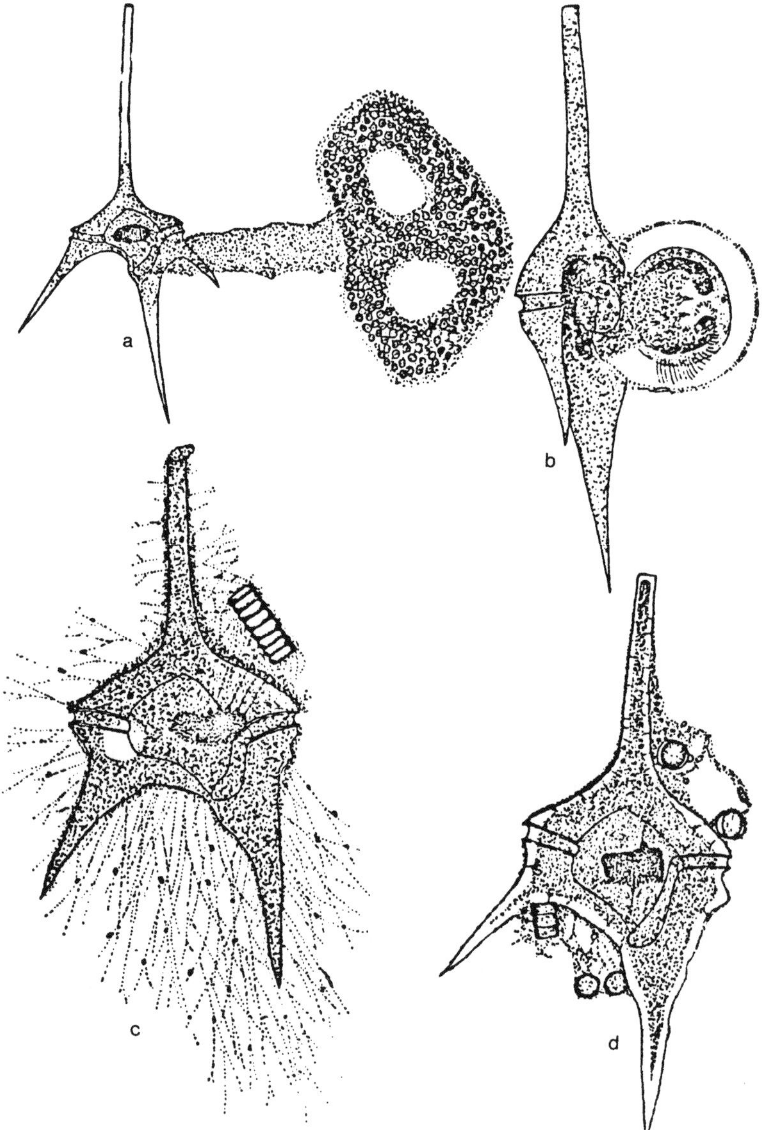

Fig. 44. *Ceratium hirundinella* holding prey; *a* a colony of *Microcystis aeruginosa* attached to a lobopodium, *b* a ciliate held by a lobopodium, *c, d* centric diatoms attached to a filopodium (after Hofeneder).

the genus *Protaspis, Katodinium fungiforme, K. woloszynskae, K. intermedium, Peridinium achromaticum* as well as by photosynthetic species such as *Gymnodinium aeruginosum, G. palustre, Katodinium montanum* and *Peridiniopsis penardii.*
Marine members of the family Blastodiniaceae are known to parasitize plants and animals. *Oodinium limneticum* has been reported to parasitize 9 different freshwater fish taxa in North America (Jakobs 1946).
Details of inorganic nutrient requirements, vitamins and trace elements are given in chapter 8.

3. Morphology

Dinophycean morphology is reported to exhibit various vegetative habits from the monadoid to filamentous forms. They have traditionally been divided among the monadoid, rhizopodial and cococcal habits. At this time we cannot confirm the capsal habit of body form because freshwater members from the genera *"Gloeodinium"* as well as the marine species *Desmocapsa gelatinosa* represent the reproductive stages of several monads. *"Gloeodinium montanum"* is the diploid stage of *Hemidinium nasutum* and *Glenodiniopsis steinii.* However, almost all stages described for *Rufusiella insignis* are very similar to those described in the life cycle of *"Gloeodinium montanum", Hemidinium nasutum,* except that *Rufusiella insignis* reproduces by means of gymnodinioid zoospores.
These palmelloid reproductive stages are not isolated reports in the Dinophyceae but are also reported for members of the genera *Gymnodinium, Woloszynskia, Peridinium* and *Peridiniopsis* as well as for *Glenodinium smreczyniense.* Thus, the importance of a mucilagenous envelope as a taxonomic character for an order is questionable. Neither does it appear important in considering body form structure in the Dinophyceae since this

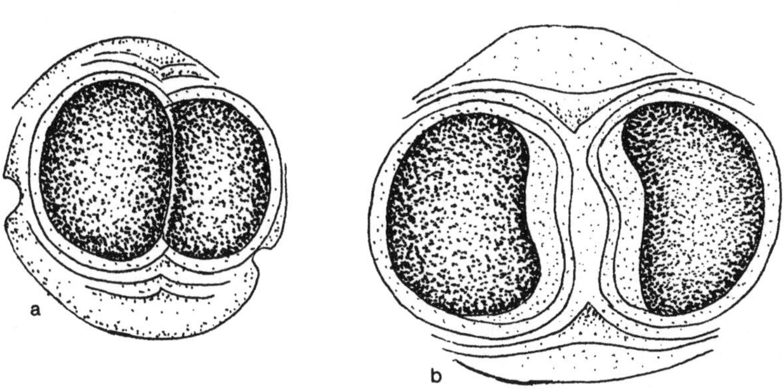

Fig. 45. *a, b* nonmotile reproductive stages of *Glenodinium smreczyniense* (nomen nudum) within a gelatinized mother cell wall *(= "Gloeodinium montanum")* (after Wołoszyńska).

may very during the life cycle. The same is true of the trichal body type. Pascher (1914) in describing the filamentous form of the marine genera *Dinothrix* and *Dinoclonium*, did not conclude that they exhibited polarity. Shussnig (1954, 1960) suggested the coccal cells formed short chains or "filaments" and named these habits parablast.
However, the classical distribution of body structure is useful, easy to follow this classification scheme this textbook.

3.1. The monadoid habit

There are freshwater monads which exist singly or very rarely in short chains such a *Gymnodinium uberrimum*, and *Ceratium hirundinella*. Monads are naked with a very fine to thick periplast (pellicle) or may be plated with cellulose (theca). The theca may be thick and contain chitin or sporopollenin. If the theca or amphiesma is thin it appears structureless at the light microscope; if it is thick it may be sculptured. In some genera such as *Hemidinium*, *Woloszynskia* and *Glenodiniopsis*, the theca is delicate with very fine structure. The theca is thick and structured in such genera as *Peridinium*, *Peridiniopsis*, *Sphaerodinium*, *Diplopsalis*, *Dinosphaera*, *Gonyaulax*, *Thompsodinium*, and *Ceratium*. Naked monads may be metabolic when engesting solid particles changing their shape and cell dimension. Many are capable of forming various types of pseudopodia such as rhizopodia, lobopodia, and a tentacle-like haustorium, which serve as means of preying on other algae and protists. There are two flagella which differ from each other in morphology and function. Cell shape as viewed from the ventral side is either spherical, heart-shaped, oval or double-conical. From the lateral or apical view the dinoflagellate body is flattened. *Desmomastix globosa* has apically positioned flagellae and is spherical to heart-shaped. Members of the Protaspidaceae are elongate oval cells with a flattened ventral side. Members of the Protaspidaceae have only a longitudinal furrow which is located on the ventral side and in which the longitudinal flagellum is located. The second flagellum is inserted subapically and, stretched forward in the direction of the cell axis. *Spiromonas perforans* morphologically is very similar to that of the colorless flagellate *Bodo*. The body is pear-shaped with a sharp rostrum, the furrows are absent. In differs from *Bodo* in having two, different flagella. The base of one is placed in a longitudinal axis and the second fagellum in a transverse axis. Members of the order Peridiniales have a dorsiventral cell orientation with their cells divided into an epicone and hypocone by a transverse equatorial furrow called the cingulum or girdle. The sulcus or longitudinal furrow is located on the ventral side of the cell. The sulcus originates in the junction of both furrows and sometimes reaches the antapex. It may extend into the epicone slightly. The cingulum is girdle shaped of slightly spiral. It usually spirals to the left, rarely to the right.

In *Gymnodinium*, *Woloszynskia*, *Glenodiniopsis*, *Peridinium* and *Peridiniopsis* the cingulum divides the cell into approximately equal parts. It does not spiral more than 1/5 of the cell lenght. In the Gymnodiniales the taxonomic and diacritical feature is the position of the cingulum. Im *Amphidinium* the epicone is small, i.e., a maximum of 1/3 of the cell volume, and the cingulum is located near the apex. In the genus *Katodinium* the epicone is large, i.e., a minimum of 2/3 of the cell volume, and the cingulum is located near the antapex. In the genus *Gyrodinium* the cingulum is more that 1/5 the lenght of the cell. In a few rare instances the cingulum does not encircle the cell such as in the genera *Hemidinium* and *Bernardinium* (fig. 46).

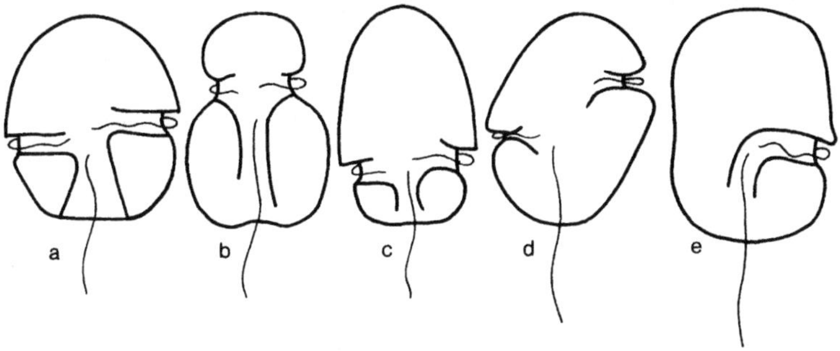

Fig. 46. The position of the cingulum is the main diacritical feature between genera: *a Gymnodinium (Woloszynskia), b Amphidinium (Amphidiniopsis), c Katodinium, d Gyrodinium, e Hemidinium (Bernardinium)* (diagrammatic).

In the Peridiniales which have a distinct theca the cell is more or less rounded, double-conical to elongated double-conical in shape. The cells appear slightly to strongly flattened from the apical or lateral view. In the genus *Ceratium* the epicone projects into the apical horn which has an apical pore. The hypocone possess 2 to 3 horns, one of which lies in the longitudinal cell axis and is called the antapical horn; the other two are termed the postequatorial right and left horns.

Cell walls of some dinoflagellates exhibit single dots *(Woloszynskia)* or rows of dots *(Hemidinium)*. Plates forming the cell wall are sculptured as seen with the light microscope. Sculpturing may consist of irregularly arranged dots (papillae), or rows of dots, or the walls be covered by aerolae. Plate margins on the hypotheca may be ornamented with one to several teeth-like projections or spines. These bold or delicate teeth-like structures may cover the entire plate surface. Margins of the cingulum and sulcus may have hyaline wings. These wing-like structures may also be present on some of the sutures found on the epi- and hypothecae as in *Peridinium cinctum* and *P. bipes*. The rhomboidal plate designated 1′ (or "r" plate), has 5 angles except in *P. striolatum* where the 1′ plate is 6 angled, and may be elongated to the apex and terminated by an apical pore. The genus *Peridinium* is divided into two subgenera on the basis of the presence or absence of the apical pore. Those *Peridinium* species which lack the apical pore are classified in the subgenus *Cleistoperidinium*. Those possessing the pore are placed in the subgenus *Poroperidinium*. Other armored genera such as *Ceratium* possess an apical pore covered by a hyaline coronnet.

Other plates present in armored dinoflagellates are polygonal and slightly to heavily irregular. Plates on the epivalve may be symmetrically arranged. The number, shape and arrangement of the plates forming the "armor" are the dominant taxonomic and diacritical features for armored genera.

Monads reproduce by division, cyst formation, amoeboid forms, and sexual stages.

All monads are not known to produce cysts. Presumably, non-motile dinoflagellates enveloped by a thick cell wall or by thick multilayered mucilage are resting cysts (dormant cysts) capable of surviving unfavorable environmental conditions. Naked monads lose their flagella, round up and develop a thick, sometimes multilayered wall during cysts formation. Cysts are often enveloped by a thick "gall-like" covering and are usually spherical somewhat oval or egg-shaped. The cyst wall may be covered with short spines or bristles as in *Gymnodinium palustre*. Cysts in armored taxa are formed within the theca. In some taxa such as *Hemidinium nasutum ("Gloeodinium montanum")*, *Peridinium cinctum*, and *Spiromonas perforans* (Brugerolle and Mignot 1979) the non-motile, sexual and vegetative reproductive stages are described as thick walled and temporary cysts, respectively.

3.2. The rhizopodial habit

Until recently the only dinoflagellates known to have amoeboid habits were the marine species *Dinamoebidium varians* and the freshwater taxon *D. coloradense*. We now know that many, possibly most, dinoflagellates exhibit amoeboid stages or tendencies such as the formation of pseudopodia during phagocytosis at some stage in their life histories. Several members of the Blastodiniales are known to have amoeboid habits during sexual reproduction wherein short lobopodia are formed. Extremely large amoeboid forms have occasionaly been observed in cultures of *Peridinium cinctum* [UTEX 1336] (Pfiester, unpublished observations). It is difficult to distinguish between motile and non-motile dinoflagellates as some taxa once thougth to be non-motile are now know to change their shape and move very slowly. These movements may take hours. Such stages are now known in the life histories of the coccal dinoflagellates *Stylodinium globosum*, *Cystodinedria inermis*, and *Cytodinium bataviense (Dinococcus oedogonii)*. In some of these taxa the rhizopodial habits are a means of vegetative reproduction and are indistinguishable from described protozoan taxa such as *Vampyrella lateritia*, *Actinophrys sol*, *Amoeba verrucosa*, and other such as *Vampyrella lateritia*, *Actinophrys sol*, *Amoeba verrucosa*, and other suctorians. The rhizopodial habit of *Chlamydomyxa labyrinthuloides* (Chlorochytriaceae) (Popovský 1979) may be a stage in the life history of a dinococcalian.
Dinamoebidium reproduces vegetatively by dividing and forming either two gymnodinoid zoospores or two amoebae. The previously mentioned rhizopodial habits of dinococcalian taxa and *Chlamydomyxa* appear to function as independent trophozoids able to reproduce in kind during some stages of their life cycles. Cysts are described only for *Dinamoebidium coloradense*. Detailed descriptions of the thickness of the cyst membrane and thickness of the cell wall are lacking. See chapter 3.3. for detailed descriptions of *Cystodinedria inermis'* and *Stylodinium globosum's* cysts.

3.3. The coccoid habit

Vegetative coccoid habits lack flagella and furrows. Cell morphology varies. Described classically dinococcalians have global, ovoid or spindle-shaped cells with projected ends or a tetrad shape with horns. Two larger or smaller horns may be present on the cell surface. Rarely, delicate bristles (fine pseudopodia) can be observed on the horns in genera such as *Cystodinium*. Stalk morphology has been used as an important taxonomic character in classifying sessile forms. Little interest was shown in the substrate to which these taxa were attached (see chapter 2.4.).
Sexual reproduction has been reported to occur by production of 2–16 gymnodinioid zoospores or by autospores produced from a vegetative cell. Zoospores may become

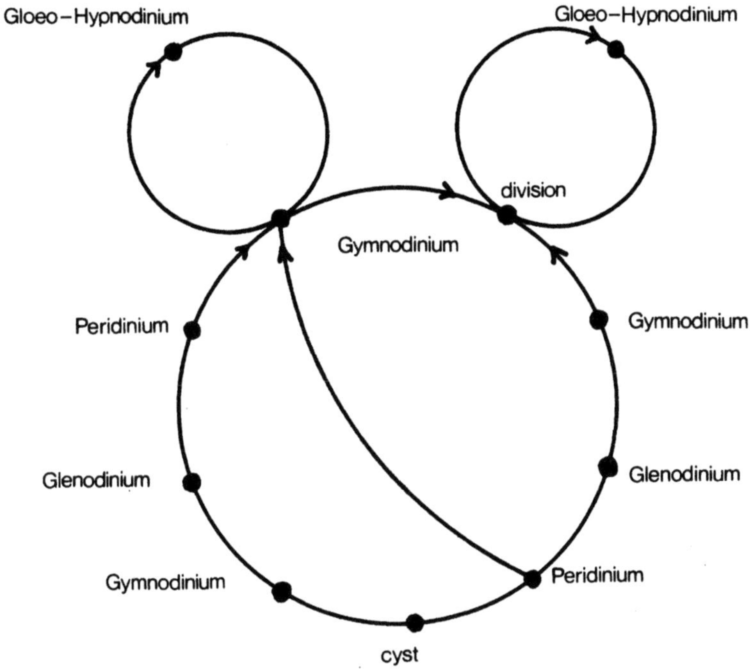

Fig. 47. A diagrammatical life history of *Peridiniopsis borgei* with armored and unarmored, motile and nonmotile stages (after Entz).

attached to a substrate after losing their flagella by forming a fine, thick layer of mucilage.

Dormant cysts have been reported in the genus *Stylodinium*. Cyst shape is similar to that of vegetative cells. Their cell membranes appear thickened and may bear short projections. A number of dinococcalian cysts appear similar to those of other dinophyceans. The resemblance, however, is only superficial. Differences exist between taxa in the thicknesses of their cyst walls and protoplast structure. It is highly probable that some described genera such as *Hypnodinium* and *Phytodinium* reported to reproduce by autospores are merely stages in the life histories of other dinophyceans (fig. 47). *Hypnodinium* has recently been shown to be a stage in the life history of *Cystodinium bataviense (Dinococcus oedogonii)*. We have partially described the life histories of *Cystodinedria inermis, Stylodinium globosum* and *Cystodinium bataviense* in earlier chapters. These life cycles could be described as the life histories of protozoans. Since it is highly probable that their complete life histories have not yet been described the stage with which we beginning our description of their life cycles is irrelevant. The vegetative cells of these taxa are motile. *Cystodinedra inermis* moves at a rate of 10 $\mu m \cdot sec^{-1}$. Stalks or disc described for these genera are in fact tentacles or haustoria used for preying, phagocytosis, and movement. Thus, the taxonomic characters formerly used such as cell size and shape are meaningless. Further the "coccal" stages are reported to develop into the free-swimming stage and vice versa. That is, *Cystodinium* produces zoospores which may form the *Dinococcus* form which develops into the *Cystodinium* form or the zoospores may form the *Cystodinium* stage immediately. Also *Cystodinium* has been reported in the life cycle of *Cystodinedria*. Their complicated life cycles (see Chapter 6) include amoeboid stages which exhibit characteristics common to helizoans, rhizopods and suctorians. The gymnodinioid zoospores described as the vegetative reproductive forms are probably gametes and the dormant cysts, zygotes. All these stages with the exception of the dormant cysts are motile along a substrate or in water. Motile stages except zoospores and gametes prey upon green filamentous algae and *Oedogonium* zoospores. Amoeboid stages, trophozoids, *(Vampyrella lateritia, Amoeba verrucosa,* and *Actinophrys sol)* have rhizopodia, lobopodia and silicious spines or typical axopodia. Small trophozoids also have rhizopodia which rapidly disappear as the cell transforms itself into a small *Cystodinedria*. The sporocytes are round and their cell wall bears a few to many spines. Trophozoids, which develop into small *Cystodinedria*, develop from the sporocytes. In some rhizopodial stages *(Vampyrella lateritia)* rhizopodia have been reported to function as flagella. In others *(Amoeba verrucosa)* it has not been possible to determine the manner by which they swim or move.

The fate of described zoospores is unknown except for those of *Cystodinedria inermis* and *Cystodinium bataviense*. Formation of motile cells though to be gametes has been reported for *Stylodinium globosum*. Two morphologically distinct "gametes" are produced per cell. One member of the pair is gymnodinioid in shape, the second gyrodinioid. Gametic fusion has not been observed although pairs of gametes have been reported dancing around each other. Vegetative reproduction in *Cystodinedria* and *Stylodinium* occurs by formation of 2–4 trophozoids similar to *Vampyrella lateritia* and *Actinophrys sol* which escape from the parent cell wall. The formation of many, small trophozoids resembling *Vampyrella lateritia* has been reported from a state in *Cystodinedria inermis* life cycle referred to as *Amoeba verrucosa*. Numerous sporocytes are formed in the trophozoid-heliozoid stage. Doflein and Reichenow (1949–52) have noted the ability of Vampyrellidae to change from a helizoan-like morphology to that of an amoeboid form. Each sporocyte is capable of developing into a trophozoid such as *Vampyrella lateritia* which rapidly changes into a small *Cystodinedria*.

Fusion of motile thecate gametes following ecdysis has been reported to occur in *Cystodinium bataviense (Dinococcus oedogonii)*. The development of the zygote has not been followed but large, round, thick-walled cells which may be the zygotic stage have been observed in cultures. These resemble *Hypnodinium*. *Cystodinedria inermis* and *Stylodinium globosum* are known to produce dormant cysts. That of *Cystodinedria inermis* is covered by a thick membrane with numerous, short spines while the membrane of *Stylodinium globosum's* cyst is thick and smooth. Both cysts have a shape similar to that of their respective vegetative cells.

The following previously described alga taxa have been reported to be but a stage in the life history of *Cystodinedria inermis*: *C. maxima, C. setosa,* several *C.* spp. and *Cystodinium phaseolus*. The are distinguishable only by cell shape and dimensions, or by the presence of spines. *Stylodinium longipes, St. truncatum, St. phaseolus* and *St.* sp. have been identified as stages in the life history of *Stylodinium globosum*. *Dinococcus oedogonii* and *Hypnodinium* sp. have been observed as stages in the life history of *Cystodinium bataviense* (Popovský and Pfiester 1982, Pfiester and Lynch 1980).

4. Reproduction

4.1. Asexual reproduction

There are 5 known types of asexual reproduction in freshwater taxa: zoospores, autospores, trophozoids, and sporocysts. (1) In *Peridinium* and *Peridiniopsis* the protoplasm divides while the cell is in a non-motile stage (temporary cyst). The flagella disappear and the dinokaryon divides together with cytoplasma into two equal parts. The plane of the division is vertical or oblique. After a short period the dorsal plates on the parent theca separate or the epi- and hypotheca separate freeing the gymnodinioid daughter cells which are gymnodinioid-like. Following growth, these cells lose their motility and develop into the "Gloeo-Hypnodinium" stage (fig. 47). (2) *Ceratium cornutum* divides in the motile stage. After division of the dinokaryon and cytoplasm the theca divides obliquely. Each daughter cell forms a complete new theca. Then that half received from the parent cell is sloughed off (Durr 1979). (3) In several *Peridinium* species it is well known that the cyst, covered with mucilage, divides in the spring producing up to four daughter cells which eventually escape the parent theca after it ruptures and water is imbibed. The cyst's protoplasm may not divide and develope directly into a gymnodinioid form (Kiselev 1954). A somewhat similar complex method of reproduction in mucilage has been detected in *Hemidinium nasutum* and *Glenodiniopsis steinii* (Popovský 1971 a,b). (4) The multiplication by means of trophozoids and sporocytes has been observed in *Cystodinedria inermis, Stylodinium globosum* and *Spiromonas perforans* (Alexeieff 1929, Pfiester and Popovský 1979, Popovský and Pfiester 1982). For details see chapter 5. Bursa (1970) described reproduction in *Dinamoebidium coloradense* by means of zoospores and trophozoids following longitudinal division of the cell. (5) The reproduction means of 2 autospores is common in coccal dinoflagellates (Starmach 1974).

4.2. Sexual reproduction

Dinoflagellate sexual reproduction has been a controversial subject for many years. Joseph's (1879) early mention of fusion of the motile cells in *Peridinium stygum* is the first reference found in the literature to sexual reproduction in this class. Pouchet (1883) was the first to describe gametic union and nuclear rotation now referred to as "cyclose nucleaire" shown (v. Stosch 1972 and paper presented at Hexrose Conference, 1981) to be associated with meiosis. Zederbauer's (1904) description of sexual reproduction in *Ceratium hirundinella* was later shown to be incorrect (v. Stosch 1964, 1965). Diwald's (1937) report of sexual reproduction in *Peridiniopsis lubieniensiforme* is similar to descriptions of dinoflagellate sexual reproduction in several more recent paper (v. Stosch 1973, Pfiester 1975) but was not generally accepted at the time. V. Stosch (1964, 1965) has described sexual reproduction of *Ceratium horridum* and *Ceratium cornutum* in detail. In 1967, Cao Vien first described sexual fusion in *Amphidium carteri* and then, in 1968 zygotic division. Because of these later reports and subsequent findings by researchers, which will be presented in detail in this chapter, sexual reproduction in dinoflagellates is an accepted fact. The ability to culture dinoflagellates, to induce the sexual cycle and follow it closely is leading to an explosion of knowledge in this area. Presently the sexual life histories of 29 of the approximately 1 500 dinoflagellate species known have been described. Many students of dinoflagellates believe that many, possibly all, reproduce sexually.

The detailed sexual life cycles of only a few dinoflagellates will be presented here. A pattern of sexual reproduction for the division as a whole is beginning to emerge as more reports of sexual reproduction within the group are being published.

The sexual life history of *Peridinium cinctum* is the first such life history to be studied at both the light, scanning and transmission electron levels (Pfiester 1975, Pfiester and Skvarla 1980, Spector et al. 1981) and appears to be typical of dinoflagellates life histories studied.

Sexual reproduction in *Peridinium cinctum* is induced when exponentially growing cells are inoculated into nitrogen deficient medium (Pfiester 1975). This appears to be a common induction method in laboratory cultures of dinoflagellates (v. Stosch 1973, Turpin et al. 1978, Anderson 1980, Spector et al. 1981). Under nitrogen deficient conditions smaller vegetative cells divide producing two daugther cells which, if kept in a nitrogen deficient environenment, will act as gametes, but if immediately placed in a nitrogen rich environment will produce a vegetative population which can later be induced to reproduce sexually. Thus vegetative populations are haploid and homothallic. At the light microscopical level gametes appear to be naked, i.e., without a cell wall. At the transmission electron microscopical level both naked and thecate fusing gametes have been observed (Spector et al. 1981). There are several possible explanations for this phenomenon. One, gametes may fuse naked or thecate. The unanswered question then become what happens to the thecae as the two fuse. V. Stosch (personal communication) has reported on fusion of thecate gametes in *Ceratium* in which the thecae are enzymatically dissolved. Thecate gametes were also observed in *Peridinium gatunense* and were thougth to dissolve as fusion progressed (Pfiester 1977). However *Peridinium cinctum* thecate gametes that were observed fusing at the light microscopical level were never able to complete fusion (Pfiester 1975). The theca develops in young cells trough time. Thus, it may be that gametes must begin fusing early before the theca becomes well developed or fusion will not be completed. *Peridinium cinctum* gametes have few chloroplasts (Spector et al. 1981) or are said to be depauperate (Pfiester 1975). This is true of a number of dinoflagellate gametes (v. Stosch 1972, 1973). They have numerous membrane bound storage bodies, many starch grains, and chromosomes which appear slightly unwound as viewed with transmission electron microscopy (Spector et al. 1981). In nitrogen deficient medium during the dark phase of the light-dark cycle *Peridinium cinctum* gametes approach each other and appear to "dance". There must be some as yet undescribed hormonal attraction.

At the light microscopical level fusion appears to be lateral but when viewed with transmission electron microscopy it appears to occur in an area of the sulcus devoid of reticulate thecal plates at or adjancent to the flagellar pores. Spector et al. (1981) reported the formation of a fertilization tube which widens along the sulcus. Fusion takes about 45 minutes to complete. Fusing cells may remain motile throughout the fusion process. The resulting planozygote at first resembles the typical vegetative cell with the exception of having two trailing flagellae, a condition common to all dinoflagellate planozygotes studiend thus far. Spector et al. (1981) have found that karyogamy occurs within the fertilization tube before plasmogamy is complete. This is contrary to Pfiester's first report (1975) of a binucleate "planozygote" presumed to represent the dikaryotic pre-fusion condition. However, it now appears that the first meiotic division occurs immediately, following syngamy long before the second such division. V. Stosch (1972, 1973) has also reported on dinoflagellate meiotic divisions separeted in time. The *P. cinctum* planozygote remains motile for two weeks, approximately doubling its size 38 × 35 µm to

72 × 68–70 µm. The theca which it develops shortly after fusion has the usual *Peridinium cinctum* plate tabulation. It accomodates protoplast enlargement by a tremendous widening and striation of the intercalary bands which join the thecal plates. At the ligth microscopical level the planozygote appears to be lumpy or warty as it grows. At the scanning electron microscopy level this can be seen to be caused by an unequal widening of these bands (Pfiester and Skvarla 1980). Most dinoflagellates studied thus far enlarge greatly during the planozygotic stage. Since this stage is of long duration it is not unusual that the size range and nature of intercalary bands given in some taxonomic descriptions are indeed descriptions of the planozygote rather than the vegetative stage as the original authors supposed. The planozygote develops a secondary wall internal to the thecal plates that is closely appressed to the protoplast. This secondary wall contains many finger-like projections which also allow for the increase in size which takes place over a two week period. At the end of two weeks when the planozygote loses its motility, its protoplast contracts, one or more large red "accumulation bodies" develop and the zygote develops into a stage referred to as the hypnozygote with an extremely thick outer wall. The large red accumulation bodies are thougth to be lipid in nature. Under culture conditions at 20°C the hypnozygote will remain dormant for about 3 months. Some have remained viable for five months at 4°C in light and in dark. If dinoflagellate hypnozygotes are treated to cold for several weeks, dormancy can be broken (v. Stosch 1972). This long dormant period is true of all fresh-water dinoflagellates studied thus far with the exception of *Peridinium gatunense* (Pfiester 1977). Few marine species require such a dormant state. However many dormant cysts known from the fossil record may prove to be hypnozygotes (Wall and Dale 1968). The hypnozygote can be shown to have a 3-layered cell wall: (1) a thick outer wall referred to as the exospore, (2) a thin middle wall referred to as the mesospore, and (3) a thick inner wall referred to as the endospore. The latter must be laid down sometime during formation of the hypnozygote. Hypnozygote walls are resistant to acids and enzymes and are reported to contain chitin (Pfiester 1975) and/or sporopollenin (Spector et al. 1981) As such it has been difficult to penetrate the hypnozygote with nuclear stains to determine meiotic figures. Thus meiosis has not been fully described in dinoflagellates until recently (v. Stosch, personal communication, Coats et al. 1984, Pfiester et al. 1984).

The first meiotic in *Peridinium cinctum* must occur shortly after syngamy as nuclear fusion occurs prior to cytoplasmatic fusion and the resulting planozygote has two nuclei (Spector et al. 1981). The second meiotic division is long separated from the first in *Peridinium cinctum* (Pfiester 1975). V. Stosch (1972) has also observed that in some dinoflagellate species the second meiotic division is long separated from the first. One is able to stain the nuclei in hypnozygotes shortly before germination observing the presence of 4 nuclei. As only one post-zygotic cell emerges from the hypnozygote it is presumed that at least two of the four nuclei abort. Post-zygotic cells retain the large red accumulation body until after their first division which occurs within 24 hours of their emergence from the hypnozygote. V. Stosch (1972) has shown that the nuclear phenomenon referred to as "cyclose nucleaire" first reported by Pouchet (1883, 1885) and later by Biecheler (1952) is associated with meiosis. Pouchet and Biecheler both observed that the nuclei of some large dinoflagellage cells with thick walls would become very large and rotate or swirl. Biecheler observed that the cells would divide sometime after the nuclei had come to rest. Biecheler suspected that this nuclear enlargement and swirling might be related to meiosis but was unable to prove it since only one division followed. V. Stosch has observed "cyclose nucleaire" in those dinoflagellates whose sexual life

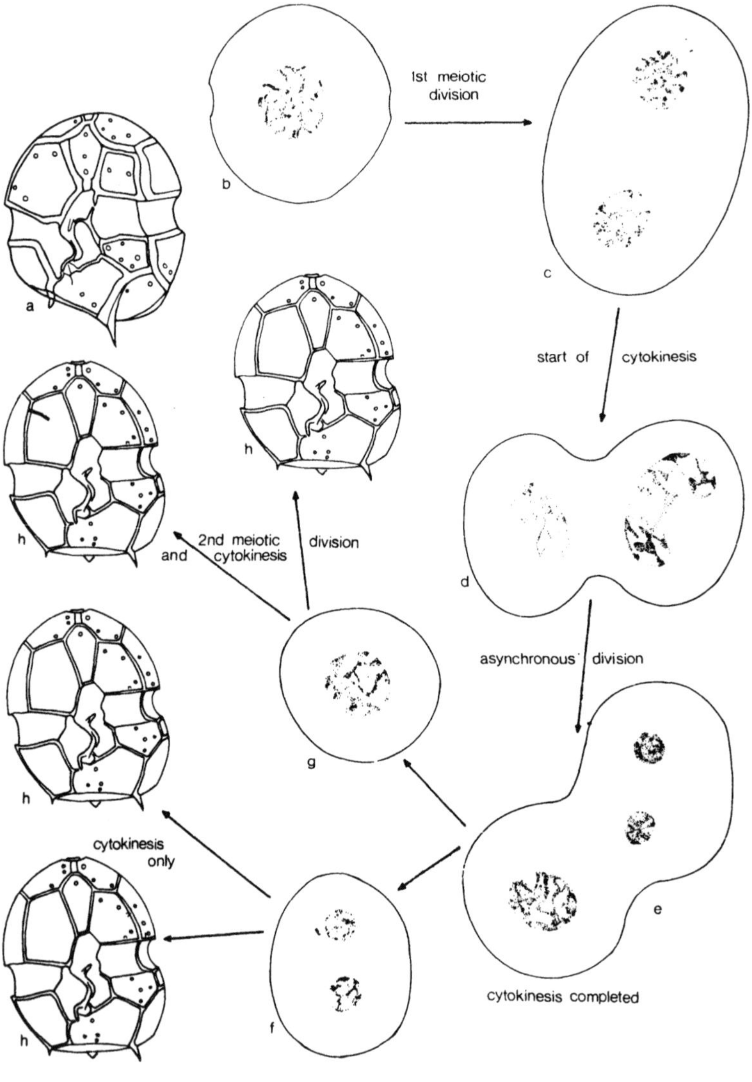

Fig. 48. Diagram of meiosis in *Peridinium umbonatum*, *a* plated zygote, *b* zygote following ecdysis, *c* first binucleate stage, *d* peanut-shaped cell with two C-shaped nuclei, *e* trinucleate stage, *f* second binucleate stage resulting from cytokinesis of a trinucleate cell, *g* cell with an arrested meiosis product, *h* four motile haploid cells. Note the processes involved to produce these cells (after Pfiester et al.).

histories he has elucidated and correlated it to the first meiotic division showing that meiosis in those taxa is a two-step meiosis (v. Stosch 1972, 1973).
V. Stosch presented details of meiosis in *Ceratium cornutum* at the Hexrose Conference in Tübingen (1982) in which by analysis of ordered tetrads he showed that meiosis is indeed a two-step process. Cyclose nucleaire has not been observed in *Peridinium cinctum*. However, the two meiotic divisions, though long separated in time, are thougth to occur very rapidly in this species and it is possible that this phenomenon may occur but not have been observed. Pfiester et al. (1984) reported sexual reproduction in a homothallic strain of *Peridinium umbonatum (P. inconspicuum)* in which protoplasmic fusion begins between two thecate gametes; but zygote formation is completed in a space outside the fusing pair. This diploid cell can form a plated theca which is shed as the cell enlarges. The resulting spherical zygote then forms a new non-plated theca. The process of ecdysis and the formation of a new non-plated theca is repeated several times. During this process the zygote gradually elongates and by cytoplasmic infurrowing becomes peanut-shaped. Eventually two cells are formed. The first and second meiotic divisions are greatly separated in time. The first meiotic division occurs in the peanut-shaped zygote before it completes cytokinesis. The second meiotic division may not be synchronous, occasionally resulting in a trinucleate stage. Eventually four flagellated, haploid products are produced.
The sexual life histories of most of the dinoflagellates studied thus far follow the pattern of *Peridinium cinctum* with variations occurring in the presence or absence of thecate gametes, in the time of nuclear fusion, in the number of accumulation bodies in the zygote, in the period of zygote dormancy and in the number of products released from the hypnozygote. There are however, some notable exceptions. Sexual reproduction in *Ceratium cornutum* is anisogamous with the smaller male gamete being engulfed or swallowed by the female (v. Stosch 1964 and personal communication). Plates in the female sulcal area appear to be "hinged" in such a way as to open as the male gamete approaches and closes after the male gamete is engulfed.
The sexual life history of freshwater dinoflagellates studied thus far with few exceptions (Pfiester 1977, Pfiester et al. 1984) exhibit many stages in common: (1) fusion of gametes at the light microscopical level appears to be lateral, (2) planozygotes have two trailing flagellae and enlarge during a long motile period, and (3) non-motile 3 wall-layered hypnozygotes with one or more red „accumulation bodies" form which have long dormant periods. Many of the marine dinoflagellates exhibit the same stages in their sexual life histories with the exception that, in most, the hypnozygotic stage is short-lived. Two marine taxa that differ from this pattern are *Crypthecodinium cohnii* and *Noctiluca miliaris*. The sexual life history of *Crypthecodinium cohnii* was discovered and reported on independently by both Beam and Himes (1974) and Tuttle and Loeblich (1974, 1975). *Crypthecodinium cohnii* is homothallic (Beam et al. 1977, Beam and Himes 1977, 1980, Himes and Beam 1978b). Both Beam and Himes (1980) and Tuttle and Loeblich (1974) have been able to develop mutants which have made tetrad analysis possible. Fusion is lateral, zygotes encyst and undergo meiosis within 24–36 hours. The zygotic cyst wall softens following meiosis and 2–8 progency are produced. On the basis of tetrad analysis Beam and Himes (1980) have proposed a one-step meiosis in contrast to the two-step meiosis proposed by v. Stosch at the Hexrose Conference in 1981.
Noctiluca miliaris is at present the only dinoflagellate known in which meiosis is gametic rather than zygotic. Thus, it is diploid in the vegetative stage. Hofker (1930) presented evidence that his strain was heterothallic while Zingmark's strains were homothallic.

Reports of dinoflagellate sexual reproduction.

Species	Gamy	Citation	Habitat (marine or freshwater)	Thally
×*Amphidinium carteri*	ISO	Cao Vien (1967, 1968)	m	homo
×*Ceratium cornutum*	ANISO	v. Stosch (1965)	f	hetero
Ceratium horridum	ANISO	v. Stosch (1964)	m	homo
Cystodinium bataviense		Pfiester and Lynch (1980)	f	
×*Crypthecodinium cohnii*	ISO	Beam and Himes (1974)	m	homo
		Tuttle and Loeblich (1974)		
Fragilidium sp.		Owen and Norris (1985)	m	
×*Gonyaulax monilata*	ISO	Walker et Steidinger (1979)	m	
×*Gonyaulax tamarensis*	ANISO	Turpin et al. (1978)	m	
		Anderson and Wall (1978)		
×*Gymnodinium paradoxum*	ISO	v. Stosch (1972)	f	hetero
×*Gyrodinium uncatenum*	ANISO	Coats et al. (1984)	m	
	ISO	Anderson et al. (1984)		
○*Helgolandinium subglobosum*	ANISO	v. Stosch (1972)	m	homo
×*Katodinium fungiforme*	ISO	Spero and Morée (1981)	m	
○*Noctiluca miliaris*	ISO	Zingmark (1970)	m	homo
		Hofker (1930)		(hetero)
Oxyrrhis marina	ISO	v. Stosch (1972)	m	homo
×*Peridiniopsis lubieniensiforme*	ISO	Diwald (1937)	f	hetero
×*Peridinium cinctum* (UTEX 1336)	ISO	Pfiester (1975)	f	homo
Peridinium cunningtonii	ISO	Sako et al. (1984)	f	homo
○*Peridinium gatunense*	ISO	Pfiester (1977)	f	homo
×*Peridinium limbatum*	ISO	Pfiester and Skvarla (1980)	f	homo

×*Peridinium umbonatum (P. inconspicuum)*	ISO	Pfiester and Skvarla (1984)	f	homo
×*Peridinium volzii*	ISO	Pfiester and Skvarla (1980)	f	hetero
×*Peridinium willei*	ISO	Pfiester (1976)	f	homo
×*Polykrikos kofoidii*		Morey-Gaines and Ruse (1980)	m	
×*Protogonyaulax catenella*		Yoshimatsu (1981)	m	hetero
Ptychodiscus brevis		Walker (1982)	m	hetero
Scrippsiella trochoidea		Watanabe (1982)	m	
×*Woloszynskia apiculata*	ISO	v. Stosch (1973)		hetero
×*Woloszynskia pseudopalustris*		v. Stosch (1973)	f	
×*(Gymnodinium excavatum)*	ANISO	v. Stosch (1972)	m	homo

× – dormant cysts
○ – temporary cysts

Zingmark (1970) observed vegetative cells which underwent a division which he interpreted as meiosis. This was followed by several to many synchronous mitotic divisions which resulted in the production of up to 1024 mature uniflagellated isogametes attached on the surface of the gametocyte mother cell body. These isogametes appeared to be formed and released from the gametocyte by some type of a budding process. He later observed these gametes, which had morphological affinities with dinoflagellates, swimming in the medium fusing with their flattened sides at 90° angles. As the gametes fused their flagella shortened and were eventually absorbed into the zygote. Most zygotes formed under culture conditions died, but Zingmark was able to observe partial growth and differentiation to the mature vegetative size in one. That zygote enlarged from a 25 μm diameter to 200 μm diameter and developed a coiled tentacle before dying. Further Zingmark (1970) showed that the *Noctiluca* vegetative cell nucleus is eukaryotic rather than mesokaryotic as in most dinoflagellates. Thus on the basis of its diploid vegetative state and eukaryotic nucleus he has questioned its dinoflagellate affinities. Jollos (1910) assigned it to the Dinophyceae on the basis of morphologial affinities between its small, uniflagellated swimming stage (now known to be the gametic stage) and the gymnodiniid-like structure of other dinoflagellates. The question still remains, "Is this resemblance merely a superficial one?"

5. Resting stages

Resting stages have been documented in only a few freshwater dinoflagellates. They probably play a role on sexual cycles, survival during unfavorable conditions, propagation, and dispersion (Dale 1982). Three basic types of these resting stages have been described: (1) a palmelloid or mucilage stage in which the monad loses its flagella and becomes enveloped in multilayered mucilage which is commonly thick and brown in color. Some palmelloid stages serve a reproductive function in that the cell divides, sometimes unevenly, and the 2 or 3 daughter cells produced develop their own multilayered mucilage layer. Later, in response to some presently unknown stimulus or trigger these cells will produce zoospores or autospores. *Glenodinium smreczyniense* (fig. 45) and *Peridinium cinctum* are examples of freshwater dinoflagellates that may form this palmelloid type of resting stage. *Hemidinium nasutum* (figs. 144–148) and *Glenodiniopsis steinii* (fig. 151) also form a palmelloid stage that apparently has a reproductive function by means of autospores, zoospores or possibly akinetes. The exact mechanism by which this non-motile stage develops and its specific role is at present unknown (Popovský 1971). (2) zoospores and vegetative cells of *Cystodinedria inermis* also become enveloped by a thick, fine homogenous mucilage (Popovský 1967) (fig. 55) which probably serves in enabling the zoospore to attach to the substrate. Vegetative cells form a mucilage layer before developing into the rhizopodial-trophozoid stage (Pfiester and Popovský 1978). Monads of *Gymnodinium palustre* and *Woloszynskia limnetica* (fig. 141) become enveloped in thick, fine mucilage. (3) another type of non-motile cell found in dinoflagellate life cycles is a dormant resting stage referred to as a thick or thin walled cyst. Both armored and unarmored forms can form these cysts characterized by a thick wall. In germinating some characteristically dislodge specific plates thus forming a pore referred to as an archeopyle. In other taxa the cyst wall breaks into 2–3 parts as the protoplast emerges (fig. 49). The cyst surface may be somewhath smooth or rugged of ornamented by spines or small teeth; furrows are never presented. The protoplast is usually opaque at the light miroscopical level. Many storage products are present. Viable *Ceratium hirundinella* cysts have been found in the mud 5 or more years after their formation. (Huber and Nipkow 1922, 1923). *Peridinium* thick walled cysts are undoubtedly hypnozygotes which develop from planozygotes. Since the sexual life histories of only 29 species have been documented to date far more research needs to be conducted in order to document the number of described cysts which are indeed zygotes. Apparently a period of low temperatures (4 °C) is required for germination for some cysts (zygotes). Once water temperatures increase the cytoplasm appears to circulate, the nucleus enlarges and swirls (cyclose nucleaire). Meiosis is thougth to occur at this time. One to four gymnodinioid zoospores then germinate from the cyst. Daughter cell products from these cells resemble a typical vegetative cell of their respective taxa (v. Stosch 1972, Dale 1982, Pfiester 1985a).

Thin-walled cysts are formed under conditions unfavorable for the dinoflagellate vegetative population. Under those conditions the protoplasts rounds up and forms a smooth wall around itself. In armored forms the epi -and hypothecae separates releasing the thin-walled cysts. These cysts can survive 4 °C temperature for 4 or more months and germinate within 24 hours when brougth to room temperature (Biechler 1952, Pfiester 1975).

Some dinoflagellates have temporary resting stages. Some amoeboid forms lose their flagella. Digestion of engested food takes place during this stage and the cell may enlarge and change its shape as *Gymnodinium helveticum* does (fig. 86). This stage has incorrectly been referred to as a "digestive cysts" or "food cysts".

The study of living dinoflagellate cysts especially that of marine forms is emerging as an important field of research for both micropaleontologists and phytoplankton biologists. For the paleontologist, living cysts offer the exciting possibility for interpretation of the fossil record based on a better understanding of the organism involved (Dale 1982).

6. Life history and cell growth

The life history of most freshwater, amoeboid and free-living dinoflagellates can be divided into a vegetative and reproductive phase. During the vegetative phase the young cell grows and differentiates into the mature adult cell which is then capable of dividing and producing vegetative cells or gametes. At the end of the growing season a resting (dormant) stage referred to as a cyst is usually produced. The cyst may, in time, when conditions become favorable divide producing vegetative cells. Detailed life cycles are known for some monads such as *Katodinium fungiforme, Spiromonas perforans, Peridiniopsis lubieniensiforme, Ceratium hirundinella, Ceratium cornutum* and a number of *Peridinium* species. A detailed, but a present still incomplete life history of *Hemidinium nasutum* has been reported. Extremely complex life cycles have been described for some of the dinococcalians.

6.1. Monads

Vegetative cells of *Peridiniopsis lubieniensiforme* (Diwald 1938) (fig. 49) are armored and divide along the cingulum (equatorially) into two daughter cells while motile.

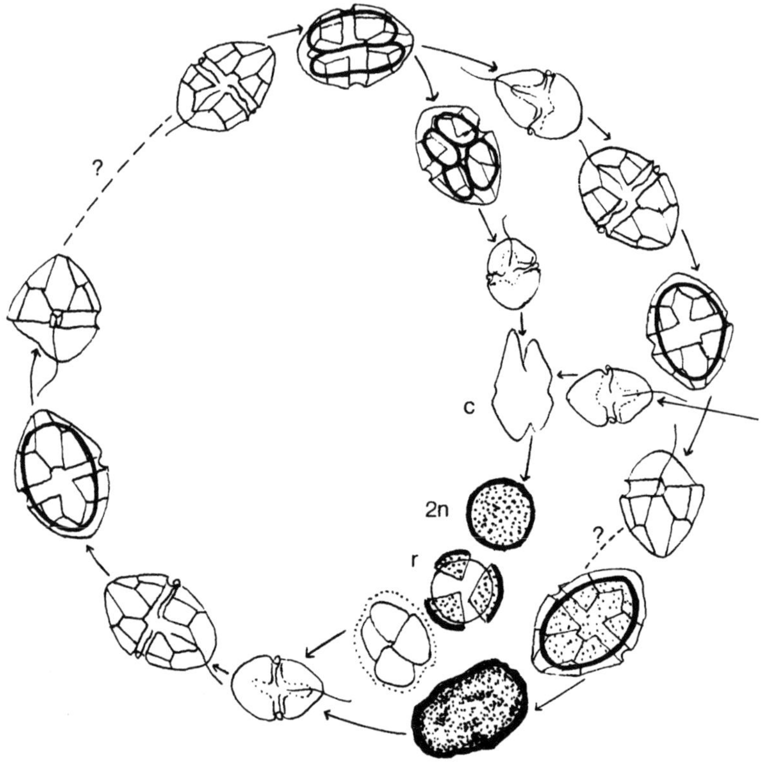

Fig. 49. A life history of *Peridiniopsis lubieniensiforme;* c copulation, 2n a diploid zygote, r reduction (after Diwald).

Division requires 1 hour with each daughter cell retaining one half the parent cell wall. The other half of the cell wall is formed during the first hour following division. The new cell wall appears to lack sutures at first. These form as the cell increases in size. After 6 or 7 such vegetative divisons the protoplast divides logitudinally into two gymnodinioid zoospores which turn 90° within the parent wall before escaping from the parent wall. Approximately 30 hours later they reach the size dimensions of the vegetative cell. Cell wall development begins in the cingulum area and continues irregularly over the entire cell. Four gymnodinioid isogametes are formed in sexual reproduction by longitudinal division of this protoplast, although one or two gametes are usually aborted. Gametes are released when the cell wall breaks equatorially. Fusion is ventral and requires approximately 30 to 40 minutes. The flagella disappear and the resulting zygote is round and becomes surrounded by an unequally thickened wall. Ten days after fusion its wall breaks into three pieces and becomes surrounded by a thick gelatinous mass which develops around the zygote. The protoplast divides producing 4 gymnodinioid zoospores. Each zoospore produces a cell wall within 24 hours at which time is has reached maximum vegetative cell size. Two or 3 of the zoospores usually abort. Aplanospores with a thick, homogenous wall develop periodically in the life history of *Peridiniopsis lubieniensiforme*. These aplanospores produce two zoospores, only one of which normally survives.

The zygote produced by members of the genus *Peridinium* and *Peridiniopsis* remains motile, for 3 to 14 days. During this time it enlarges with cell wall enlargement taking place in the suture areas. Sutures may increase tremendously – becoming as large as some of the plates forming the cell wall. Sutures on plate 1′ and the adjoining plates 1′′ and 7′′ do not enlarge as greatly as those of the other plates. Once a maximum size is reached, the planozygote loses motility and develops a thick wall. At this stage it is referred to as a

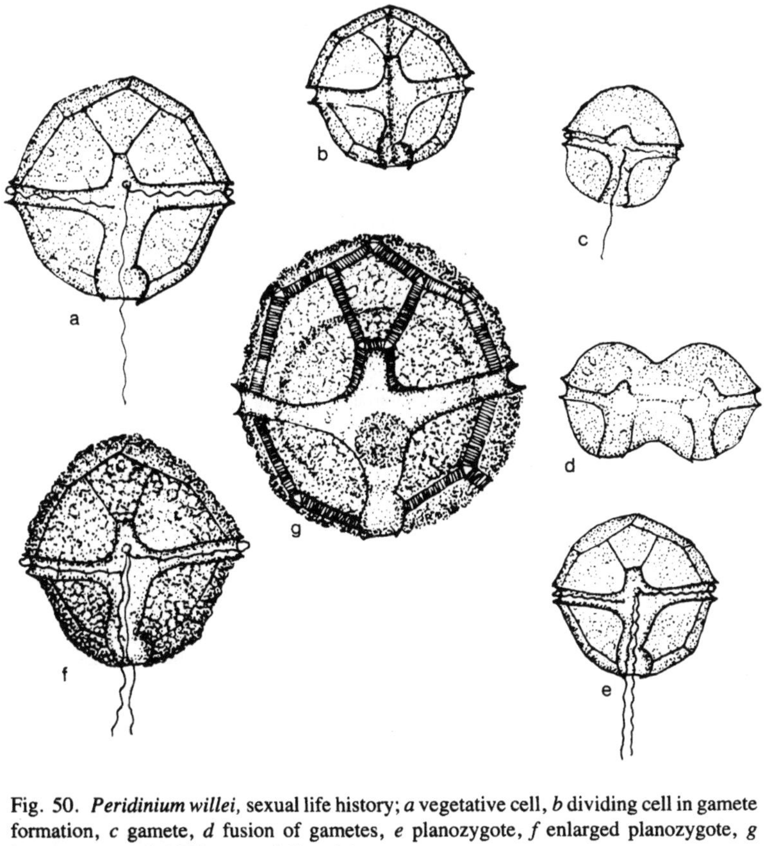

Fig. 50. ***Peridinium willei***, sexual life history; *a* vegetative cell, *b* dividing cell in gamete formation, *c* gamete, *d* fusion of gametes, *e* planozygote, *f* enlarged planozygote, *g* hypnozygote (after Pfiester and Skvarla).

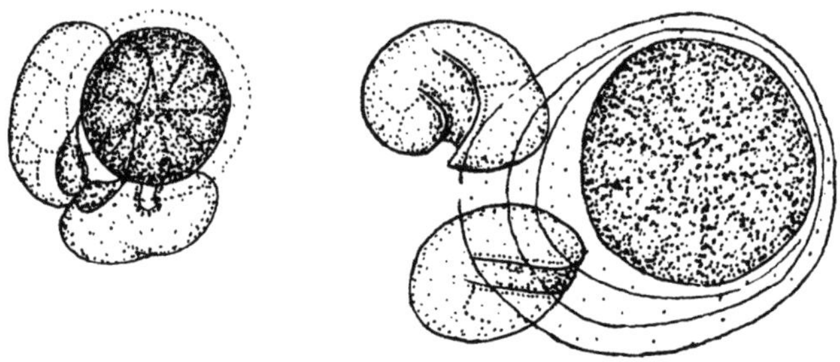

Fig. 51. Two empty cell walls of *Hemidinium nasutum* attached to immobile stage (*"Gloeodinium montanum"*) (after Popovský).

hypnozygote. These germinate following a dormant period (Pfiester 1975, 1976, 1977, Pfiester and Skvarla 1979, 1980) (fig. 50).

The life histories of *Hemidinium* and *Woloszynskia* have not been fully documented. It is highly probably that described dinocapsal genera such as *"Gloeodinium"* and *Rufusiella* are sexual and vegetative reproductive stages of other described taxa. Two empty thecae of small *Hemidinium nasutum* cells have been observed attached to *"Gloeodinium montanum"* (fig. 51). Perhaps *Hemidinium nasutum* represents the gametic phase and *"Gloeodinium montanum"* the zygotic phase in the life cycle. Further development and germination of the cysts or zygotes has not been observed (Popovský 1971).

The life history of *Ceratium* has been well documented and is anisogamous. Male gametes are smaller and thinner than female gametes. The female gamete engulfs the male gamete through an opening in the sulcal area. Aplanozygotes are formed which enlarge. Thecal enlargement again takes place by increased suture width. Hypnozygotes are eventually formed which are irregularly shaped like a tetraedron or cone. Two or 3 shorter horns may be present. Hypnozygotes germinate following several weeks of dormancy at 4 °C when the temperature is raised. Four haploid zoospores are produced which have been referred to as the "gymnoceratium" stage. Horns are visible two hours after germination. At this stage the cells are referred to as "preceratium". This stage develops into the mature vegetative cell within 6 hours. It is capable of vegetative division and develops aplanospores at the end of the growing season. Aplanospores are released from the parent cell wall and settle on the bottom of the pond or lake where they were produced (v. Stosch 1964, 1965).

The life history of marine *Katodinium fungiforme* grown phagotrophically in cultures of *Dunaliella salina* has been documented (fig. 52). Mitotic division and cytokinesis occur within a non-motile cell or zoosporangium (asexual division cyst). This state always follows feeding and is initiated when the vegetative cell settles to the bottom, sloughs its flagella, rounds off its outer cell surface and develops a cleavage furrow (zoosporangium) which divides the cytoplasm into two new cells. The isogamous pairs of gametes are similar in appearance to new cells released from a zoosporagium. The gametes orient 90° one another and fuse such that the anterior apex of one is attached to the sulcal region of the other. All fused gametes observed were seen revolving as they swam through the water. Motile planozygotes with dual longitudinal flagella were also seen. Following the

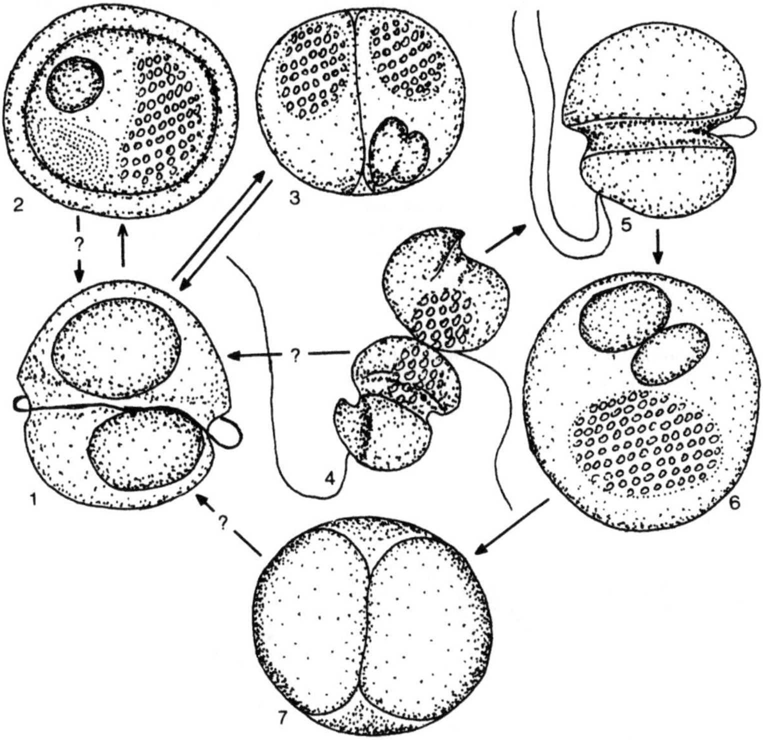

Fig. 52. Known and postulated stages in the life cycle of *Katodinium fungiforme*. Asexual phase: 1 swimming cell, 2 long-term resting cyst, 3 zoosporangium (mitotic division). Sexual phase: 4 gametes, 5 planozygote, 6 hypnozygote, 7 hypnozygote (meiosis) (adapted from Spero and Morée).

feeding period, during which the planozygote expands to the maximum dimensions, it settles to the bottom, sloughs its flagella, and develops into the hypnozygote with a thickened outer wall. Nuclear cyclosis was observed in several cells following hypnozygote formation. At this time, the nucleus could be seen rotating within the cell. Dividing hypnozygotes and post-division hypnozygotes were observed and these always appeared to have only two new cells the release of which was not observed (Spero and Morée 1981).

Spiromonas perforans is a predator of *Chilomonas paramaecium*. It is free-swimming in its vegetative habit (trophozoid). Eventually it anchors its rostrum close to a flagellar base of the *Chilomonas*. After several hours it penetrates into the *Chilomonas* cell. It engulfs the *Chilomonas* cytoplasm very quickly. The cell then enlarges, looses its flagella, separates from the *Spiromonas,* rounds up and becomes non-motile. After some period the nucleus and cytoplasm divide producing up to four gametes. Sexual fusion takes place within the parent wall. Trophozoid release was not recorded (Brugerolle and Mignot 1979) (fig. 60).

6.2. Coccoids

Dinococcalian life cycles have thus far been found to be the most complex in the Dinophyceae. Heterothallism and metamorphism are well developed. The term "coccal" is used in the traditional sense here (see chapter 1). Apparently all dinococcalians are capable of predation. Dinococcalians are motile in all stages of their life histories except the cysts and sporocyte. Several stages in their life histories are identical to those of named protozoan taxa.

Cystodinium bataviense (fig. 53). These cells have been described to have horn-like ends. However clones have produced rounded cells identical to the genus *Hypnodinium.* Rounded cells collected in nature have also later developed into the horned *Cystodinium*

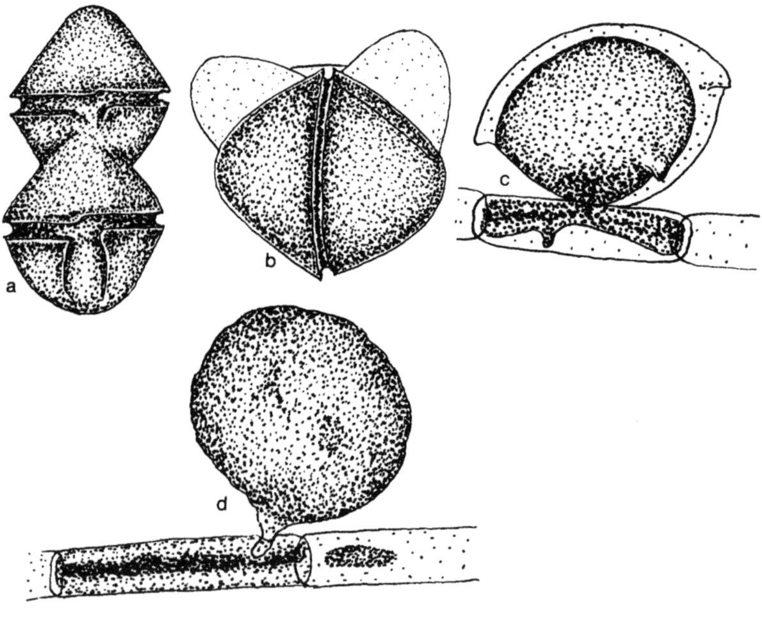

Fig. 53. *a* sexual fusion of *Cystodinium bataviense* gametes, *b* a resulting zygote, *c Cystodinium (Dinococcus)* cell phagocytizing *Oedogonium, d* an amoeboid stage with haustorium phagocytizing *Mougeotia* (after Pfiester and Lynch).

(Pfiester, unpublished observation). Cell division has been observed that produced 2 to 3 unequal division products. Division may result in the production of thecate zoospores which are released from the parent cell wall. These zoospores are motile for a short period, lose their cell wall and develop into the typical vegetative cell. The cell wall remains attached to the non-motile protoplast for some time. Isogametes are formed by the same type of division process observed in the vegetative cells. Fusion is rapid (4 to 5 minutes). Thecae are shed by fusing gametes but remain attached to eighter side of the zygote. Zygotes are round and possess a thick cell wall. *Cystodinium bataviense* may also reproduce vegetatively by the formation of amoebae. At this time the vegetative cell protoplasm becomes frothy and very slowly (8–10 hours) cleaves into numerous, small light-brown amoebae with axopods. These cells resemble the amoebae *Vampyrella lateritia* or *Actinophrys sol*. They prey upon *Oedogonium*, *Spirogyra* and *Mougeotia* cells eventually developing the *Cystodinium* form. Bright green to brown *Cystodinium* "vegetative" cells have been observed to slowly (10–12 hours) prey upon *Oedogonium* cells. This is accomplished by the extension of a short pseudopodium (haustorium or stalk) which engests the *Oedogonium* protoplast. This stage has been identified as *Dinococcus oedogonii* (Pfiester and Lynch 1980).

Stylodinium globosum (fig. 54). Vegetative reproduction by means of 2–4 zoospores from a sessile stage has long thougth to be the only means of reproduction. Vegetative cells varying in color from bright green to brown to bright orange have recently been reported to rupture liberating two orange amoeboid-trophozoids resembling *Vampyrella lateritia*. Orange trophozoids so produced possess two flagella-like filopodia, which shorten within seconds as a trophozoid settles on an intact *Oedogonium* cell. Within seconds it penetrates (enzymatically?) the *Oedogonium* cell wall and "swells" as it engulfs the *Oedogonium* protoplast. The trophozoid may develop into the typical *Stylodinium* "sessile" stage at this point or move to the next intact *Oedogonium* cell and repeat the predatory process or it may fuse with another *Stylodinium* cell (Timpano and Pfiester 1986). It has been recorded to prey upon five to seven *Oedogonium* cell within 30 minutes. At any time during this process it may develop into the typical, sessile *Stylodinium* form within a 10–15 sec period. A few minutes after developing the sessile form (10–30 min) the protoplast may divide into 4–8 small gymnodinioid zoospores. A few minutes after forming the *Stylodinium* phase a large central food vacuole is visible, but chromatophores and nuclei cannot be observed at the light microscopical level. Some amoebae have been observed to have a suctorian-like tentacle by means of which they puncture the *Oedogonium* cell wall. This tentacle rapidly develops into the stalk characteristic of the sessile form. Silicious, discoidal spicules have been observed on some cells. *Stylodinium globosum (St. cerasiforme)* appears to be related to the protozoan genus *Raphidiocystis*. Its method of predation resembles that described for the genus. Formation of probable gametes has been observed. Several pairs of "gametes" have been observed dancing around each other. One member of the pair was slightly larger and gymnodinioid shaped than the other which was gyrodinium-like. Fusion has not been observed.

A similar but much more complex life history has been observed in *Cystodinedria inermis* (fig. 55). *Cystodinedria inermis* has been reported to prey on from 1 to 8 cells in a row before resting on an intact *Spirogyra* or *Oedogonium* cell. It changes its shape and enlarges in size as it moves from one cell to another. Thus the species *Cystodinedria maxima* was probably described from a *Cystodinedria inermis* cell that had preyed upon a large number of *Spirogyra* or *Oedogonium* cells. *Cystodinedria inermis* reproduces by zoospores. A free swimming zoospores loses motility and becomes surrounded by a thick mucilage. The sulcus and girdle disappear. The cell attaches itself to the substrate by means of this mucilage. Once attached the mucilage disappears. Four other means of reproduction have been reported. One alternative means of reproduction is by the release of two trophozoids similar to *Actinophrys sol* following protoplast division. For details see the *Stylodinium* cycle since the trophozoids *Vampyrella lateritia* and *Actinophrys sol* are identical in both life histories. Thus it is possible that *Stylodinium globosum* and *Cystodinedria inermis* represent different stages in the life history of one organism. This trophozoid may assume a *Cystodinium phaseolus* form; in such instances its fate is unknown. Usually a trophozoid will move across an intact *Oedogonium* or *Spirogyra* cell. Axopods shorten and the trophozoid settles saddle-like on the filament. Within 5 seconds the filamentous cell will be penetrated enzymatically; the protoplast will become distorted and engested within 10 minutes. At this stage the *Cystodinedria* cell may change in eighter of two ways: (1) it may settle on a cell, release a small microspore and then leave the filament. It will then immediately fuse with what resembles a microspore (autogamy?). (2) The trophozoid resembling *Vampyrella lateritia* in its non-heliozoan stage enzymatically penetrates a *Spirogyra* or *Oedogonium* cell by means of a filopodium and preys upon the cell content. This free swimming amoeba shortens its filopodia and develops into a typical *Cystodinedria* vegetative cell. A second alternative means of reproduction is that in which a suctorian trophozoid resembling *Trichophrya* sp. absorbs its tentacles and develops into a *Cystodinedria* cell which continues preying on several filamentous cell. It will secrete a fine mucilagenous covering and within 10 minutes leaves the filament. At this time the mucilage is indetectable at the light microscopical level. This trophozoid has lobopodia, resembles *Amoeba verrucosa* and moves extremely slowly. *Oedogonium* zoospores have been observed to approach the cell and lose their motility (dead?). These inert *Oedogonium* zoospores were slowly engulfed by the dinoflagellate. At the same time the dinoflagellate was seen to release a small bit of protoplasm with short filopodia which changed into the non-heliozoan *Vampyrella lateritia* stage. These cells preyed upon *Oedogonium* or *Spirogyra* cells, suddenly lost motility and developed the *Cystodinedria* form. Four motile gymnodinioid zoospores were produced at this stage; their fate is unknown. A third method of reproduction by trophozoids has been observed in which the trophozoid resembling *Actinophrys sol* is first produced which in some ways resembles the heliozoan taxon *Dinamoeba mirabilis*. Some speciemens have been observed to shed their covering and settle *Cystodinedria* style on a filament. These shelled trophozoids have also been observed to move slowly and selectively engulf inert *Oedogonium* zoospores leaving behind inert ciliates. Again, following engestion of 1 or 2 *Oedogonium* zoospores a small piece of protoplasm, a sporocyte, has been reported to leave the dinoflagellate and quickly develop into two small *Vampyrella lateritia* like trophozoids which then prey upon *Oedogonium* or *Spirogyra* filaments. Several hours later they lost motility and developed into the *Cystodinedria* stage. The fourth and final means of reproduction begins with a *Vampyrella lateritia-Dinamoeba mirabilis* stage which develops into an attached stage similar to *Actinophrys sol*. This stage preys upon 1 to 4

Fig. 54. A life history of *Stylodinium globosum; a, e, h, i, m, o, p, r, t, u, v* typical dinococcal forms, *c, d, g* trophozoids *"Vampyrella lateritia"*, *n, s* trophozoids *"Actinophrys sol"*, *k, l* motile cells, *f* zoospores, *y* heterothallic gametes. For more detailed explanation see the text, pages 230 and 231 (after Popovský and Pfiester).

Fig. 54

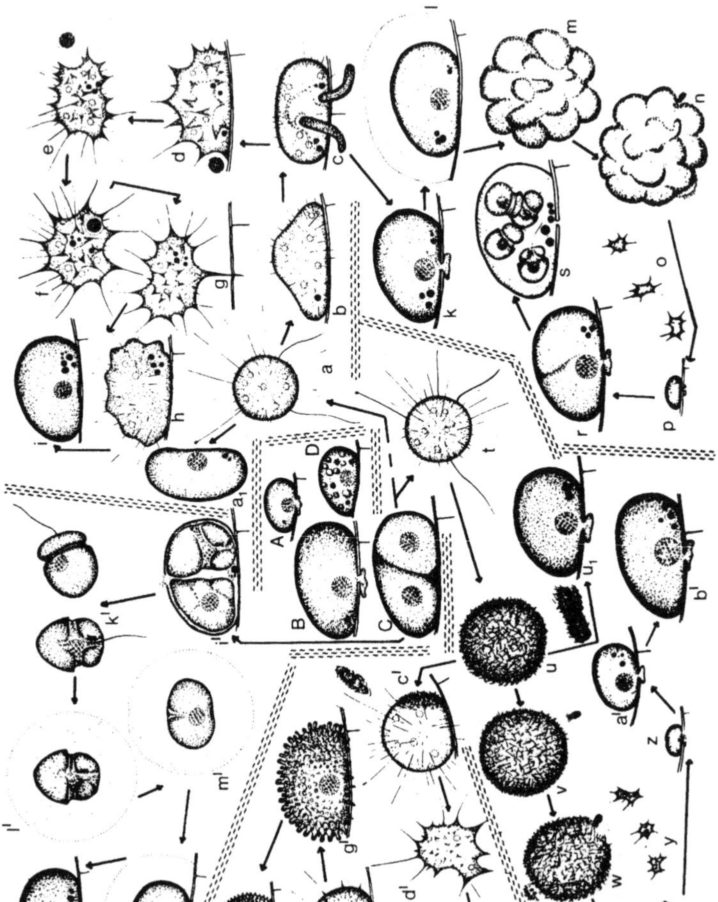

Fig. 55

Oedogonium or *Spirogyra* cells in succession. Following predation it suddenly develops into a trophozoid and leaves the filament. The trophozoid swims for a short period then settles on an *Oedogonium* or *Spirogyra* filament and begins to prey upon several cells. Cell shape is more constant during this process and resembles a typical *Cystodinedria* vegetative cell. The axopods are absorbed and spicules, characteristic of *Cystodinedria setosa* are visible. Within a period of an hour these spicules are either absorbed or develop into short, rapidly moving lobopodia. Approximately two hours later the cell wall becomes rigid, somewhat wrinkled and nontransparent. Presumably the cell is in a resting stage, i.e. dormant cyst (Pfiester and Popovský 1979, Popovský and Pfiester 1982).

One is able to conclude from the descriptions given above of the life histories of *Cystodinium bataviense, Stylodinium globosum,* and *Cystodinedria inermis* that this knowledge is not yet complete. Little is known of the exact relationship between the various modes of reproduction observed in *Cystodinedria.* Further, the similarity of the trophozoids observed in the life histories of the three taxa are puzzling. We presume that these organisms may be able to remain in one stage or phase of the life cycle indefinitely. That is, each phase in the life history appears to be capable of reproducing its own kind for a long period of time. The stimulus which triggers the next phase is unknown. Physiologically speaking it is important to note here that several stages have been observed which have previously been described as separate taxa. They are differentiated only by cell shape, presence or absence of motility, color and size. The amoeboid, free-swimming stage of the green alga *Chlamydomyxa labyrinthuloides* may belong in the Dinococcales and be a species of *Tetradinium* (Popovský 1979). Pearlmutter and Timpano (1984) state that it is more closely related to the chrysophytes.

The taxonomy and systematics of the Dinophyceae is at this time unclear, primarily because of a lack of knowledge concerning the organisms within the class. Thus, we do not propose any principle changes following the International Code of Botanical Nomenclature.

6.3. Cell growth

Two types of cell growth exist in the Dinophyceae: vegetative and cell enlargement by predation.

Vegetative growth of both armored and unarmored species is well documented. Unarmored dinoflagellates appear at times to be metabolic. Their cell covering is flexible. Vegetative growth of armored taxa is by the formation of naked zoospores. In *Peridiniopsis lubieniensiforme* the zoospores reach their maximum size within approxima-

---

Fig. 55. A life history of *Cystodinedria inermis;* A *C. inermis,* B *C. maxima,* f′ *C. setosa, i, k, l, p, r, s,* u_l, *z, a′, b′, f′, g′, i′, m′, n′, o′,* typical "coccal" forms, *d, e, f, g, o, y, d′, e′,* trophozoids *"Vampyrella lateritia", a, t, c′,* trophozoids *"Actinophrys sol", m, n,* sporozoids *"Amoeba verrucosa", u, v, w,* sporozoids *"Vampyrella lateritia"* in a heliozoan form, *h′,* dormant cyst, *s, i′, k′, l′,* zoospores, gametes?, a_l *Cystodinium phaseolus.* For more detailed explanation see the text pages 227, 228, 232, and 233 (after Popovský and Pfiester).

tely 30 hours of release. Their cell wall developes in an irregular pattern over the entire cell surface. Once the theca is formed little cell enlargement takes place. Each of the two zoospores formed by *Peridiniopsis lubieniensiforme* maintain one-half of the parent theca. Each regenerates the missing half within 30–60 minutes. This regeneration begins in the cingulum in the epitheca and surrounding the sulcus in the hypotheca. Within minutes thecal regeneration reaches the epi- or hypocones respectively. At that time the theca with plates is clearly visible (Diwald 1938).
However, during sexual reproduction the planozygotic theca grows as the protoplast enlarges. The rigid theca accommodates protoplast enlargement through increase in the intercalary or suture bands rather than in enlargement of cell plates. Intercalary bands in vegetative cells are narrow and uniform in size. As the planozygotes enlarge, these bands enlarge greatly and usually become striated. The vertical sutures of the plate 1' enlarge only slightly (Pfiester and Skvarla 1979). Planozygotes of *Woloszynskia* species appear to split at the carina and along the cingulum in reponse to protoplast enlargement as intercalary bands are lacking. This splitting occurs prior to cell division in *Woloszynskia reticulata* (Pfiester et al. 1980).

7. Occurrence and ecology

The majority of Dinophyceae are found in marine and brackish waters. Freshwater dinoflagellates are found in such varied habitats as streams, ponds, heavily polluted wastes, and swamps. The dinoflagellates seem to be cosmopolitan with only a few taxa specific to a given geographical location. Dinoflagellates are usually dominant in summer months. They sometimes become so numerous that they discolor the standing waters. Some species appear to be stenothermic obligate cryophiles and for this reason they are collected in the surface layers during winter months in high mountain lakes, in circumpolar waters in the spring, or during the summer in deep lake waters (Pechlaner 1971). Some are known which color snow red. Some species are thought to be halophobic and sphagnophilic. A common observation is that during blooms of green algae or of planktonic diatoms the dinoflagellates remain low in number or are not present. Contrary to these reports Skuja (1938), Wołoszyńska (1935), and Huber-Pestalozzi (1950) have reported the presence of dinoflagellates along with blue green algae and diatoms. The senior author has recorded the presence of a number of dinoflagellate species together with diatoms and blooms of blue green algae in several reservoirs in Central Europe. Evidently the relationship between blue green algae, diatoms, and dinoflagellates is not simple. It must depend on competition for food and light, on the phase of the growth cycle and on the vertical distribution of the phytoplankton in reservoirs and lakes.
We can make some ecological statements concerning dinoflagellate growth and occurrence following Höll's (1928) observations on some physical and chemical variables affecting dinoflagellate growth. From his study we can conclude that *p*H, temperature, and the concentration of organic matter and ions such as calcium, chlorine, nitrogen, and phosphorus are dominant factors regulating the occurrence of dinoflagellates in nature. Other ecological factors which influence the occurrence of photosynthetic dinoflagellates are light intensity and the changing spectrum in relation to the water column and color as well as the concentration of dissolved oxygen. Some species are more efficient in utilizing various light components including ultraviolet light (Pechlaner 1971) than others. Some

dinophyceans such as *Ceratium hirundinella* are tolerant to low concentrations of oxygen in the water. The combination of light spectrum and oxygen are factors which effect the daily vertical migration of populations of planktonic dinoflagellates. These migrations may reach up to several meters per day. Pechlaner (1967, 1971) showed that once the snow cover disappears from high-mountain lakes more of the high light intensities can penetrate. Under these conditions *Gymnodinium uberrimum* and *Peridinium umbonatum* migrate downward to a depth of 20–25 meters thus avoiding radiation. *Katodinium bohemicum* is found in the lower levels of the phototrophic zone in the Černé Lake (Fott 1938) as is *Amphidinium elenkinii* in the High Tatra mountain Lakes (syn. *Amphidinium tatrae* and its fo. *achromaticum* Wołoszyńska 1936). When these taxa or members of a population are found below the phototrophic zone they usually do not have chromatophores. Many dinoflagellates migrate vertically daily because of changes in light intensities but also because of the availability of food used in heterotrophic nutrition. Changes in water temperature during the year is an important factor influencing dinophycean occurrence but also appears to cause temporary morphological variation such as have been reported in *Woloszynskia leopoliensis*. Huber (1905, 1914), Huber and Nipkow (1922) and Huber-Pestalozzi (1923) have studied the effects of temperature on morphological changes in *Ceratium hirundinella*. They showed that *Ceratium hirundinella* is capable of living at a minimum temperature of 5°C, an optimum temperature of 15 to 20°C and a minimum temperature of 28 to 30°C. Thus temperature appears to be a major factor influencing the occurrence of *Ceratium hirundinella*. As temperatures rise from the maximum, individuals with 3 to 4 horns are present. Those possessing two horns are not found until the water temperature has surpassed 15°C. They also showed that temperature affects cell length in *C. hirundinella*. Knowledge of this cyclomorphosis is valuable in phytogeographical studies of *Ceratium hirundinella*: this species is not present in polar areas and it reaches a maximum during the summer months in temperature zones and during cold periods in trophical areas. Thus, its polymorphism can be explained by changes in water temperature during the season and by vertical migration. Peters (1934) reported that marine species of *Ceratium* had shorter and thicker cell walls than those in tropical marine waters.

However, we cannot compare the few ecological studies conducted on freshwater dinophyceans to those of other freshwater algae because the present knowledge of their biology is too incomplete. Further, they are often overlooked or misidentified in samples. Thus, it is not often possible to correlate their occurrence with physical and chemical factors in aquatic studies of algal communities.

The use of increasing amounts of fertilizers in agriculture, forestry, and aquaculture is resulting in eutrophication and increased salinity in freshwater habitats. Salting roads around metropolitan areas during the winter has also been a major factor in increasing the salinity of freshwater systems. Thus it is highly possible that some "marine" Dinophyceae will be found in freshwater localities.

The red tide phenomenon has been reported to occur in warm continental shelf waters of America and West Europe. It results from a bloom of marine dinoflagellates (10^6 to 2.10^7 cells.liter^{-1}), in which ocean patches up to several square kilometers in exent are colored red at the surface, or reddish-brown or yellow. Toxic red tides are caused by species of *Gymnodinium, Gonyaulax,* or *Ptychodiscus*. These algae produce saxitoxin which causes fish poisoning and concentrates in shellfish. These shellfish then become toxic for human consumption. Kristiansen (1974) recorded an eel kill in farm tanks in Horn Lake (Denmark) during a red tide caused by brackish dinoflagellate *Oxyrrhis marina*. The lake opens to the North Sea and has a salinity of 24‰. Hashimoto and Okaichi (1968) isolated an ichthyotoxic substance named "glenodinine" from freshwater *Peridiniopsis polonicum*.

Dinoflagellates, in the form of zooxanthellae, live as symbionts within some marine invertebrates. These coccoid cells are surrounded by a thin periplast; older cells have a thicker wall and appear cyst-like. Vegetative cells normally divide into two equal products. Single, naked gymnodinium-like zoospores can also be formed within cells. Genera known to form these symbolic relationships are *Symbiodinium* and *Amphidinium* (syn. *Endodinium*) (Bold and Wynne 1978, 1985).

Kryptoperidinium foliaceum and *Peridiniopsis balticum* are two marine species known to have a second, eukaryotic nucleus. This represents a different type of symbiotic relationship in the Dinophyceae (see chapter 2.3.). The marine species *Ornithocercus* and *Histioneis* form a symbiotic relationship with blue-green algae which live in a hollow chamber formed by a cingular wing. These blue-green algae may fix nitrogen which is a limiting factor in tropical oceanic waters (Fott 1971).

Fossil Dinophyceae are known from the Carboniferous as primitive members of the Peridiniales. Artzner et al. (1979) listed 276 genera of fossil Dinophyceae from the Peridiniales, Dinophysiales and Gymnodiniales. Fossil dinoflagellates similar in appearance to extant taxa have been discovered in the Triassic and Lower Crecateous.

Parasitism is a well-developed mode of nutrition in the marine Dinophyceae. Parasitic species are often exceedingly specialized in their structure and reveal their correct placement in this division either by their gymnodinioid motile stages or by their persistent chromosomes. Although many of these genera have mesokaryotic nuclei typical for the class, other genera lack such a nucleus and may be classified here incorrectly. Both endoparasites and ectoparasites are known (Bold and Wynne 1978). Blastodiniales are attached parasitic forms on invertebrates and fish. *Ooidinium limneticum* has been reported to parasitize 9 different freshwater fish taxa in North American (Jakobs 1946). A list of parasitic genera has been published (Sournia et al. 1975), and a compilation of their hosts and of the mode of attachment has also appeared (Chatton 1952).

8. Sampling, identification and culture methods

Sampling methods used for dinoflagellates do not differ from those used in collecting other protists. In order to concentrate planktonic species one may centrifuge them at ca. 1500 cycles $\cdot$ min^{-1} for 5 to 10 minutes or they may be filtered onto a membrane. Larger taxa may be collected in a plankton net having a pore size of 30–40 µm. Litoral species may be collected with a pipette or by squeezing Sphagnum and other moss samples into a bottle. These samples should be stored at a low temperature (8° to 10°C) and transferred into a petri dish prior to microscopic examination.

Preservatives. Armored species and those with thick membranes are preserved with 4% IKI after Utermöhl, or with Transeau's preservative 6:3:1 (water, 95% ethanol and formalin); five ml of glycerol may be added to 100 ml of this solution. Gluteraldehyde is used to preserve species for electron microscopy.

Dinoflagellates should be observed with the light microscope as soon as possible following collections since monads with a delicate cell covering are sensitive to minute environmental changes. Some are especially sensitive to the glass used in cover slips and slides. This sensitivity is a rough feature that is useful in differentiating naked species from those with a thick covering. Once the delicate cell covering is broken in naked cells the protoplast becomes amorphous. Examples of such sensitive species are found in the following genera: *Amphidinium, Katodinum, Bernardinium, Protaspis* and many species of *Gymnodinium*. Those genera which have a thick pellicle or a delicate cell wall maintain their cell shape for some time. Such genera are: *Woloszynskia, Glenodiniopsis, Sphaerodinium, Hemidinium,* young *Peridinium* and *Peridiniopsis* cells, and some species of *Gymnodinium*.

In order to identify dinoflagellates one should first study their overall shape and behavior. This is best accomplished by means of an immersion objective. One should determine at higher magnification the cell length, width, thickness, flagellar length, and cell shape from all angles and cross-sections of the epi- and hypocones, as well as the furrow, the presence and shape of pseudopodia or cell projections when present. In studying cell content one should determine the number, shape, color and position of the chloroplasts, the dimension, shape and position of the dinokaryon, eyespot and other organelles such as vacuoles, contractile vacuoles, storage products and remnants, etc.

Some samples can be kept at low temperatures for weeks and life cycles studied from them. The color of the chloroplasts, the position, shape and presence of the nucleus often depends on the phase of the life cycle as does flagellar length and its presence or absence.

If one were to list the single most important factor which has led to the tremendous increase of knowledge of dinoflagellates that has occurred in the past ten years, it would be the ability to culture them. Dinoflagellates are morphologically complex. In many instances their sexual life histories are long and the organism may pass through several distinct stages. It is not surprising that in the past some of these stages have been named as separate taxa by phycologists forced to work with field samples alone, many of which were preserved. The ability to culture both freshwater and marine dinoflagellates has enabled phycologists to study their biochemistry, ultrastructure, vegetative and sexual life histories, and mineral and vitamin requirements. As Pringsheim (1967) so aptly stated, "It is by combining field and culture work that researchers are able to best understand (know) the organisms with they work."

Freshwater growth media

Soil Water Medium (Pringsheim 1946, Starr 1964, Nichols 1973)

Use of soil water medium when first attempting to isolate a dinoflagellate should result in a high degree of success. This is especially true if the medium is made using soil taken from the collection site. Use of soil water medium usually results in "normal" algal growth forms. Soils with high clay or humus content should not be used nor should recently fertilized soil.

(a) Add a pinch of $CaCO_3$.
(b) Fill a test tube 1/10 to 1/8 full of soil.
(c) Fill the test tube 3/4 full of distilled water.
(d) Cover test tube and steam at 100°C for 2 hours.
(e) Place test tubes in the light for 24 hours.
(f) Rapeat stages (d) and (e) at least two times.

Carefoot's Medium (Carefoot 1968)

This medium has been used successfully to culture freshwater members of the genus *Peridinium*. The junior author has modified it by adding vitamins, as at least one freshwater *Peridinium, P. limbatum,* requires thiamin and *Peridiniopsis polonicum* requires B_{12} for growth (Holt and Pfiester 1981).

(1) Stock Solutions

$MgSO_4 \cdot 7H_2O$	0.098 g in 200 ml H_2O
NaCl	0.33 g in 200 ml H_2O
KH_2PO_4	0.45 g in 200 ml H_2O
K_2HPO_4	0.194 g in 200 ml H_2O
$CaCl_2$	0.33 g in 200 ml H_2O
$NaNO_3$	5.0 g in 200 ml H_2O

(2) PIV Metal Solution

To 500 ml glass-distilled water add 0.750 g of Na EDTA. After this dissolves, add the following.

$FeCl_3 \cdot 6H_2O$	0.097 g
$MnCl_2 \cdot 4H_2O$	0.041 g
$ZnCl_2$	0.005 g
$CoCl_2 \cdot 6H_2O$	0.002 g
$Na_2MoO_4 \cdot 2H_2O$	0.0049 g

To 935 ml water add 10 ml each of the above (1) stock solutions and 5 ml of the (2) PIV solution.

(3) Vitamin Solution

This solution should be kept frozen between uses. These vitamins may be autoclaved. To 500 ml distilled water add.

Thiamine	5.0 g
B_{12}	5.5 mg
Biotin	5.5 mg

add 0.1 ml vitamin solution (3) per liter of media.
Do not titrate for *p*H.

NR_{12} Media (Linstrøm and Rodhe 1978)

This medium was developed for *Peridinium cinctum* (fo. *westii)* from Lake Kinneret (Israel). It contains selenium and has been used to culture a number of other freshwater dinoflagellates. This author has modified it by adding the vitamin mix listed above.

(1) Stock Solutions

$Na_2 SeO_3 \cdot 9H_2O$	0.057 g · 500 ml H_2O
$NaNO_3$	0.635 g · 500 ml H_2O
$NaHCO_3$	6.72 g · 500 ml H_2O
$MgSO_4 \cdot 7H_2O$	0.615 g · 500 ml H_2O
K_2HPO_4	0.087 g · 500 ml H_2O
$CaCl_2 \cdot 2H_2O$	0.735 g · 500 ml H_2O

(2) Metal Mix Stock Solutions

Add 1 ml of each of following to 500 ml H_2O to make matal mix solution.

$FeCl_3 \cdot 6H_2O$	0.6759 g · 125 ml H_2O
H_2SeO_3	0.0006 g · 125 ml H_2O
$ZnCl_2$	0.452 g · 125 ml H_2O
$CoCl_2 \cdot 6H_2O$	0.059 g · 125 ml H_2O
Na EDTA	3.4898 g · 125 ml H_2O
$MnCl_2 \cdot 4H_2O$	0.0625 g · 125 ml H_2O
$Na_2MoO_4 \cdot 2H_2O$	0.3 g · 125 ml H_2O

(3) Vitamin Mix

Thiamine	5.0 g · 500 ml H_2O
B_{12}	5.5 mg · 500 ml H_2O
Biotin	5.5 mg · 500 ml H_2O

NR_{12} Modified Media

Add (1) 10ml of each stock solution
(2) 10 ml of metal mix solution
(3) and 0.1 ml vitamin mix solution to water – bring to 1 liter. So not tritrate for *p*H.

Modified Chu-Solution (v. Stosch 1973)

This medium has been used to culture freshwater *Gymnodinium* taxa.

	Additions per liter distilled water
$Ca(NO_3)_2 \cdot 4H_2O$	70.80 mg
K_2HPO_4	8.71 mg
$MgSO_4 \cdot 7H_2O$	24.60 mg
$NaHCO_3$	33.60 mg
NaCl	0.58 mg
Na EDTA	0.74 mg
$FeSO_4 \cdot 4H_2O$	278.00 µg
$MnCl_2 \cdot 4H_2O$	19.80 µg
$ZnSO_4 \cdot 7H_2O$	0.29 µg
$CuSO_4 \cdot 5H_2O$	0.25 µg
$CoSO_4 \cdot 7H_2O$	0.30 µg
$Na_2MoO_4 \cdot 2H_2O$	0.24 µg
$Na_2B_4O_7 \cdot 10H_2O$	95.35 µg

For more detailed description of isolation techniques the reader is referred to the Handbook of Phycological Methods (Stein 1973).

Unialgal Clonal Cultures

Several methods can be used to obtain unialgal cultures from a mixed sample:

(a) Dilution Technique

(1) Place 10 drops of liquid medium into a watch glass or disposable plastic petri dish. Add approximately 5 drops of the natural sample from which the dinoflagellate is to

be isolated. Number of drops of liquid medium and natural should be varied depending on the sample density.

(2) Place ten separate drops of media into a petri dish on a circular fashion.

(3) Take a sterilized pasteur pipette and heat the tip over a flame until soft. Remove the tip from the flame and quickly draw the tip out to a very fine point with forceps. Break off the tip of the capillary pipette near the point where the glass bends under its own weight. The opening should be smooth and several times the diameter of the algal units being isolated. Attach a rubber tube of suitable lenght to the large end of the pipette so that the other end can be placed in one's mouth and algal cells from the sample be drawn into the fine capillary tip by suction.

(4) Locate the desired dinoflagellate with a stereomicroscope with 2 × magnifier. Place capillary tip near the cell and draw it in with gentle suction.

(5) Place the dinoflagellate cell into a drop of medium (see step 2).

(6) Take a clean pipette and draw it out as in step (3). Attach rubber tube and move the dinoflagellate from drop 1 to drop 2. Repeat this procedure until the cell been "rinsed" in ten drops of medium.

(7) Place cell into test tube containing approximately 15 ml media.

(8) Culture at approximately 20 °C on a 12–12 light-dark regime. Light intensity should be 450–500 lux.

(b) Spray Plating

(1) Add 15 g to 1 liter of desired medium. Autoclave and pour petri plates. Allow to cool.

(2) Fasten a pasteur-type pipette onto a table so that the thin end is free and the other end is attached to an air source by means of a rubber tube. This will be used to create a jet spray. Collect a few drops of the natural sample into a pasteur-type pipette with rubber bulb. Hold an agar petri plate about 20 cm perpendicular to the pasteur pipette attached to the air source. Place the tip of the pipette containing the natural sample in front of the air jet and by squeezing on the bulb release a drop of sample. The sample will atomize and spray onto the agar surface. The degree of separation of algal cells can be controlled by either varying the diameter of the pasteur-type pipette or by varying of air flow. Culture the sprayed plates for 24 hours (20°C, 12 L–D, 450 lux). Using a stereomicroscope equipped with a 2× magnifier and a pasteur-type pipette prepared as in (3) pick up isolated, healthy looking cells and place in liquid medium.

Axenic Cultures

Axenic cultures are defined as cultures free of all other living organisms other than the one being cultured. One must first obtain a unialgal culture. Then:

(a) By repeated washing and transfer of cells with drawn out pasteur pipettes obtain an axenic culture.

(b) Centrifugation – Ultrasonic Method

(1) Centrifuge a dinoflagellate aliquot from a unialgal culture. Decant, add fresh sterilized media to the dinoflagellate pellet obtained. Mix.

(2) Ultrasonicate mixture for 5–10 seconds. This will knock bacteria free from the dinoflagellate cell.

(3) Repeat steps 1 and 2 approximately four times. Bacteria released from dinoflagellate surfaces should be lighter than dinoflagellate cells and thus removed by decanting following centrifugation. Most, but not all bacteria will be removed in this manner.

(4) Spray algal cells onto petri dishes as described in section on spray plating. Culture agar plates for 4–5 days under 20°C, 12 L–D at 400–500 lux.
(5) Aseptically remove cells from agar that do not appear to have bacteria associated with them. Place one cell in test tube of liquid media. Culture. Repeat until desired number of axenic clones are set up.
(6) Prepare slant agar test tubes by adding 15 g agar to 1 liter media, autoclaving and pouring approximately 15 ml into sterile text tubes which are slanted while cooling. Cap with cotton or sterile caps. After two weeks aseptically take an aliquot from axenic culture and inoculate slanted text tube. Place in warm dark place for 72 hours. Check to see if bacterial colonies and/or fungi have grown aseptically. Check for presence of dinoflagellates in 3–4 weeks. Axenic dinoflagellate cultures grow more slowly than those containing bacteria.

Antibiotic Treatment

Antibiotics may be applied singly or in combination to kill strongly attached contaminants. Antibiotics may be placed in liquid or agar medium and should be prepared as follows:

(a) Dissolve 100 mg penicillin G (K or Na salt) and 50 mg streptomycin SO_4 in 10 ml distilled H_2O; add 10 mg chloramphenicol in 1 ml 95% ethanol to the penicillin-streptomycin solution and mix well.
(b) Filter solution quickly with a membrane or Seitz-filter apparatus.
(c) Place 1 ml of dinoflagellate suspension in each of six 125 ml Erlenmeyer flasks containing 50 ml sterile medium.
(d) Add one of the following volumes of antibiotic solution to each of the flasks:
    3.0 ml
    2.0 ml
    1.0 ml
    0.5 ml
    0.25 ml
    0.125 ml
(e) Culture.
(f) Transfer some cells after 24, 48, and 72 hours to flasks of sterile medium. Make triplicate cultures.
(g) Culture.
(h) After several weeks check for bacteria as described in section on obtaining axenic cultures.

Culture collections

(1) University of Texas Culture Collection (UTEX)
Department of Botany
University of Texas
Austin, Texas, USA

(2) American Type Culture Collection (ATCC)
12301 Parklawn Drive
Rockville, Maryland 20852, USA

(3) Culture Collection of Algae (WHOI)
Woods Hole Oceanographic Institution
Woods Hole, Massachusetts 02543, USA

(4) Culture Collection of Thermophilic Bluegreen Algae
c/o Dr. R. W. Castenholz
Department of Biology
University of Oregon
Eugene, Oregon 97403, USA

(5) Carolina Biological Supply Company
Burlington, North Carolina 27215, USA

(6) Culture Centre of Algae and Protozoa (Cambridge)
The Director
36 Storeys Way
Cambridge, CB3, ODT, United Kingdom

(7) Freshwater Biological Association
Supply Department
Windermere Laboratory
The Ferry House
Ambleside, Cumbria, LA 22 OLP, United Kingdom

(8) Sammlung von Algenkulturen
Universität Göttingen
Nikolausbergweg 18
D-3400 Göttingen, Federal Republic of Germany

(9) Algal Culture Collection
Institute of Applied Microbiology
University of Tokyo
Bunkyo-ku
Tokyo, Japan 113

9. Classification

Class Dinophyceae

Subclass Adinophycidae
  Order Desmomastigales
  + Family Desmomastigaceae
  + Family Protaspidaceae
    Family Adinimonadaceae
    Family Desmocapsaceae
Subclass Dinophycidae
  Order Dinophysiales
    Family Amphisoleniaceae
    Family Dinophysiaceae

Order Peridiniales
+ Family Gymnodiniaceae
+ Family Actiniscaceae
Family Entomosigmaceae
Family Polykrikaceae
Family Warnowiaceae
Family Kofodiniaceae
Family Leptodiscaceae
Family Noctilucaceae
Family Archaeosphaerodiniopsidaceae
Family Crypthecodiniaceae
+ Family Woloszynskiaceae
+ Family Hemidiniaceae
+ Family Glenodiniopsidaceae
+ Family Peridiniaceae
+ Family Dinosphaeraceae
+ Family Gonyaulacaceae
Family Ceratocoryaceae
Family Ceratoperidiniaceae
Family Cladopyxidaceae
Family Heteraulacaceae
Family Heterodiniaceae
+ Family Lophodiniaceae
Family Osteropsidaceae
Family Oxytoxacaceae
Family Podolampaceae
Family Pyrophaceae
Family Thecadiniaceae
+ Family Ceratiaceae
Order Dinococcales (incl. Dinamoebidiales)
+ Family Dinococcaceae
Order Prorocentrales
Family Prorocentraceae
Order Pyrocystales
Family Pyrocystaceae
Order Zooxanthellales
Family Zooxanthellaceae
Order Blastodiniales
Family Apodiniaceae
Family Blastodiniaceae
Family Cachonellaceae
Family Haplozoaceae
+ Family Oodiniaceae
Family Protoodiniaceae
Order Brachydiniales
Family Brachydiniaceae
Order Chytriodiniales
Family Chytriodiniaceae

Subclass Ebriophycidae
  Order Ebriales
    Family Ebriaceae
    Family Hermesinaceae
Subclass Ellobiophycidae
  Order Thalassomycetales
    Family Thalassomycetaceae
Subclass Syndiniophycidae
  Order Syndiniales
    Family Amoebophryaceae
    Family Coccidiniaceae
    Family Duboscquellaceae
    Family Sphaeriparaceae
  + Family Spiromonadaceae
    Family Syndiniaceae

(+ families with freshwater species)

Note on the zoological system

For completeness we cite the protozoological classification of the dinoflagellates (Kudo 1971).

Phylum Protozoa
Subphylum Plasmodroma
Class Mastigophora
Subclass Phytomastigina
Order Dinoflagellida
Suborder Prorocentrina
  Peridiniina
  Cystoflagellina

II. Special part

Dinophyceae Pascher 1914

Peridineae Ehrenberg 1830, *Cilioflagellata* Claparède and Lachmann 1858, 1861. Arthrodele Flagellaten Stein 1883, *Dinoflagellata* Bütschli 1885, Kofoid and Swezy 1921, Lebour 1925, *Dinoflagellatae* Schilling 1913, *Peridineae* Lindemann 1928
Important references: Pascher 1914, 1927, 1931, Schiller 1933, Bourrelly 1970.

Key to the subclasses

1a. Vegetative cells possess two different, apically inserted flagella: the cingulum is absent, the sulcus may be present . . . . **1. Adinophycidae** (p. 83)
1b. Vegetative cells, or zoospores and gametes possess two different, ventrally inserted flagella; the cingulum and sulcus are always present . . **2. Dinophycidae** (p. 88)

1. Subclass **Adinophycidae** Bourrelly 1970
Desmokontae Pascher 1914, *Adiniferae* Lindemann 1928

Most members are unicellular. Cells may be planktonic or parasitic. Cell wall fine or in two valves similar to that of a bivalve. Plates absent. Palmelloid stages which produce biflagellate zoospores may be present. Hypnospores are known to produce 4 zoospores. Vegetative cells have two apically inserted flagella which are different in structure. A sulcus is present as are 1–2 chloroplasts, pyrenoids, stigma, pusules, pseudopodia, a rostrum, contractile vacuoles, and a food vacuole. Trichocysts may be present. The fusion of two gametes has been reported. Species are mainly marine and brackish.
Important references: Pascher 1914, Lindemann 1938, Schiller 1933, Bourrelly 1970.

1. Order **Desmomastigales** Bourrelly 1970
Desmomonadales Pascher 1914, *Athecatales* Lindemann 1928

Cells spherical, ovale, heart or drop-shaped; sometimes flattened dorsiventrally. Cell wall fine and flexible. Nutrition is autotrophic, heterotrophic, saproholophytic or parasitic. Three freshwater families are known.
Important references: Pascher 1914, Lindemann 1928, Schiller 1933, Bourrelly 1970.

Key to the families

1a. Flagella apically inserted . . . . **1. Desmomastigaceae** (p. 84)
1b. Flagella not apically inserted **2**

2a. Cells free-swimming, non parasitic . . . . **2. Protaspidaceae** (p. 84)
2b. Cells attached, parasitic . . . . **3. Spiromonadaceae** (p. 85)

1. Family **Desmomastigaceae** Christensen 1962 *Desmomonadaceae* Schiller 1933

Cells are naked, ellipsoid in shape and circular in cross section. The two disc-shaped, perforated chloroplasts are parietal. One pyrenoid and several starch droplets. The two flagella are apically inserted. One is relatively simple and prolongated in the direction of the longitudinal axis. The second flagellum is waved and sickle-shaped. Resting stages are spherical and enveloped by a cellullose membrane. Reproduction is by division of a monad or by division of the non-motile resting stage (hypnozygotes?).
Important references: Christensen 1962, Schiller 1933.

1. **Desmomastix** Pascher 1914
Important reference: Pascher 1914.

1. Desmomastix globosa Pascher 1914 (fig. 56)
Dimensions: 10–15 μm × 8–14 μm.
Occurrence: dead river meanders; Czechoslovakia.

2. Family **Protaspidaceae** Skuja 1939

Cells are somewhat metabolic; flattened dorsiventrally and ellipsoid or egg-shaped in ventral view. The apex is excavated. In lateral view the dorsal side appears more convex than the ventral. The sulcus slants along the longitudinal axis. Two flagella are inserted subapically on the ventral side. The anterior flagellum is the same length as the cell; the posterior flagellum is somewhat longer and lies within the sulcus. One or two contractile vacuoles are present in the dorsal part of the cell. Numerous refractile starch bodies are also present. Rhizipodia form from the ventral side of the cell. Nutrition is saprobic. Reproduction is by longitudinal division of the cell body while motile.
Important reference: Skuja 1939.

1. **Protaspis** Skuja 1939
Important reference: Skuja 1939.

1. Protaspis obovata Skuja 1948 (fig. 57).
Cells are egg or pear-shaped; rounded anteriorly and posteriorly elongated and beveled. The sulcus widens in the posterior part. One contractile vacuole is present.
Dimensions: 26–40 μm × 14–25 μm.
Occurrence: freshwater pools; Sweden.

Appendix

Protaspis metarhiza Skuja 1939 (fig. 58) has been collected from brackish waters. Its cells are egg-shaped or oval, with a small sinus on the apex. The antapex is usually apiculate. No contractile vacuoles are observed. Rhizopodia have been reported.

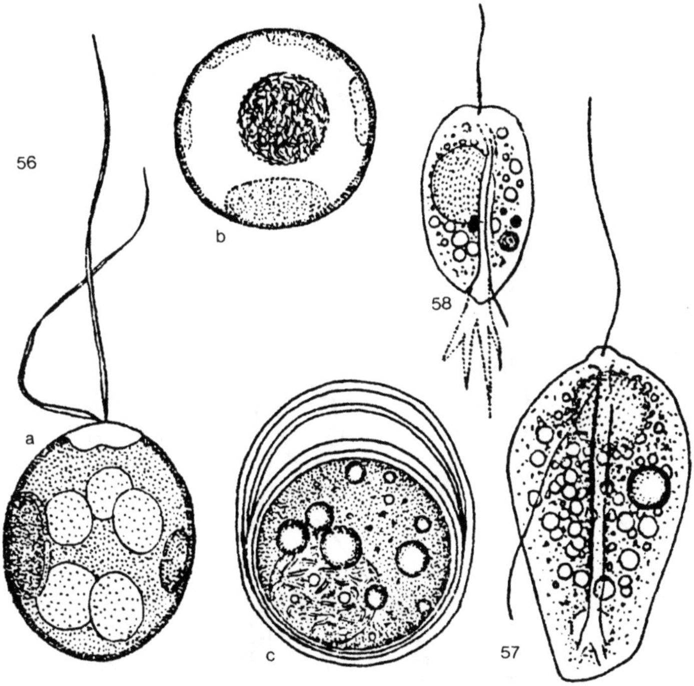

Fig. 56. *Desmomastix globosa* Pascher, *a* ventral view, *b* cross section, *c* resting stage (after Pascher).

Fig. 57. *Protaspis obovata* Skuja, ventral view (after Skuja).

Fig. 58. *Protaspis metarhiza* Skuja, ventral view (after Skuja).

Dimensions: 30–36 μm × 16–25 μm.
Occurrence: puddle with brackish water, seashore of the Baltic Sea; USSR.

3. Family **Spiromonadaceae** Brugerolle and Mignot 1979

Parasitic monads with elongated, drop-shaped cells with an apiculate apex (rostrum). The cell membrane is flexible. Two different flagella are inserted subapically on the ventral side at the base of the rostrum. A contractile vacuole is located at the base of the flagella and a dictyosome lies beside it. The dinokaryotic nucleus, numerous polysaccharide bodies and a food vacuole are also present in the cell. Trichocysts have been reported. The monad attaches onto the apical end of a prey with its rostrum and quickly absorbes the cell content into its food vacuole. Once the food is digested the cell then forms a resting stage.

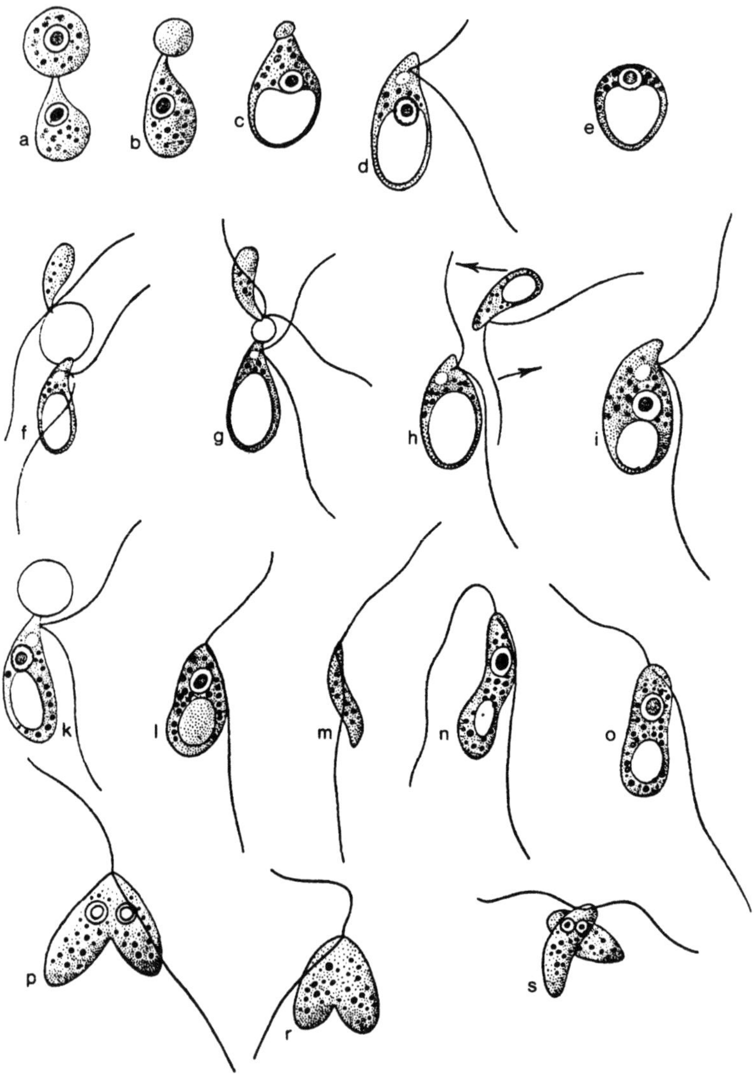

Fig. 59. *Spiromonas angusta* (Dujardin) Alexeieff, *a–d, k* different phases of phagocytosis, *e, i, l* cells red colored, *f–h* two predators with one prey, *m* thin, starved cell, *p, r, s* stages of sexual fusion (after Alexeieff).

The hypnospore is spherical and enveloped by a thick membrane. Four trophozoids are formed by two succesive cell divisions. The fusion of two gametes has been recorded. There is a striking similarity to the Sporozoa.
Important references: Brugerolle and Mignot 1979.

1. Spiromonas Dujardin 1841

Important references: Grassé 1952, Hollande 1938, Brugerolle and Mignot 1979.

Key to the species

1a. Parasitic on freshwater protists . . . . **1. S. angusta**
1b. Parasitic on *Chilomonas paramaecium* . . . . **2. S. perforans**

1. Spiromonas angusta Dujardin 1841, Alexeieff 1929 (fig. 59)
Parasitic on freshwater protists.

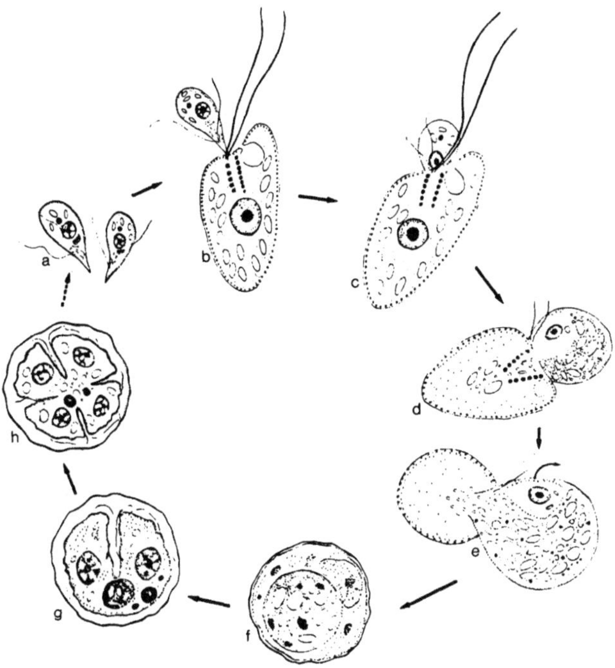

Fig. 60. *Spiromonas perforans* (Hollande) Brugerolle and Mignot, *a* trophozoids, *b* attached to *Chilomonas, c* penetration, *d* endocytosis, *e* predator rotation with separation following, *f* resting stage, *g* 1st division, *h* 2nd division (after Brugerolle and Mignot).

2. Spiromonas perforans (Hollande) Brugerolle and Mignot 1979 (fig. 60)
Bodo perforans Hollande 1938
Parasitic on *Chilomonas paramaecium* Ehrenberg 1839
Dimensions: approx. 18 × 8 µm.

2. Subclass **Dinophycidae** Bourrelly 1970
Dinoflagellatae Schilling 1913, Pascher 1914 sensu stricto, *Diniferae* Lindemann 1928

Motile vegetative cells have a cingulum and a sulcus and two different types of flagella. Amoeboid forms lack a cingulum and sulcus. Quadraflagellate reproductive stages with cingulum and sulcus are known.
Important references: Pascher 1914, Schiller 1933, 1937, Bourrelly 1970.

Key to the orders

1a. Vegetative cells are mostly free-schwimming, furrows are always present . . . . **1. Peridiniales** (p. 88)
1b. Vegetative cells are attached to a substrate or are planktonic; furrows are absent . . . . **2. Dinococcales,** incl. **Dinamoebidiales** (p. 218)

1. Order **Peridiniales** Schütt 1896

Chloroplasts, when present, vary from bright green (cyanelles) to yellow, to dark brown in color. Pyrenoids, stigma, trichocysts and pusules may be present. Both unarmored and armored species are present. The majority of species are marine.
Important references: Schiller 1930, Bourrelly 1970.

Key to the families

1a. Cells unarmored . . . . . **1. Gymnodiniaceae** (p. 89)
1b. Cells armored, with plates **2**
2a. Cells with one apical and 1–3 antapical horns . . . . . **6. Ceratiaceae** (p. 205)
2b. Cells without distinct horns **3**
3a. Cingulum is incomplete . . . . . **3. Hemidiniaceae** (p. 151)
3b. Cingulum is complete **4**
4a. Numerous plates, small and similar in size . . . . **2. Woloszynskiaceae** (p. 139)
4b. Plates not numerous and dissimilar in size **5**
5a. One antapical plate (1'''') . . . . . **8. Dinosphaeraceae** (p. 213)
5b. Two antapical plates (2''''), occasionally only one with an accessory plate present **6**
6a. One antapical plate (1'''') and one posterior accessory plate (1p) present . . . . . **7. Gonyaulacaceae** (p. 211)
6b. The two antapical plates (2'''') are equal in size **7**

7a. Five postcingular plates (5''') . . . . **5. Peridiniaceae** (p. 164)
7b. Six or 7 postcingular plates (6'''–7''') . . . . **4. Glenodiniopsidaceae** (p. 156)

1. Family **Gymnodiniaceae** (Bergh) Schütt 1896
Gymnodinida Bergh 1881b, *Gymnodiniaceaė* Lemmermann 1900, *Kyrtodiniaceae* Schilling 1913, *Kryptoperidiniaceae* Lindemann 1925/26

Unarmored cells that are naked or with a flexible periplast (pellicle) which may be ornamented with small dots (papillae) in rows. Chloroplasts are often lacking. Nutrition is autotrophic, saprobic or holophytic.
Important references: Schütt 1896, Schiller 1933, Bourrelly 1970.

Key to the genera

1a. Cingulum equatorial 2
1b. Cingulum not equatorial 4
2a. Cingulum incomplete . . . . **5. Bernardinium** (p. 138)
2b. Cingulum complete 3
3a. Cingulum circumscribes the cell . . . . **2. Gymnodinium** (p. 98)
3b. Cingulum spirals distinctly . . . . **4. Gyrodinium** (p. 135)
4a. Cingulum placed forward; the epicone is less than 1/3 the length of the cell . . . . **1. Amphidinium** (p. 89)
4b. Cingulum posteriorly placed; the epicone occupies more than 2/3 the length of the cell . . . . **3. Katodinium** (p. 125)

1. **Amphidinium** Claparède and Lachmann 1858/61 (fig. 46b)

Cells egg or broadly egg-shaped; sometimes dorsiventrally flattened. Cingulum placed anteriorly forward, with little or no displacement (thus, the epicone may be very minute). Sulcus usually extends from the cingulum (rarely from apex to antapex) without torsion. The cell wall is thin and structureless. The cytoplasm is hyaline, with or without chloroplast or cyanells. A stigma is usually present in zoospores. Rhizopodia have been reported in motile cells. Nonmotile monads embedded in fine mucilage have been reported. Resting stages are spherical and enveloped by a thick gelatinous layer. Some species are know to be capable of inhabiting both fresh and saline waters. Cells are autotrophic, saprobic, or holozoic.
Important references: Kofoid and Swezy 1921, Schiller 1933, Huber-Pestalozzi 1950, Javornický 1957, 1967.

Key to the species

1a. Cell length $< 20\,\mu m$ 4
1b. Cell length $> 20\,\mu m$ 2
2a. Epicone $< 1/5$ the lenght of the cell . . . . **4. A. lacustre**
2b. Epicone $> 1/5$ the lenght of the cell 3
3a. Chloroplasts radially arranged . . . . **6. A. sphagnicola**
3b. Blue-green chloroplasts not radially arranged 5
4a. Cells elongated and spindle-shaped . . . . **5. A. mucicolum**

4b. Cells not elongated or spindle-shaped . . . . **2. A. elenkinii**
5a. Sulcus extends into the epicone . . . . **1. A. amphidinioides**
5b. Sulcus does not extend into the epicone . . . . **3. A. lacunarum**

1. Amphidinium amphidinioides (Geitler) Schiller 1933 (fig. 61)
Gymnodinium amphidinioides Geitler 1924, *Amphidinium wigrense* Wołoszyńska 1925, *A. geitleri* Huber-Pestalozzi 1950, *A. bourrellyi* Wawrik 1982
Cells are ovoid and slightly flattened dorsiventrally. The epicone may be rounded or bluntly conical or variable in shape. The shape of the hypocone also varies. It may be broadly rounded or narrowly rounded or cylindrical with a rounded bottom. The deeply incised cingulum is wide and angled to the left. The sulcus often extends into the epicone and may reach to the antapex in the hypocone. A nose-shaped projection may be present at the junction of the cingulum and sulcus. The numerous chloroplasts are oval or irregularly ribbon-shaped, scattered throughout the cytoplasm and blue-green (cyanells?) in color. Several refractile bodies are also present in the hyaline cytoplasm. The presence of one or two red droplets, erroneously identified as the stigma, have been recorded. Nutrition varies from autotrophic to phagocytic. Spherical resting stages are surrounded by a thick membrane. The non-motile stage of a monad has been observed embedded in the fine, thick layer of mucilage which surrounds the resting stages.
Dimensions: 18–30 μm × 11–20 μm.
Occurrence: ponds, peat bogs, puddles; Central Europe, Sweden, S. America.

2. Amphidinium elenkinii Skvorcov 1925 (fig. 63)
A. larvale Lindemann 1928, *A. hyalinum* Entz 1930, *A. tatrae* Wołoszyńska 1936, *A. tatrae* fo. *achromaticum* Wołoszyńska 1936, *A. luteum* Skuja 1939, *A. gyrinum* Harris 1939/40, *A. tenagodes* Harris 1939/40, *A. turicense* Huber-Pestalozzi 1950, *A. lohamari* Skuja 1956, *A. skujae* Christen 1958, *Gymnodinium rarum* Litvinenko 1963.
Cells are rounded or oval-shaped with a highly variable epicone. The epicone varies from button shaped to broadly conical; the hypocone is oval. The cingulum and sulcus are distinct and relatively deep: the sulcus reaches the antapex in most cells. The longitudinal flagellum is usually double the cell's length. The cytoplasm is hyaline and colorless and includes small refractive bodies. Chloroplasts, when present, are of various sizes, oval shaped, and from 1 to 7 in number. A stigma has sometimes been observed in vegetative cells.
Dimensions: 6.4–16 μm × 4.8–15 μm.
Occurrence: plankton of oligo- and eutrophic rivers, pools, ponds and lakes.

3. Amphidinium lacunarum Skuja 1964 (fig. 62)
Cells are ellipsoid and somewhat dorsiventrally flattened. The epicone is rounded, the hypocone is elongated and rounded; the antapex is sometimes excavated. The deep cingulum circumscribes the cell, converges on the ventral side and is "V" shaped. The sulcus sometimes reaches to the antapex. Refractile bodies are present in the cytoplasm. Six to ten blue green (cyanells) chloroplasts are present.
Dimensions: 25–30 μm × 17–20 μm.
Occurrence: puddles; Sweden.

4. Amphidinium lacustre Stein 1883 (fig. 64)
Cells are egg-shaped; the epicone is button shaped. The length of the epicone is 0.13 that of the cell and its width 0.52 that of the width of the cell. The short hypocone is heart-

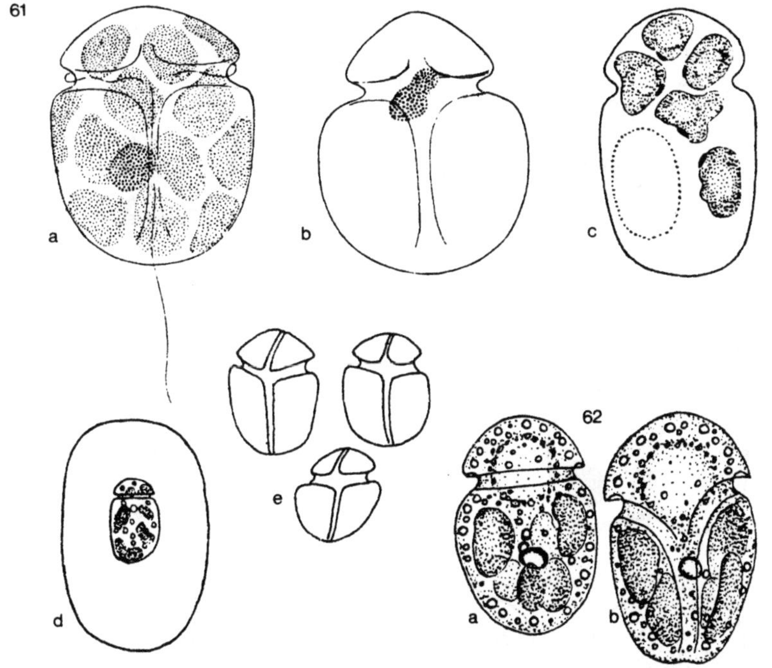

Fig. 61. *Amphidinium amphidinioides* (Geitler) Schiller, *a, b* ventral view, *c* chloroplasts with pseudostigmas, *d* resting stage, *e* ventral view (*a, b* after Javornický, *c, d* after Geitler, *e* after Wołoszyńska).

Fig. 62. *Amphidinium lacunarum* Skuja, *a* dorsal view, *b* ventral view (after Skuja).

shaped. The sulcus is wide at the cingulum and narrow at its end near the antapex. Several large, pariental chloroplasts are present. A stigma is absent.
Dimensions: 23 × 18 µm.
Occurrence: fresh and saline waters; Central Europe, S. America.

5. Amphidinium mucicolum Conrad 1941 (fig. 65)
Cells are elongated, spindle-shaped, and they are not dorsiventrally flattened. The epicone is broadly egg-shaped or pentagonal, and obliquely cut. The hypocone is elongated and elliptical or egg-shaped. It is sometimes conical and rounded at the antapex. The cingulum circumscribes the cell; the sulcus is strongly curved. Refractive bodies and food vacuoles are present in the colorless protoplasm.
Dimensions: 4–18.5 µm × 5–6.5 µm.
Occurrence: inside or outside the gelatinous matrix found with *Batrachospermum* filaments; Belgium.

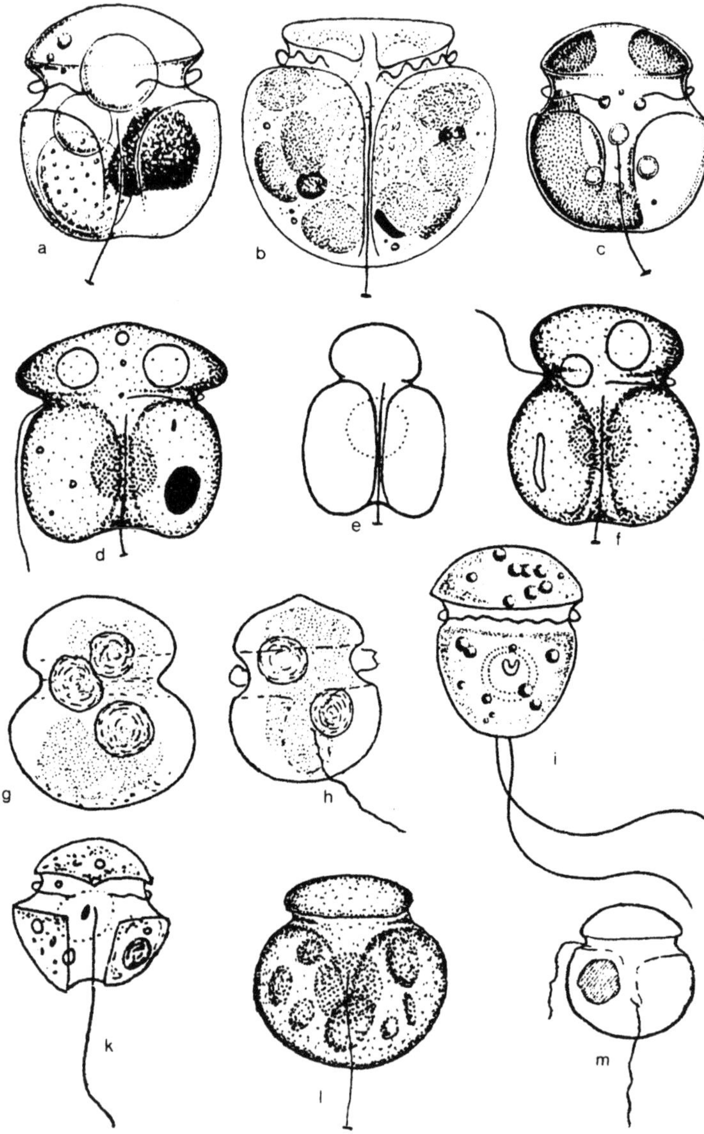

Fig. 63. *Amphidinium elenkinii* Skvorcov *a–l* variability of the cell shape, *i* planozygote, *a, b, c, g, h, k* specimens with chloroplasts (*a, c* after Javornický, *b* after Ettl, *d–f, l* after Popovský, *g, h, m* after Harris, *i* after Litvinenko, *k* after Entz).

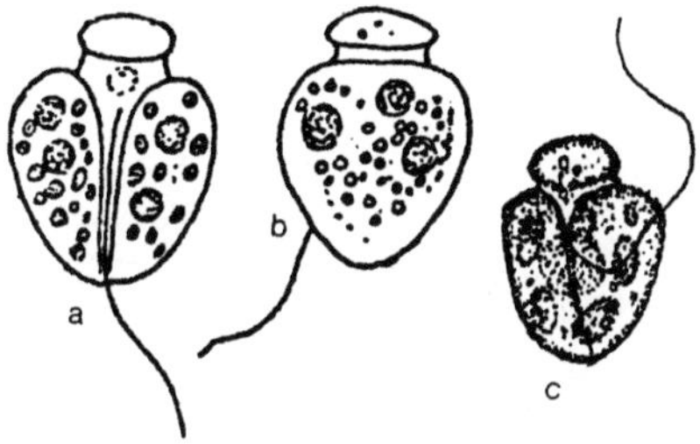

Fig. 64. *Amphidinium lacustre* Stein (*a, b* after Stein, *c* after Bicudo and Skvortzov).

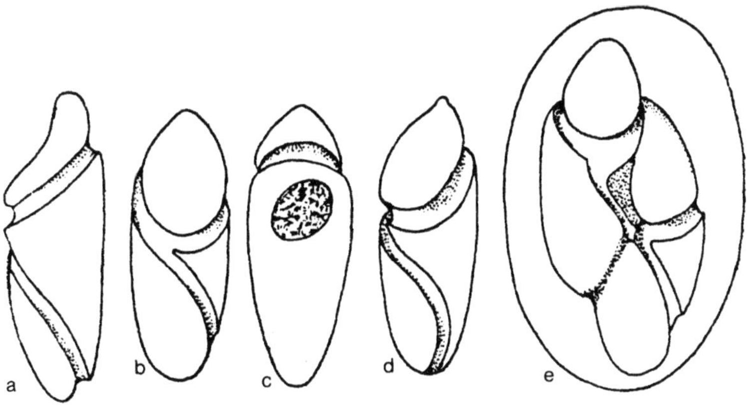

Fig. 65. *Amphidinium mucicolum* Conrad, *a–d* vegetative cells, *e* division in nonmotile stage (after Conrad).

6. Amphidinium sphagnicola Conrad 1946 (fig. 66)
Amphidinium radiatum Javornický 1957

Cells are nearly spherical or widely ovoid and may be slightly flattened dorsiventrally. The epicone is rounded or bluntly conical; the hypocone is rounded. The cingulum is wide and spirals sligthly to the left. The sulcus is poorly defined and does not reach the antapex. The chloroplasts are discoidal and club-shaped. They are usually radially arranged in the epicone and are often net-like in the hypocone. Minute rod-like bodies are scattered in the cytoplasm. A stigma is lacking.

Dimensions: 11–33 µm × 9–27 µm.

Occurrence: dam reservoir in high mountains, peat bogs; Belgium, Czechoslovakia.

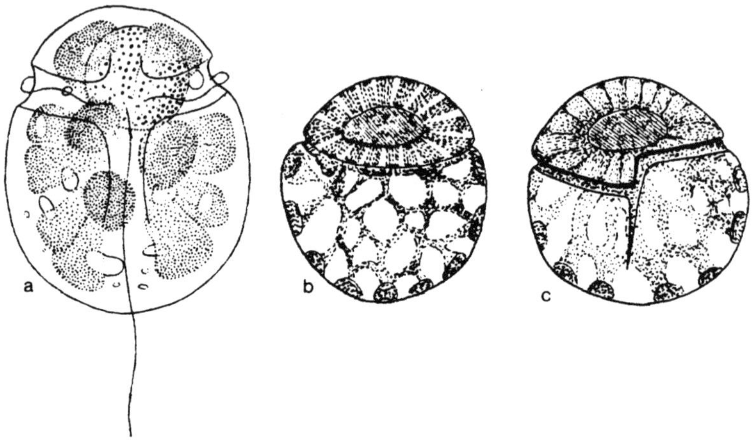

Fig. 66. *Amphidinium sphagnicola* Conrad (*a* after Javornický, *b, c* after Conrad).

Appendix

1. Amphidinium aeschrum Harris 1939/40 is in fact probably a dividing *Amphidinium* or *Gymnodinium* monad (fig. 67).

2. *Amphidinium cryophilum* Wedemayer, Wilcox and Graham 1982 (fig. 68)

Cells ovoid, epicone acute, more commonly bluntly cone shaped. Hypocone subcylindrical. The cingulum is premedian, deep, and displaced one time its width due to an abrupt turn near its terminal junction with the sulcus. The deep sulcus is narrower than the cingulum, and nearly reaches the apex but veers to the left of it. In the hypocone the sulcus reaches about halfway to the antapex. Chloroplasts are numerous and peripheral. Chloroplasts shape varies from spherical to elongate to somewhat irregular. A stigma is lacking.
Dimensions: 33 × 22 µm.
Occurrence: euplankter, winter; USA.
Note: In many respects resembles *A. amphidinioides* (Geitler) Schiller and *Woloszynskia pascheri* (Suchlandt) v. Stosch.

3. Amphidinium kesslitzii Schiller 1928 (fig. 69)

Cells somewhat ovoid and slightly flattened dorsiventrally. The epi- and hypocones are subspherical. The cingulum encircles the cell and the sulcus extending into the hypocone where it reaches the antapex. Chloroplasts are disc-shaped and numerous.
Dimensions: 7–14 µm × 4.5–7 µm.
Occurrence: Mediterranean sea and in freshwater temporary pools in S. America.

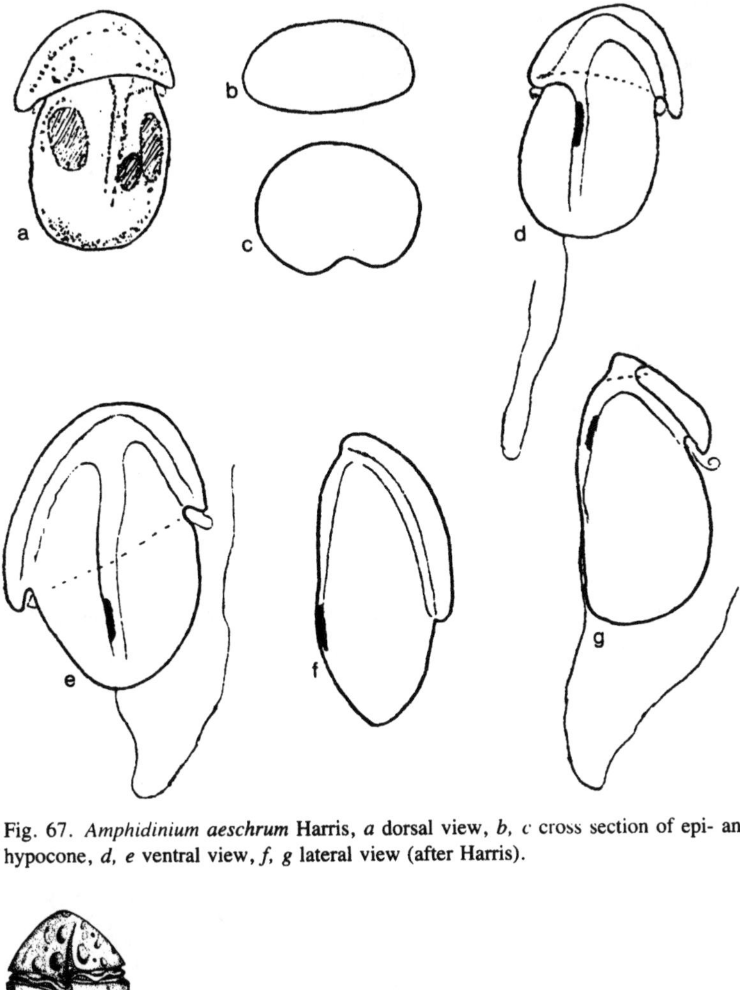

Fig. 67. *Amphidinium aeschrum* Harris, *a* dorsal view, *b, c* cross section of epi- and hypocone, *d, e* ventral view, *f, g* lateral view (after Harris).

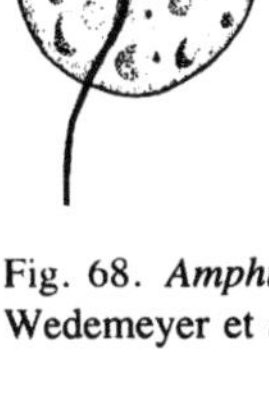

Fig. 68. *Amphidinium cryophilum* Wedemayer et al. (redrawn after Wedemeyer et al.).

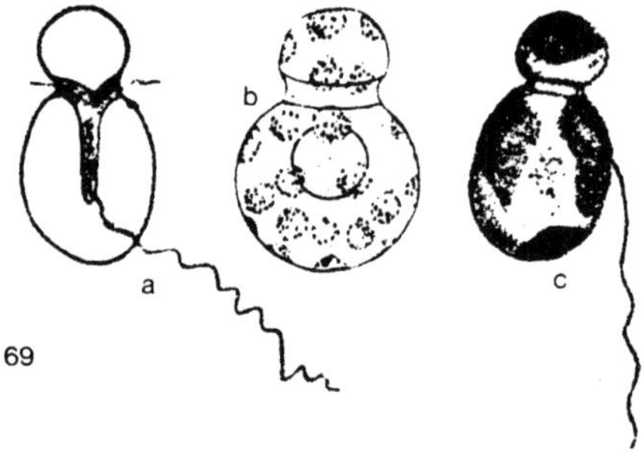

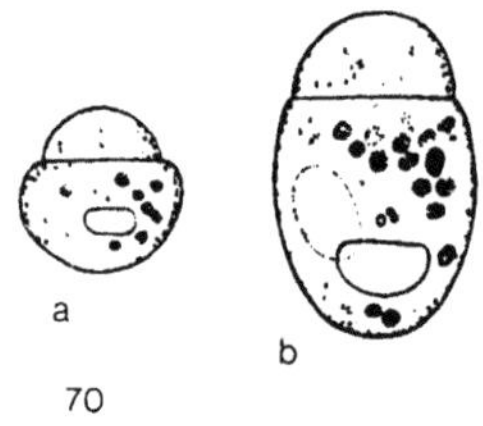

Fig. 69. *Amphidinium kesslitzii* Schiller (*a, c* after Schiller, *b* after Bicudo and Skvortzov).

Fig. 70. *Amphidinium ovoideum* Lemmermann (after Bicudo and Skvortzov).

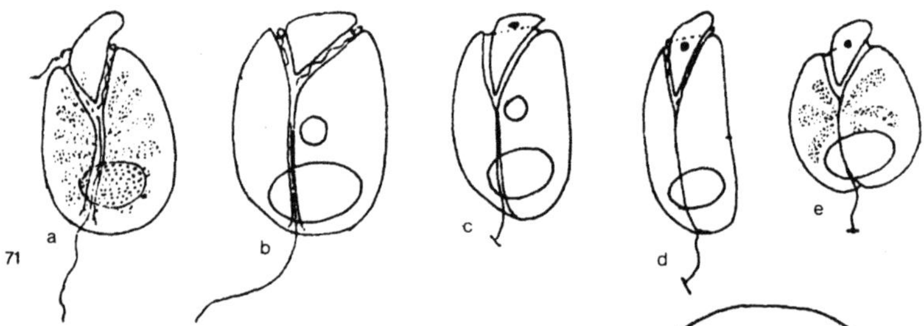

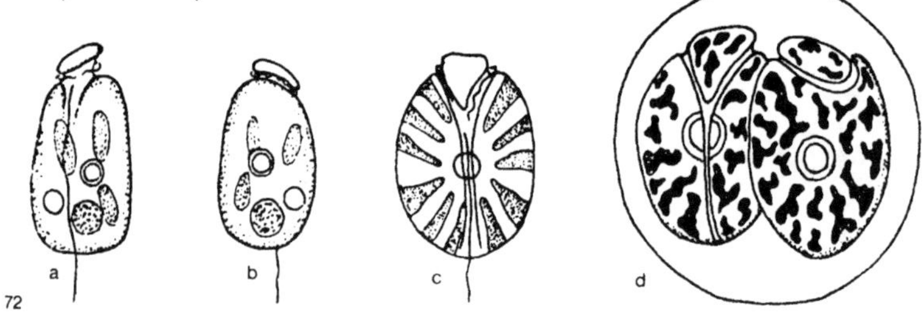

Fig. 71. *Amphidinium klebsii* Kofoid and Swezy (after Lebour).

Fig. 72. *Amphidinium operculatum* Claparède and Lachmann (*a, b* after Conrad, *c* after Calkins, *d* after Zimmermann).

4. *Amphidinium klebsii* Kofoid and Swezy 1921 (fig. 71)
A. operculatum Klebs 1884, *A. Klebsii* fo. *minus* Carter 1938, *A.* sp. Wislouch 1924, *A.* sp. Skuja 1930; non: *A. operculatum* Claparède and Lachmann 1858–1861, *A. operculatum* var. *minutum* Conrad 1926

Cells are ellipsoidal or egg-shaped and dorsiventrally flattened with a delicate pitted cell covering. The epicone is tongue-shaped and curved to the left. The hypocone is broadly elliptical and slightly asymmetric. The cingulum is narrow and deep; the sulcus is narrow and nearly reaches the antapex. Choroplasts are numerous, club-shaped and radially arranged. One central pyrenoid is present.
Dimensions: 12–36(46) µm × 7–26 µm.
Occurrence: saline waters; USSR, Great Britain.

5. *Amphidinium operculatum* Claparède and Lachmann 1858/1861 (fig. 72)
A. operculatum var. *minutum* Conrad 1926; non: *A. operculatum* Klebs 1884

Cells are oval with broadly rounded ends. The epicone is small and tongue-shaped. The hypocone is rounded. The cingulum is wide and deep; the sulcus is relatively narrow and nearly reaches the antapex. The chloroplasts are usually radially arranged.
Dimensions: 35–50 µm × 15 µm.
Occurrence: saline waters of Europe.

6. *Amphidimum ovoideum* Lemmermann 1900 (fig. 70)

Cells are sperical or broadly ovoid and somewhat dorsiventrally flattened. The cingulum encircles the cell; the sulcus is not well defined. The chloroplasts are numerous.
Dimensions: 11–24 µm × 11–21 µm.
Occurrence: brackish waters; Germany. Freshwater lakes; S. America.

7. *Amphidinium rostratum* Proškina-Lavrenko 1945 (fig. 73)

Cells are ovoid, asymmetric and dorsiventrally flattended. The small epicone is somewhat cone shaped and curved toward the dorsal side. The hypocone is rounded. The cingulum is deep; the sulcus widens toward the antapex. The chloroplasts are elongated, oval, and numerous. One pyrenoid is present.
Dimensions: 21–30 µm × 14–21 µm.
Occurrence: saline pool; Ukraine.

8. *Amphidinium rhynchocephalum* Anisimova 1926 (fig. 74)

Cells are broadly egg-shaped, and dorsiventrally flattended. The epicone is strongly conical and curved toward the ventral side. The hypocone is broadly rounded or heart-shaped. The cingulum is very deep. The sulcus is "V" shaped and does not reach the antapex. The chloroplasts are radially arranged, club-shaped. One pyrenoid is present which is surrounded by starch granules.
Dimensions: 17–23 µm × 10–12.4 µm.
Occurrence: in a mud of saline waters; USSR.
Schiller (1955) described the following species: *Amphidinium caerulescens, A. glaucovirescens, A. bidentatum, A. oculatum, A. sauerzopfi, A. ruttneri, A. eucephalum, A. multiplex, A. obliquum, A. inconstans, A. vorax, A. pusillum, A. viride*. These appear to be related to the species *A. elenkinii* Skvorcov 1925.

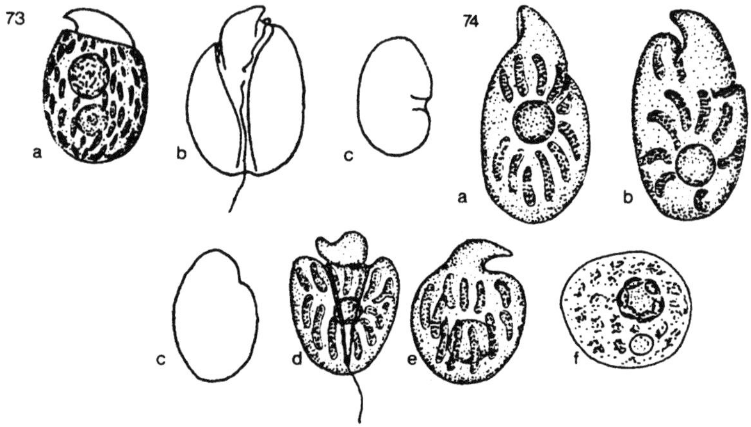

Fig. 73. *Amphidinium rostratum* Proškina-Lavrenko, *a* lateral view, *b* ventral view, *c* antapical view (after Proškina-Lavrenko).

Fig. 74. *Amphidinium rhynchocephalum* Anisimova, *a, b, d, e* vegetative cells, *c* apical view, *f* unmotile stage (after Anisimova).

2. **Gymnodinium** Stein 1878 (fig. 46a)

Peridinium Ehrenberg 1835 partim

Cells are ovoid to spherical with small projections sometimes present. The cingulum forms a descending left spiral that is displaced less than 1/5 the total length of the body and circumscribes the cell. The sulcus may extend from the apex to the antapex. The cytoplasm varies from colorless, with or without chloroplasts to colored; pigment granules and cyanells may be present. Cells are naked or may possess a pellicle. The surface may be smooth or striated. A true stigma is sometimes present in zoospores and gametes. Trichocysts have been found in some species. Motile cells usually occur singly but chain formation is known. Production of copious amounts of mucilaginous material surrounding the motile cell has been reported. Motile cells divide obliquely into two daughter cells. Autospore production has been observed in non-motile cells surrounded by concentric layers of mucilaginous material. Thick walled cells resembling 2 pyramids joined at their bases have been described as have thick-walled hexagonal cells and *Tetradinium*-like, thick walled cells. These are possible zygotes. Cells are autotrophic, holozoic or saprotrophic.

Important references: Stein 1878, Kofoid and Swezy 1921, Pascher 1927, Schiller 1933, 1937, Huber-Pestalozzi 1950, Kiselev 1954, Baumeister 1957, Cridland 1958, Popovský 1968, 1971a, 1984, Bourrelly 1970, Starmach 1974, Matvienko and Litvinenko 1977.

Key to the species

1a. Cells spherical 2
1b. Cells not spherical 17
2a. Chloroplasts present 3
2b. Chloroplasts absent 12
3a. Chloroplasts radially arranged 4
3b. Chloroplasts not radially arranged 7
4a. Sulcus extends into the epicone more than 1/3 of its length . . **20. G. uberrimum**
4b. Sulcus does not extend into the epicone 5
5a. Cells 30 µm or less in length 6
5b. Cells more than 30 µm in length . . . . **17. G. paradoxum**
6a. Cells flattened dorsiventrally; epicone broadly rounded . . . . . **18. G. thomasi**
6b. Cells slightly flattened dorsiventrally; epicone conical . . . **1. G. acuminatum**
7a. Chloroplasts bright blue-green . . . . **2. G. aeruginosum**
7b. Chloroplasts not bright blue-green or are absent 8
8a. Sulcus extends into the epicone 11
8b. Sulcus does not extend into the epicone 9
9a. Chloroplasts numerous . . . . **12. G. lacustre**
9b. Chloroplasts (or cyanells) up to 5 in number 10
10a. Chloroplasts 1 to 2 in number, hypocone rounded . . . . . . **13. G. lantzschii**
10b. Chloroplasts 1 to 5 in number, hypocone irregularly bell-shaped . . . .
. . . . **11. G. triceratium**
11a. Chloroplasts numerous . . . . **4. G. chiastosporum**
11b. Chloroplasts 1 to 2 in number . . . . **7. G. discoidale**
12a. Sulcus extends into the epicone 16
12b. Sulcus does not extend into the epicone 13
13a. Epicone pointed with spines at the apex . . . . **14. G. latum**
13b. Epicone not pointed 14
14a. Sulcus wide . . . . **15. G. mitratum**
14b. Sulcus narrow 15
15a. Epicone broadly rounded . . . . **10. G. hiemale**
15b. Epicone conical or helmet shaped . . . . **13. G. lantzschii**
16a. Sulcus deep and V-shaped . . . . **3. G. austriacum**
16b. Sulcus broad; not V-shaped . . . . **6. G. colymbeticum**
17a. Chloroplasts absent . . . . **9. G. helveticum**
17b. Chloroplasts present 18
18a. Chloroplasts parietal 19
18b. Chloroplasts radially arranged 21
19a. Cingulum does not spiral or spirals slightly 22
19b. Cingulum distinctly spiraled 20
20a. Sulcus does not reach the antapex . . . . **22. G. wawrikae**
20b. Sulcus reaches the antapex . . . . **21. G. varians**
21a. Hypocone extended into a short, cone-shaped portion . . . . **8. G. fuscum**
21b. Hypocone bell shaped . . . . **11. G. impatiens**
22a. Sulcus extends into the epicone . . . . **16. G. palustre**
22b. Sulcus does not extend into the epicone . . . . **5. G. cnecoides**

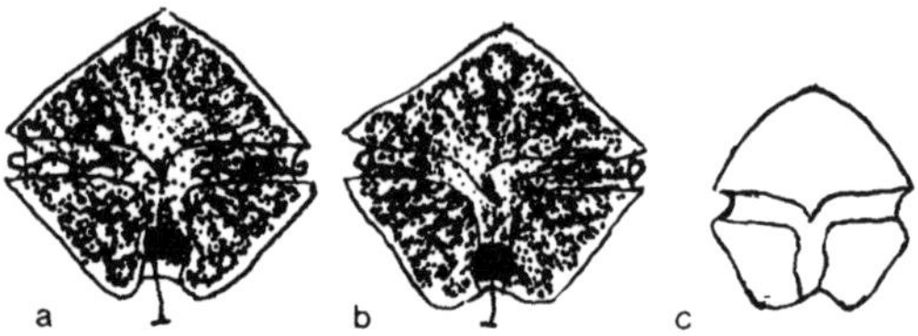

Fig. 75. *Gymnodinium acuminatum* Christen (after Christen).

1. Gymnodinium acuminatum Christen 1954 (fig. 75)
Cells are slightly flattened dorsiventrally and pentangular in ventral view. The epicone is triangular; the hypocone is large with a small sinus at its antapex. The cingulum is deep and spirals. The sulcus is narrow and reaches to the antapex. Chloroplasts are numerous, elongated and radially arranged.
Dimensions: 25–30 µm × 30–34 µm.
Occurrence: lakes; Switzerland.

2. Gymnodinium aeruginosum Stein 1883 (fig. 76)
G. P. Dohrni Wawrik 1956, *G. acidotum* Nygaard 1949, *G. viride* Penard 1891, *G. campaniforme* Popovský 1971
Cells are ovoid to rhombodial in shape and dorsiventrally flattened. The epicone is broadly hemispherical, or conical or bell-shaped. The hypocone is rounded or triangular. The cingulum is distinct; the sulcus extends from the epicone to the antapex. The numerous chloroplasts are bright blue-green. They are minute and lens or rod-shaped and parietally arranged. There is no stigma. Cysts are spherical with a fine to firm mucilaginous envelope.
Dimensions: vegetative cells 12.8–44 µm × 8.7–35 µm, cysts 35–60 µm in diameter.
Occurrence: common, cosmopolitan in ponds, lakes, bogs and rivers, from oligo- to slightly eutrophic waters.

3. Gymnodinium austriacum Schiller 1933 (fig. 77)
G. tridentatum Schiller 1933, *G. cruciatum* Thompson 1950, *G. thompsonii* (Thompson) Kiselev 1954, *G. waltzii* Baumeister 1957, *G. titubans* Christen 1958, *G. autumnale* Christen 1959
Cells are egg-shaped to oval and slightly flattened dorsiventrally. In dorsiventral view the epi- and hypocones are conical, bell-shaped, or hemispherical. The apex may have several small projections (pseudopodia?). The epicone is slightly larger than the hypocone. The wide cingulum is deeply incised and spirals slightly to the left. The narrow sulcus is deeply incised and extends approximately 3/4 of the way into the epicone. It may extend to the antapex on the hypocone where it deflects from the longitudinal axis. The lobopodia are extruded from the sulcus in the region of the hypocone. The presence of rhizopodia has also been reported. There are no chloroplasts or stigma. One to two food vacuoles are present.
Nutrition: holophytic on chlorococcal algae.
Dimensions: 17–41 µm × 13.5 × 30 µm.
Occurrence: oligo- or slightly eutrophic lakes and artifical reservoirs in Europe and N. America.

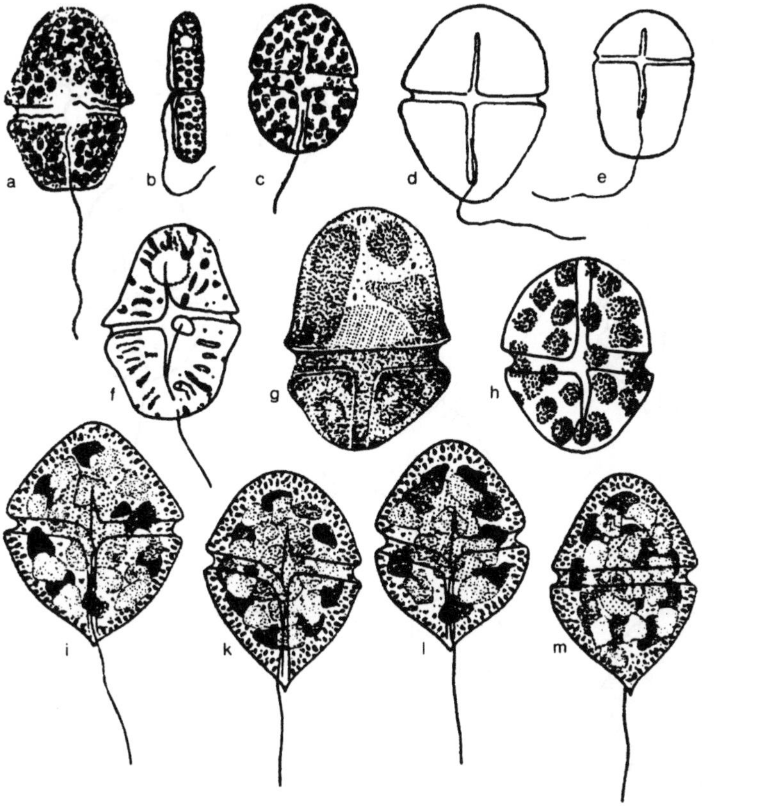

Fig. 76 *Gymnodinium aeruginosum* Stein (*a* after Schilling, *b* after Lemmermann, *c, i, k, l, m* after Nygaard, *d, e* after Harris, *f* after Penard, *g* after Popovský, *h* after Wawrik).

4. Gymnodinium chiastosporum (Harris) Cridland 1958 (figs. 78, 79)
Tetradinium chiastosporum Harris 1939/40, *Dinastridium chiastosporum* (Harris) Starmach 1974, *Gymnodinium hippocastanum* Cridland 1958, *Dinastridium sexangulare* Pascher 1927

Cells are oviod and slightly flattened dorsiventrally. The epicone and hypocone are rounded. The cingulum is wide; the sulcus extends 3/4 of the way into both the epi- and hypocones. Numerous ovoid chloroplasts are present. There is no stigma. Two or three brown corpuscles are often present in the cytoplasm. Starch is abundant in the hypocone. Free swimming individuals enlarge (to about 30 µm × 20 µm) and their girdles become indistinct. They then sink to the bottom and a cell wall and spines slowly develop

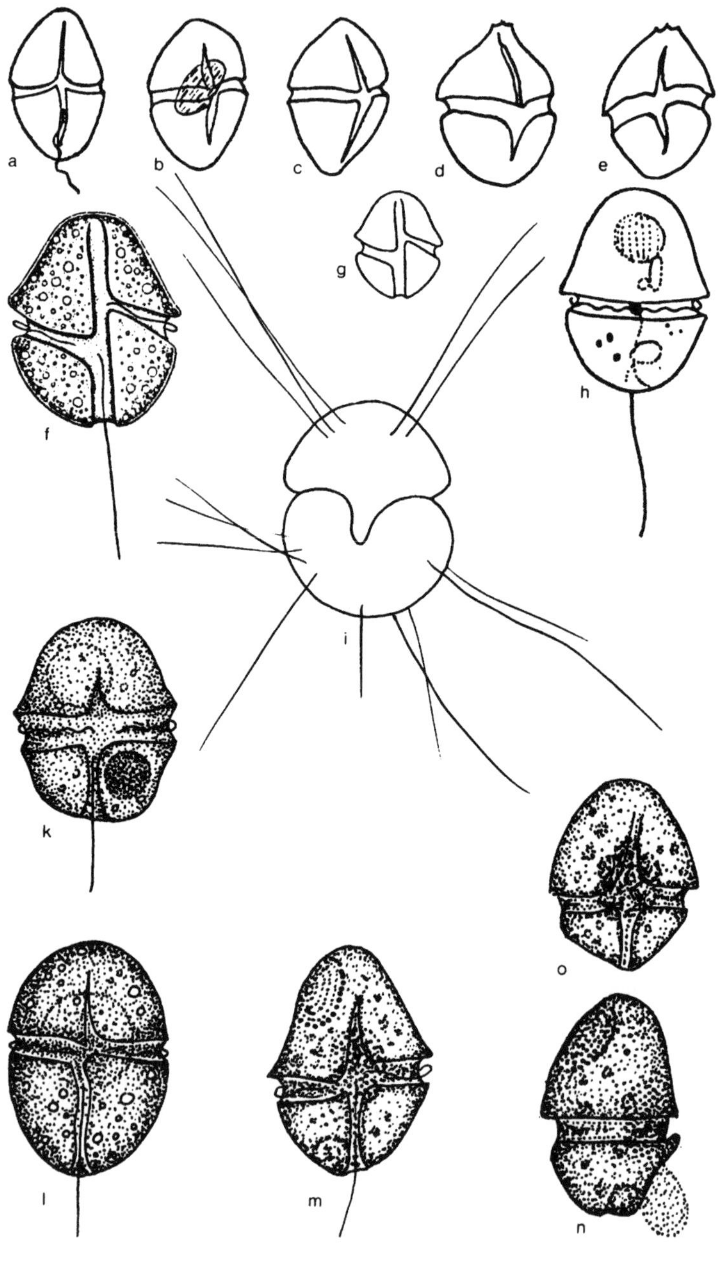
a
b
c
d
e
f
g
h
i
k
o
l
m
n

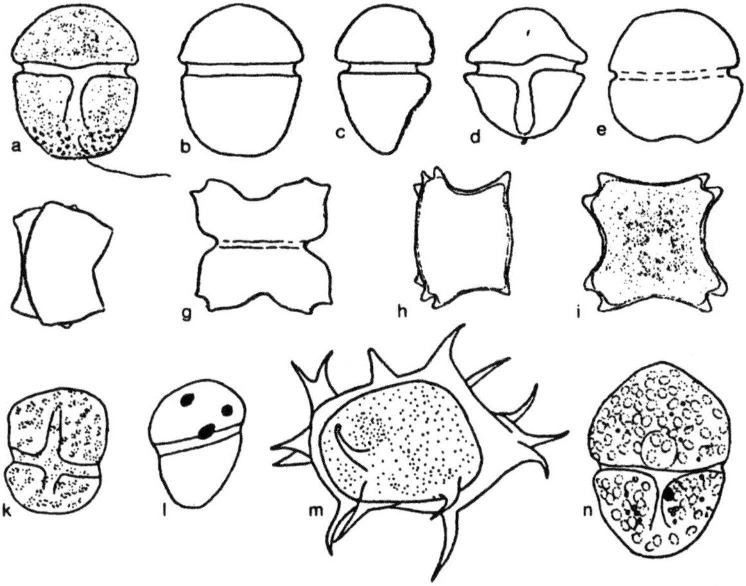

Fig. 78. *Gymnodinium chiastosporum* (Harris) Cridland, *a–e* vegetative cells, *f–i* hypnozygotes, *k, l* vegetative cells, *m* hypnozygote, *n* vegetative cell (*a–i* after Harris, *k–m* after Cridland, *n* after Pascher).

(hypnozygotes?). Reproduction by formation of non-motile daughter cell lacking furrows and flagella is known.
Dimensions: vegetative cells 12–25 µm × 10–12 µm, resting spores 13 bis 40 µm × 13–25 µm.
Occurrence: pools, ponds; Great Britain, Czechoslovakia.

5. Gymnodinium cnecoides Harris 1939/40 (fig. 80)

G. saginatum Harris 1939/40, *G. luteofaba* Javornický 1965

Cells are ovoid and slightly flattened dorsiventrally. The epicone is rounded or bluntly conical, the hypocone rounded. Both furrows are wide and deep; the sulcus almost reaches the antapex; it does not extend into the epicone. One to two chloroplasts and small droplets are visible in the cytoplasm. There is no stigma. Reproduction is by means of zoospores.

Fig. 77. *Gymnodinium austriacum* Schiller, *i* specimen with pseudopodia (*a–e* after Schiller, *f, g* after Thompson, *h, i* after Baumeister, *k, l* after Christen, *m, n, o* after Popovský).

a
b
c
d
e
f
g

Dimensions: 9–23 μm × 8–18 μm.
Occurrence: peat-bogs, ponds, lakes; Poland and Great Britain.

6. Gymnodinium colymbeticum Harris 1939/40 (fig. 81)
G. pulvisculus Klebs 1912; non: *G. pulvisculus* Pouchet 1885
Cells are spherical or broadly ovoid; the epicone and hypocone are approximately equal in size. The slightly smaller epicone is rounded or conical; the hypocone rounded. The sulcus is wide and extends into the epicone. It does not reach the antapex on the hypocone. Chloroplasts and stigma are lacking. Several large colorless bodies are present in the cytoplasm.
Dimensions: 16 × 16 μm.
Occurrence: Great Britain.
Note: *G. pulvisculus* was poorly described by Klebs (1912) and without a figure. *G. pulvisculus* Pouchet 1885 is a synonym of *Oodinium poucheti* (Lemmermann) Chatton 1912.

7. Gymnodinium discoidale Harris 1939/40 (fig. 82)
Glenodinium eurystomum Harris 1939/40
Cells are broadly rounded or obovoid and dorsiventrally flattened. The epicone is rounded or conical. The sulcus extends from the epicone to the hypocone where it does not reach the antapex. One or two large chloroplasts are present; there is no stigma. An agglomeration of refractive droplets in the epi-and hypocones.
Nutrition: autotrophic, holozoic.
Dimensions: 8–18 μm × 7–15.5 μm.
Occurrence: temporary pools, lakes, peat bots; Great Britain, Czechoslovakia, Germany.

8. Gymnodinium fuscum (Ehrenberg) Stein 1878 (figs. 83, 84)
Gymnocystodinium gessneri Baumeister 1957, *Cystodinium gessneri* (Baumeister) Bourrelly 1970, *Gymnodinium caudatum* Prescott 1944
Cells are large and ovoid. The epicone is dome-shaped; the hypocone is as broad or broader than the epicone and narrowed posteriorly forming an inverted cone with a slightly produced tip. The cingulum spirals slightly; the sulcus extends about half way into the hypocone, but barely enters the epicone. The numerous chloroplasts are ovoid, disc or rod-shaped and radially arranged. Comma-shaped trichocysts are present mainly in the hypocone. A stigma is visible in zoospores but not in vegetative cells. Mucilage production has been reported in two morphologically dissimilar non-motile stages under unfavorable conditions: 1. a spherical resting stage enveloped by a thick, somewhat irregular mucilage (Bourrelly 1970) and 2. "*Cystodinium*" stage in which the protoplasm divides into 4 zoospores. The zoospores are non-motile, however, and colorless. Their furrows disappear and the cell develops into a "*Cystodinium*" stage in which chloroplasts are again visible. The motile zoospore may divide into two daughter cells (Baumeister 1957). The zoospore monad has also been reported to lie freely in a pear-shaped mucilaginous

---

Fig. 79. *Gymnodinium chiastosporum* (Harris) Cridland, *a* two autospores escaping from hypnozygote, *b* view from above, *c* view from side, *d–f* formation of zoospores, *g* vegetative cell or zoospore (after Pascher).

80

81

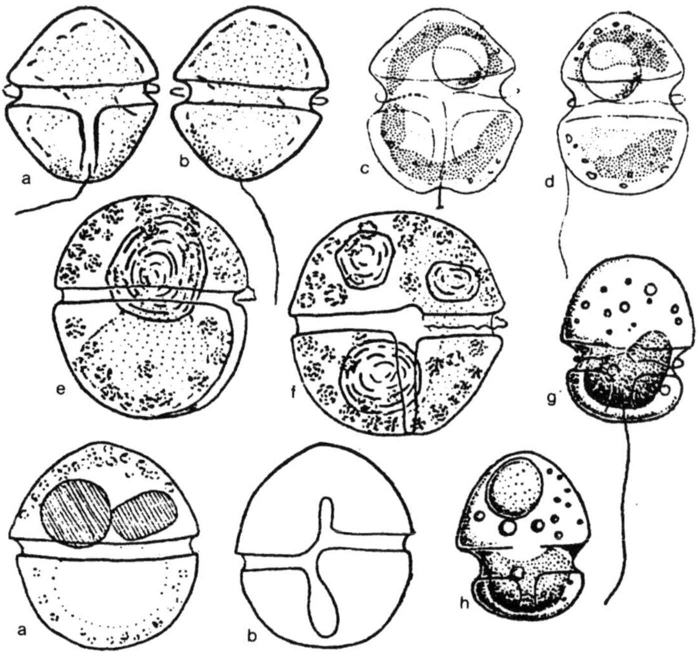

Fig. 80. *Gymnodinium cnecoides* Harris (*a, b, e, f* after Harris, *c, d, g, h* after Javornický).

Fig. 81. *Gymnodinium colymbeticum* Harris (after Harris).

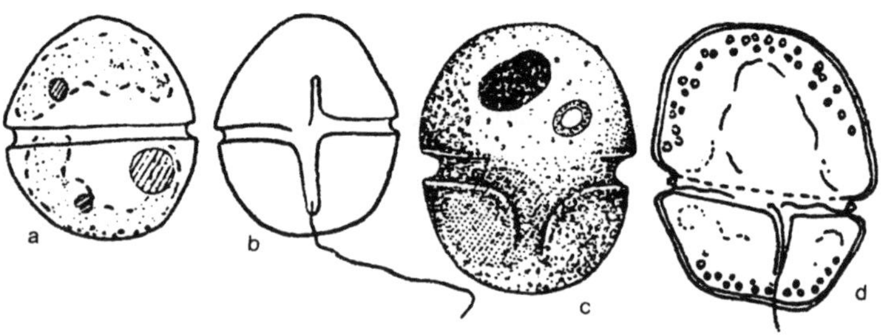

Fig. 82. *Gymnodinium discoidale* Harris (*a, b, d* after Harris, *c* after Popovský).

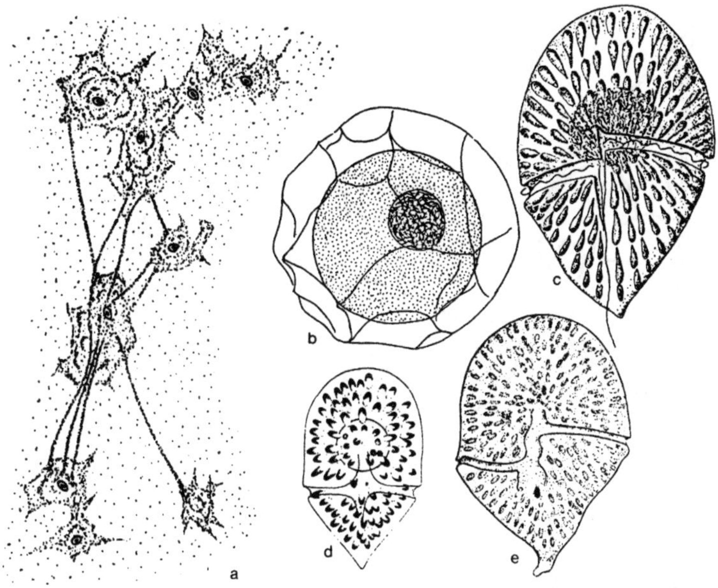

Fig. 83. *Gymnodinium fuscum* (Ehrenberg) Stein, *a* pseudocolony, monads in jelly envelopes, *b* hypnozygote, *c–e* vegetative cells (*a* after Lindemann, *b, c* after Bourrelly, *d* after Bicudo and Skvortzov), *e* (after Prescott).

envelope. Rhizopodia-like structures have been observed at the narrowed pole of the mucilaginous envelope.
Dimensions: monads 80–118 µm × 55–70 µm, zoospores 35–45 µm in length, resting stages 60–65 µm in diameter, "*Cystodinium*" stage approx. 70–80 µm × 45–55 µm.
Occurrence: peat bogs, acid oligotrophic lakes; cosmopolitan, common.

9. Gymnodinium helveticum Penard 1891 (fig. 85)

G. helveticum var. *apiculata* (Zacharias 1903) Utermöhl 1925, *Glenodinium apiculatum* Zacharias in Schilling 1913

Cells are deltoid; the epicone is somewhat conical with three teeth-like projections at the apex. The hypocone is elongated and conical with an apiculate antapex. The cingulum is wide and deeply incised; the epicone is much smaller than the hypocone (*Amphidinium*-shaped!). The sulcus is narrow, deep and extends 3/4 of the way into both the epi- and hypocones. The periplast is relatively firm but flexible. It is ornamented with fine longitudinal furrows and rows of small papillae. Chloroplasts and stigma are lacking. The cytoplasm is light pink in color. Three temporary variations (?) in cell morphology have been reported (Bachmann 1923): 1. both poles may be rounded; 2. both poles may be apiculate; and 3. the apex may be apiculate with three teeth-like projections and the hypocone rounded.

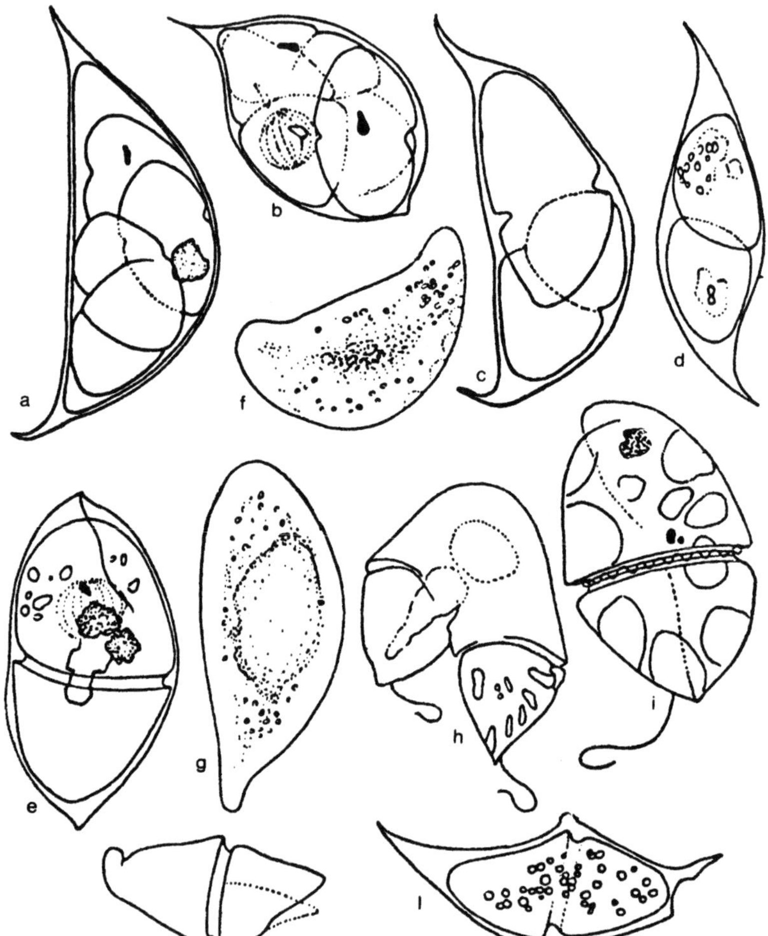

Fig. 84. *Gymnodinium fuscum* (Ehrenberg) Stein, *Cystodinium*-like reproductive cycle (after Baumeister).

Nutrition: holozoic and holophytic.
Dimensions: 50×30 µm.
Occurrence: oligotrophic lakes; Europe.

9.1. fo. *achroum* Skuja 1948 (figs. 42b, 86)
Differs in that the cones are helmet-shaped. Nutrition is holophytic and holozoic by means of a lobopodium. Hypnozygotes are rounded and *"Coelastrum"* shaped.

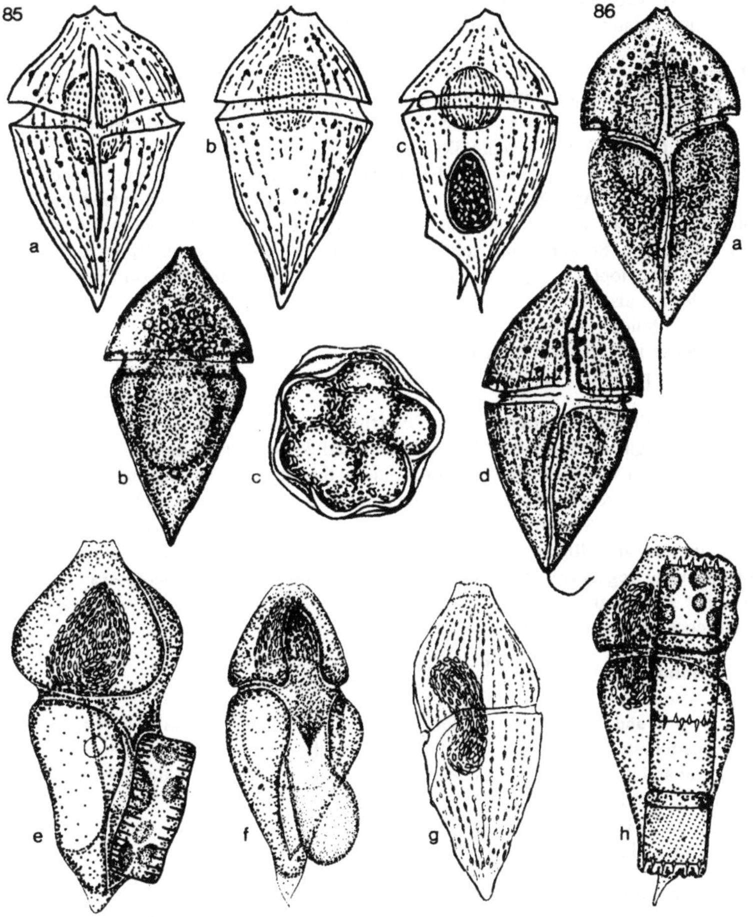

Fig. 85. *Gymnodinium helveticum* Penard (after Penard).

Fig. 86. *Gymnodinium helveticum* fo. *achroum* Skuja, *c* resting stage, *e, h* ingestion of diatoms, *f* cells with pseudopodium, food vacuole and triangularly-shaped nucleus, *g* lateral view (*a–d* after Skuja, *e–h* after Popovský).

Dimensions: vegetative cells 42–57 µm × 23–30 µm.
Occurrence: plankton of oligotrophic lakes; Europe.

10. Gymnodinium hiemale (Schiller) Popovský 1984 (fig, 87)
Massartia hiemalis Schiller 1954, *Katodinium hiemale* (Schiller) Loeblich III 1965, *Katodinium intermedium* Christen 1959; non: *Gymnodinium hiemale* Wołoszyńska 1917, *Gymnodinium hiemale* Skvorcov 1927
Cells are spherical or ovoid and slightly flattened dorsiventrally. The epi- and hypocones are rounded. The cingulum is wide and deep; the sulcus which does not extend into the epicone is narrow and does not reach the antapex. There are no chloroplasts. A stigma may be present. Several small spherical cyanelles which are parietally arranged are present in the cytoplasm. Ingested green algae may be present. Red or brown droplets and corpuscles are also present.
Nutrition is holophytic and holozoic.
Dimensions: 10–15 µm × 9–12 µm.
Occurrence: plankton; Austria, Switzerland.

11. Gymnodinium impatiens Skuja 1964 (fig. 88)
Cells are elongated ellipsoids and slightly flattened dorsiventrally. Both cones are bell-shaped and rounded. The hypocone is excavated at the antapex. The cingulum is wide and deep. The sulcus reaches the antapex. The numerous chloroplasts are radially arranged. There is no stigma. Numerous red droplets are present in the cytoplasm. The hypnospores are cylindrical and ellipsoidal. They are surrounded by a thick, colorless cell wall.
Dimensions: vegetative cells 18–21 µm × 11–14 µm, hypnospores 18 bis 20 µm × 13–14 µm.
Occurrence: pools; Sweden.

12. Gymnodinium lacustre Schiller 1933 (fig. 89)
G. profundum Schiller 1933 ad interim
Cells are oval and slightly flattened dorsiventrally. The epicone is rounded or conical; the hypocone rounded. The epicone may be larger than the epicone. The cingulum is usually equatorial and wide; the sulcus is deeply incised and "V"-shaped. It does not reach the antapex. Several red dark droplets and a stigma may be present.
Dimensions: 13–18 µm × 10–12.5 µm.
Occurrence: lakes, alpine species; Austria, Czechoslovakia.

13. Gymnodinium lantzschii Utermöhl 1925 (fig. 90)
G. minimum Lantzsch 1914, *Glenodinium minimum* (Lantzsch) Bachmann 1924, *Gymnodinium albulum* Lindemann 1928, *G. lantzschii* var. *rhinophoron* Javornický 1957, *G. rhinophoron* (Javornický) Litvinenko 1977, *G. macronucleum* Litvinenko 1963; non: *G. minimum* Klebs 1912
Cells are variable in shape; they are usually ovoid, with a conical or rounded epicone and a rounded hypocone. The cingulum is wide, sometimes barely distinct; the sulcus which is somewhat narrow and does not reach the antapex may also be indistinct. The presence of a hyaline projection has been reported at the junction of the furrows. Cells are usually colorless although 1–2 chloroplasts or cyanells may be present. There is no stigma. Colorless or red bodies are present in the cytoplasm which is sometimes "silvery." Nutrition is holozoic and holophytic.
Dimensions: 9–25 µm × 8–18 µm.

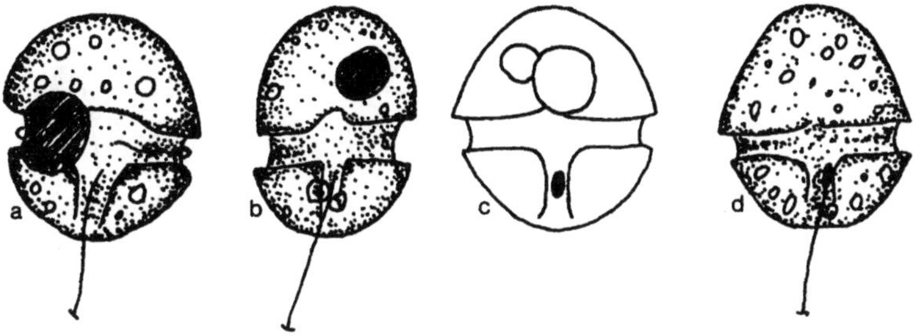

Fig. 87. *Gymnodinium hiemale* (Schiller) Popovský (*a, b* after Schiller, *c, d* after Christen).

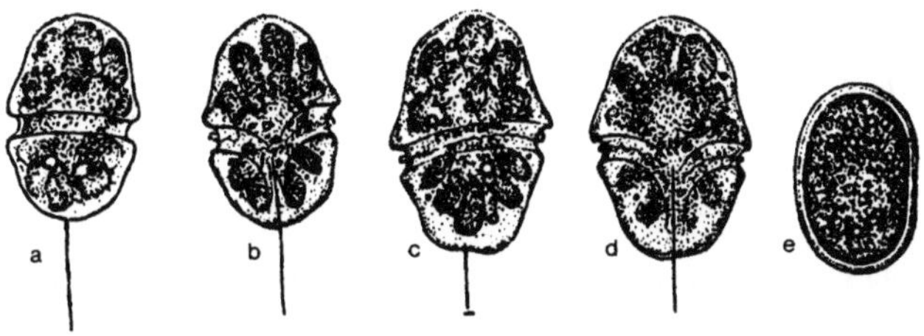

Fig. 88. *Gymnodinium impatiens* Skuja, *a, c* dorsal view, *b, d* ventral view, *e* resting stage (after Skuja).

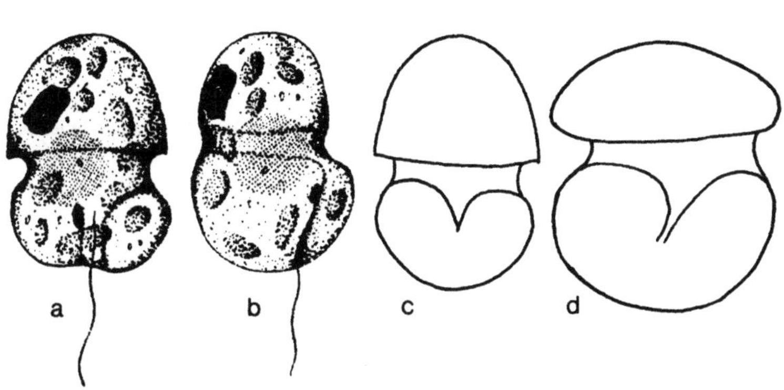

Fig. 89. *Gymnodinium lacustre* Schiller (*a, b* after Popovský, *c, d* after Schiller).

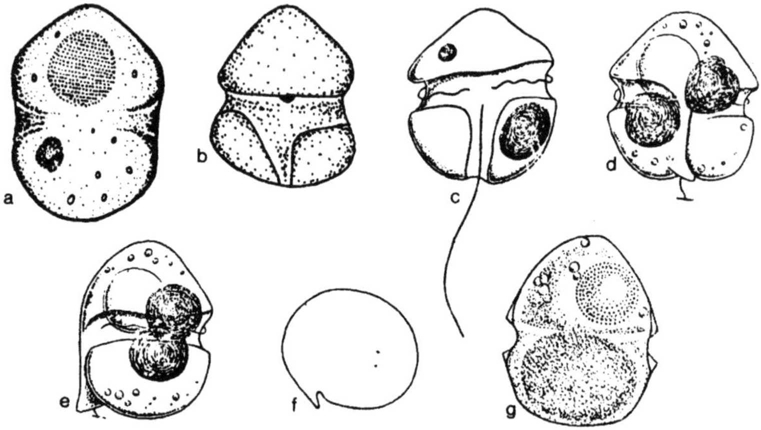

Fig. 90. *Gymnodinium lantzschii* Utermöhl, *e* lateral view, *f* cross section of epicone (*a* after Popovský, *b* after Lindemann, *c* after Schiller, *d–f* after Javornický, *g* after Litvinenko).

Occurrence: mainly a winter form in ponds, pools, lakes, and melted snow, of oligo- to polysaprobic waters.

14. Gymnodinium latum Skuja 1948 (fig. 91)
G. alsiophyllum Skuja 1964

Cells are elliptical or hexagonal and slightly flattened dorsiventrally. The epicone is conical and pointed with two spines at the apex. The hypocone is rounded or square-shaped. The wide, deep cingulum is equatorial. The sulcus almost reaches the antapex. Chloroplasts and stigma are lacking. The pink cytoplasm has refractive droplets and remains of engulfed algae (diatoms). Nutrition is holophytic and holozoic.
Dimensions: 25–45 μm × 20–34 μm.
Occurrence: pools, lakes, eutrophic waters; Sweden.
Note: There are many features by which *G. latum* resembles *G. austriacum* Schiller and *Peridinium achromaticum* Levander.

15. Gymnodinium mitratum Schiller 1933 (fig. 92)
G. eurytopum Skuja 1948, *G. simile* Skuja 1956

Cells are elliptical in ventral view and circular in apical view. The epicone is slightly larger than the hypocone; both are rounded. The wide cingulum is almost equatorial; the sulcus is wide and nearly reaches the antapex. There are no chloroplasts or stigma. Several red droplets are often present. Multiplication occurs in the motile state by the formation of two divisional products or in the non-motile state by autospore formation.
Dimensions: 13–35 μm × 9–19 μm.
Occurrence: oligotrophic lakes and eutrophic rivers; Austria, Czechoslovakia, Sweden.

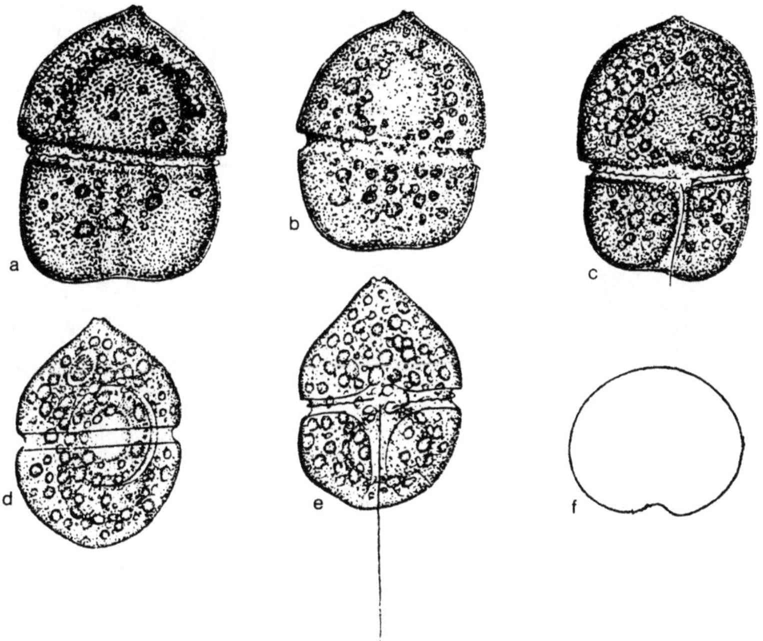

Fig. 91 *Gymnodinium latum* Skuja, *f* cross section of epicone (after Skuja).

16. Gymnodinium palustre Schilling 1891 (fig. 93)

G. carinatum Schilling 1891, *G. wigrense* Wołoszyńska 1925

Cells are ellipsoidal and elongated. The epicone is extremely rounded and somewhat cone or helmet-shaped. The hypocone is broadly rounded. The epicone may be twice the length of the hypocone. The narrow, deep cingulum does not spiral. The sulcus is narrow and extends approximately half way into the epicone. It is wider in the hypocone where it almost reaches the antapex. The numerous chloroplasts are round and parietally arranged. There is no stigma. Nutrition is autotrophic, holophytic and holozoid. A description of what appears to be lobopodia has been reported. Resting stages are spherical and enveloped in multilayered mucilage. Hypnospores are spherical with a thick membrane.

Dimensions: 35–60 μm × 22–37 μm.

Occurrence: pools, ponds, lakes; cosmopolitan.

17. Gymnodinium paradoxum Schilling 1891 (fig. 94)

G. paradoxum var. *maior* Lemmerman 1906, 1910, *G. paradoxum* fo. *astigmosa* Nygaard 1949

Cells are oval and somewhat flattened dorsiventrally. The epicone is slightly larger than the hypocone; both cones are rounded. The cingulum is poorly defined and the sulcus

92

93

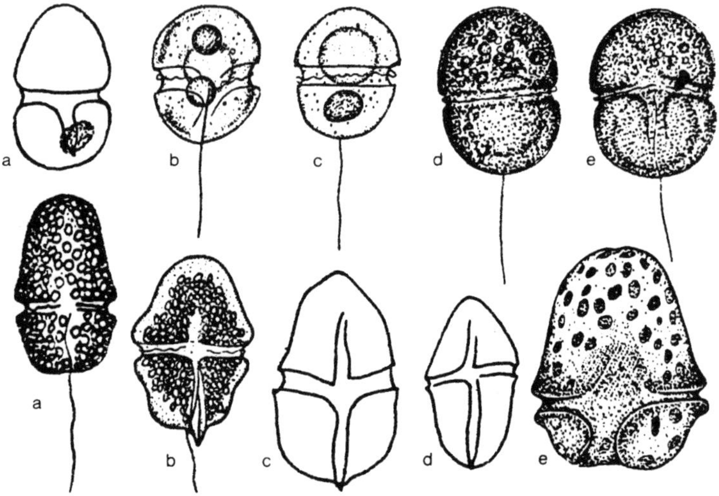

Fig. 92. *Gymnodinium mitratum* Schiller (*a* after Schiller, *b–e* after Skuja).

Fig. 93. *Gymnodinium palustre* Schilling (*a, b* after Schilling, *c, d* after Wołoszyńska, *e* after Popovský).

almost indistinct, it appears, however, to extend into both the epi- and hypocones when the cell is crenated. Chloroplasts are numerous and radially arranged. Numerous elongated bodies are present in the cytoplasm. A stigma is present in zoospores. Hypnozygotes are cube-shaped and slightly flattened with three small bladder-like structures on each corner.
Nutrition is autotrophic and holozoic.
Dimensions: vegetative cells 60–75 µm × 61–67 µm, zoospores 36 bis 40 µm × 34–35 µm, hypnozygotes 25 µm in diameter.
Occurrence: ponds, lakes; Europe.
Note: The hypnozygotes resemble some species of *Tetradinium*.

18. Gymnodinium thomasi Christen 1959 (fig. 95)
G. sp. Popovský 1968
Cells are broadly ellipsoidial and flattened dorsiventrally. The epicone is larger than the hypocone. Both cones are broadly rounded. The cingulum is wide and deep and spirals to the left. The sulcus almost reaches the antapex. The numerous chloroplasts are large and radially arranged. There are a few lightly colored droplets present in the cytoplasm. Zoospores have a stigma.
Dimensions: vegetative cells 16–17 µm × 10–11.5 µm, zoospores 13 µm × 10.5 µm.
Occurrence: peat bogs; Czechoslovakia, Switzerland.

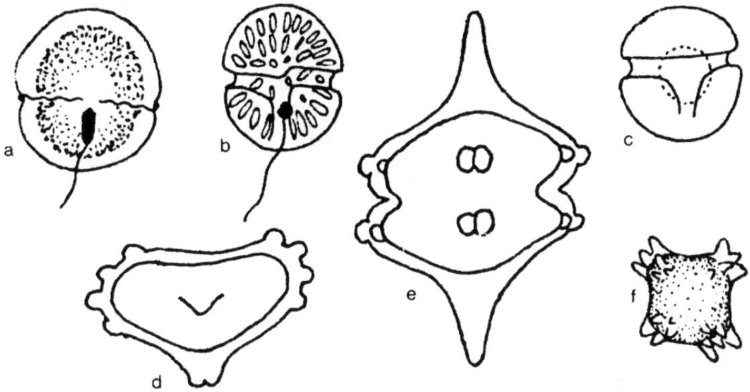

Fig. 94. *Gymnodinium paradoxum* Schilling, *d–f* hypnozygotes (*a* after Schilling, *b–f* after Nygaard).

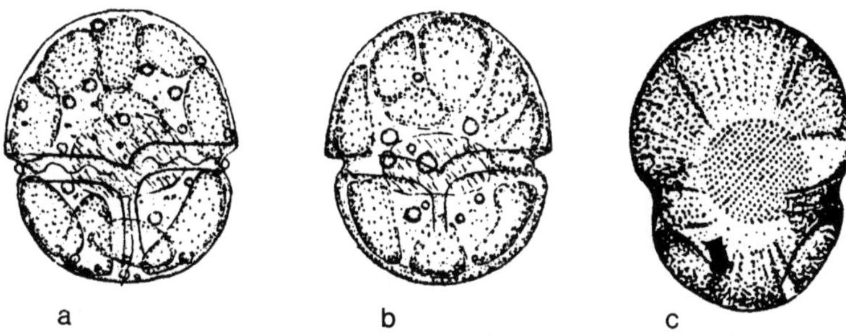

Fig. 95. *Gymnodinium thomasi* Christen (*a, b* after Christen, *c* after Popovský).

19. Gymnodinium triceratium Skuja 1939 (fig. 96)

Gyrodinium asymmetricum Wołoszyńska 1936, *Gymnodinium impar* Harris 1939/40
Cells are spherical in ventral view and slightly flattened dorsiventrally. The epicone is rounded; the hypocone is irregularly bell-shaped with one or three large projections of which the median one is the largest. The cingulum is very wide, deeply incised and spirals slightly to the left. The sulcus is very short and poorly defined. There are one to five chloroplasts which are parietally arranged and oval or kettle-shaped. There is no stigma. Several red droplets are present in the cytoplasm. Nutrition is autotrophic, holozoic, and holophytic.
Dimension: 10–20 μm × 8–20 μm.
Occurrence: peat-bogs, pools, lakes; Europe.

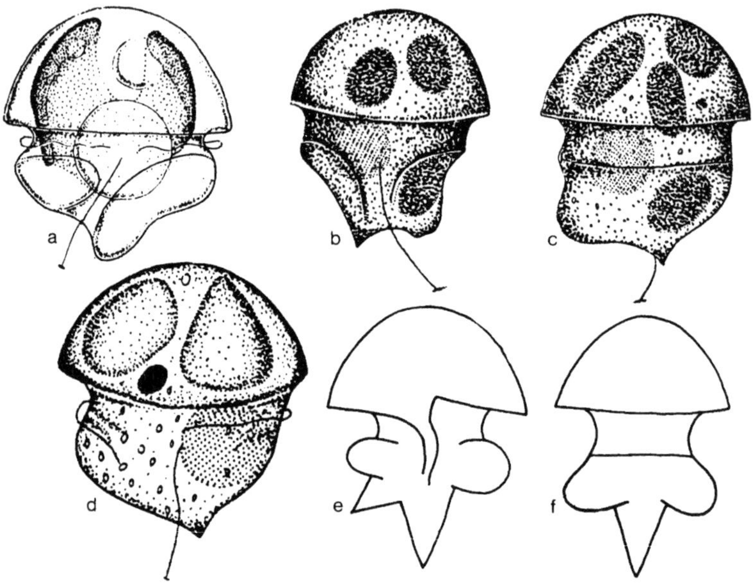

Fig. 96. *Gymnodinium triceratium* Skuja (*a* after Javornický, *b–d* after Popovský, *e, f* after Wołoszyńska).

20. Gymnodinium uberrimum (Allman) Kofoid and Swezy 1921 (fig. 97)

Peridinium uberrima Allman 1855, *Melodinium uberrimum* Kent 1880/81, *Gymnodinium mirabile* Penard 1891, *G. mirabile* var. *rufescens* Penard 1891, *G. rufescens* Lemmermann 1910, *G. limneticum* Wołoszyńska 1935, *G. irregulare* Christen 1959, *G. bogoriense* Klebs 1912, *G. obesum* Schiller 1933, *G. rotundatum* Klebs 1912, *G.* sp. Wołoszyńska 1918, *G. poculiferum* Skuja 1956, *G. limitatum* Skuja 1956, *G. uberrimum* var. *rotundatum* Popovský 1968, *G. uberrimum* fo. Skuja 1956, *Gyrodinium traunsteineri* Lindemann 1928; non: *G. limneticum* Lackey 1936

Cells are broadly spherical to ellipsoidial and flattened dorsiventrally. The epicone is rounded or bell-shaped; the hypocone rounded. The antapex is sometimes blunt. The

---

Fig. 97. *Gymnodinium uberrimum* (Allman) Kofoid and Swezy, *e* a plasmolyzed cell with trichocysts, *k* vegetative cell embedded in gelatinous material, *l* cell chain, *r* hypnozygote in gelatinous material (*a–c* after Popovský, *d* after Skuja, e, f, l after Penard, *g* after Fott, *h* after Klebs, *i, k* after Javornický, *m, n* after Lindemann, *o* after Schiller, *p, r* after Wołoszyńska).

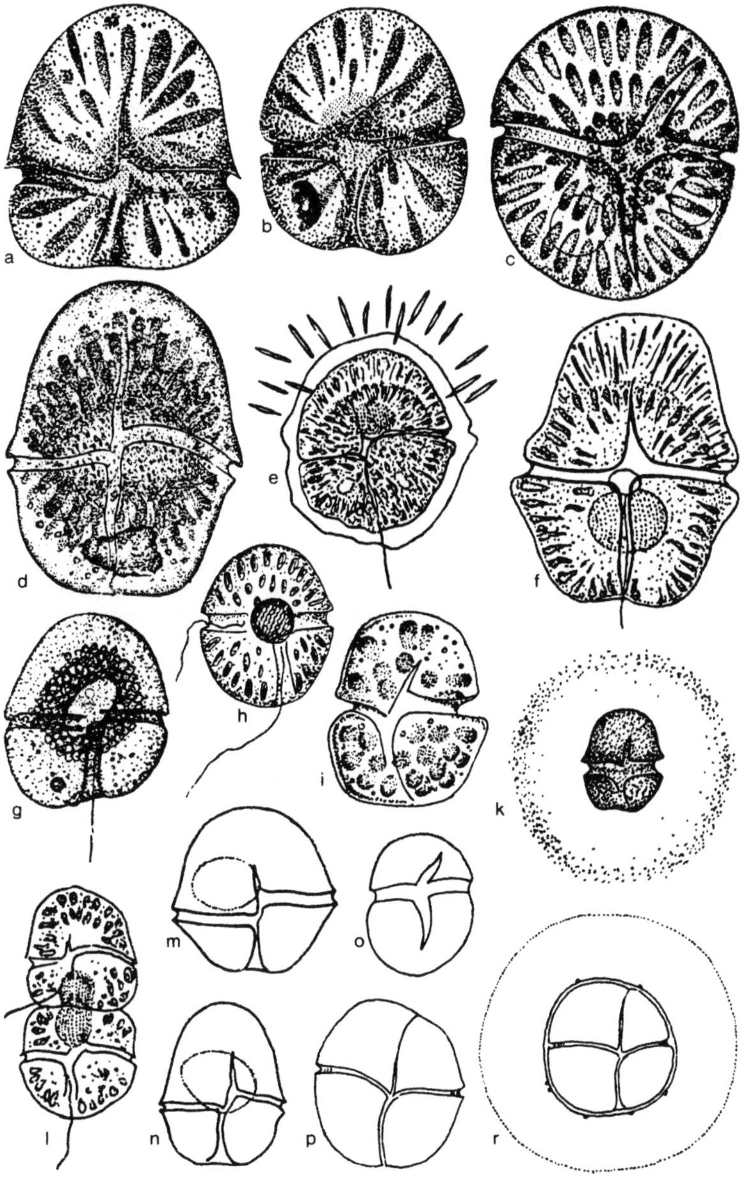
a
b
c
d
e
f
g
h
i
k
l
m
n
o
p
r

wide cingulum is incised and spirals to the left. The sulcus extends into the epicone and almost reaches the antapex on the hypocone. The deeply incised sulcus is somewhat narrow and usually pointed at each end; it deflects somewhat from the longitudinal axis. The numerous chloroplasts are wedge-shaped and radially arranged. Numerous red droplets are present in the cytoplasm. A stigma is usually lacking. Trichocysts have been observed rarely. When present they are located just below the cell membrane. A small spherical pusule is found at the proximal junction of the furrows. Pseudopodia have been reported in vegetative cells as have amoeboid stages. Vegetative cells without flagella have been observed embedded in a thick, fine mucilage. Cells divide in a motile phase or in a non-motile stage enveloped in a gelatinous matrix. Two daughter cells are produced. Motile cells will sometimes form chains – a rare phenomenon in freshwater dinoflagellates.
Dimensions: 24–90 μm × 19–75 μm.
Occurrence: lakes, oligotrophic waters; Europe.
Note: Zoospores or gametes have been described as vegetative cells. Consequently there is a large size range recorded in the literature. *Gymnodinium deformabile* Schiller 1955, invalidly described, is probably *Gymnodinium uberrimum* (Allman) Kofoid and Swezy 1921.

21. Gymnodinium varians Maskell 1887 (fig. 98)
G. minimum Klebs 1912; non: *G. minimum* Lantzsch 1914
Cells are egg or oval-shaped and somewhat flattened dorsiventrally. The epicone is rounded, the hypocone rounded or conical and sometimes obliquely cut. The wide, deep cingulum spirals slightly to the left. The sulcus is relatively narrow and nearly reaches the antapex. The chloroplasts are oval or rod-shaped and parietal. A stigma is lacking. Resting stages are spherical with a thick membrane.
Dimensions: 8–17 μm × 6–12 μm.
Occurrence: lakes; Czechoslovakia, Netherlands, Java.

22. Gymnodinium wawrikae Schiller 1955 (fig. 99)
Cells are spherical or egg-shaped and somewhat flattened dorsiventrally. The epicone is rounded conical or mitre-shaped and larger than the hypocone. The hypocone is broadly rounded. The cingulum is wide and deep, and spirals slightly to the left. The sulcus is shallow and does not reach the antapex. The numerous parietal chloroplasts are disc-shaped, large and sometimes present in clusters near the poles. A stigma and numerous corpuscles are present.
Dimensions: 16–25 μm × 15–24 μm.
Occurrence: wetlands, lakes, pools; Central Europe.

Appendix

1. Gymnodinium biciliatum Ohno 1911 (fig. 100)
Cells are broadly ellipsoidal, about 1.5 times longer than broad and not dorsiventrally flattened. The cingulum circumscribes the cell; the sulcus extends slightly into the epicone and almost reaches the antapex in the hypocone. The chloroplasts appear to be centrally massed and ellipsoidal.
Dimensions: vegetative cells 9–11 μm × 6–7 μm, planozygotes 20.6–25 μm × 12–21 μm.
Occurrence: Japan, S. America.

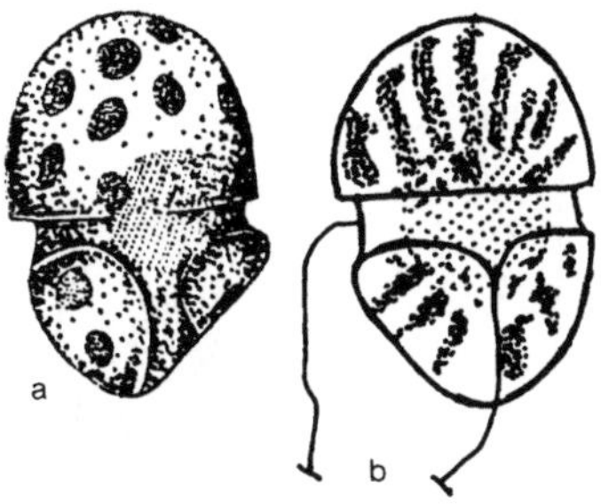

Fig. 98. *Gymnodinium varians* Maskell (*a* after Popovský, *b* after Klebs).

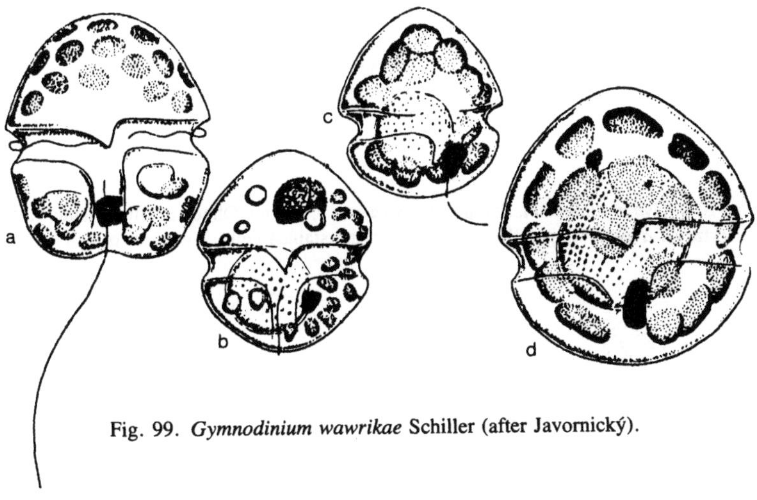

Fig. 99. *Gymnodinium wawrikae* Schiller (after Javornický).

2. *Gymnodinium indicum* Shyam and Sarma 1975 (fig. 101)
Cells are subspherical to broadly ellipsoidal and slightly flattened dorsiventrally. The prominent but shallow, wide cingulum circumscribes the cell. The sulcus does not reach the antapex. The angular chloroplasts are generally 6 to 10 in number and green to blue green in color. A stigma is absent.
Dimensions: 10.8–16.2 µm × 8.1–13.5 µm.
Occurrence: temporary pond; Varanasi, India.

3. *Gymnodinium lackeyi* (Lackey) Kiselev 1954 (fig. 102)
Gymnodinium limneticum Lackey 1936; non: *G. limneticum* Wołoszyńska 1935
Cells are oval and slightly flattened dorsiventrally. The epi- and hypocones are rounded and bell-shaped. The deep, wide cingulum circumscribes the cell. The sulcus nearly reaches the antapex. There are 8–12 discoidal chloroplasts. A stigma is absent.

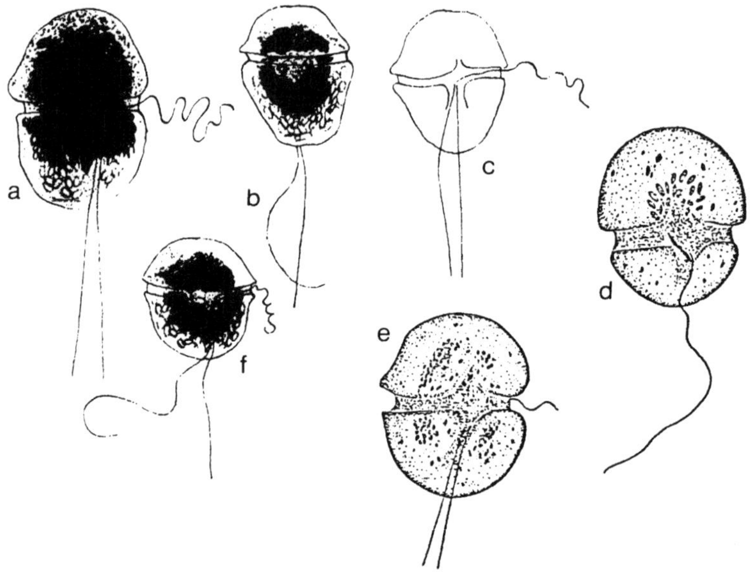

Fig. 100. *Gymnodinium biciliatum* Ohno (*a–c, e, f* after Ohno, *d* after Bicudo and Skvortzov). *a* swimming cell, *b, f* two fixed and stained cells, *c* contour of cell.

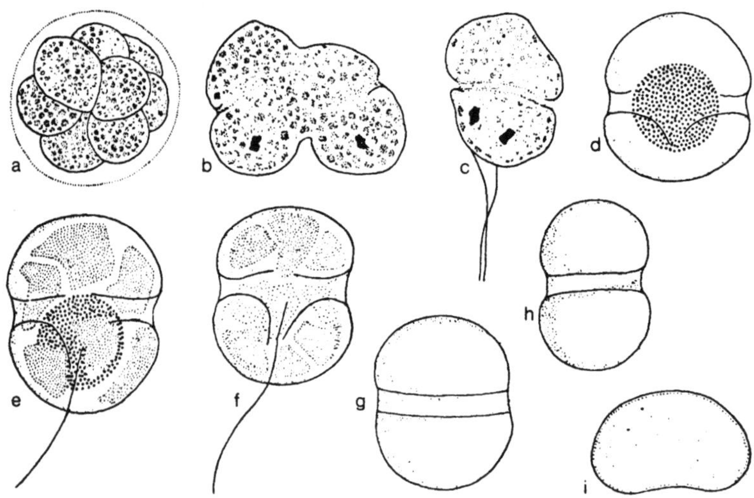

Fig. 101. *Gymnodinium indicum* Shyam and Sarma (after Shyam and Sarma).

102

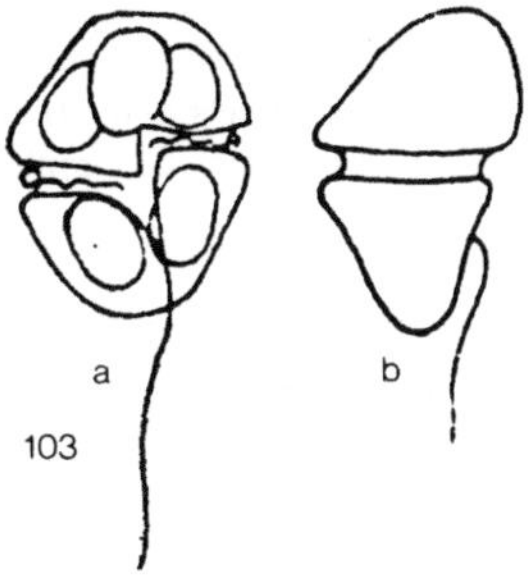

103

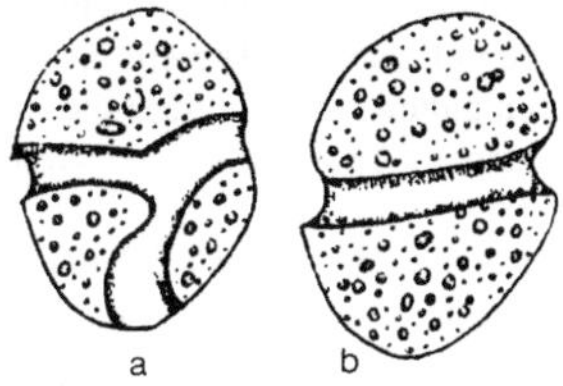

104

Fig. 102. *Gymnodinium lackeyi* (Lackey) Kiselev (after Lackey).

Fig. 103. *Gymnodinium marylandicum* Thompson (after Thompson).

Fig. 104. *Gymnodinium najadeum* Schiller (after Schiller).

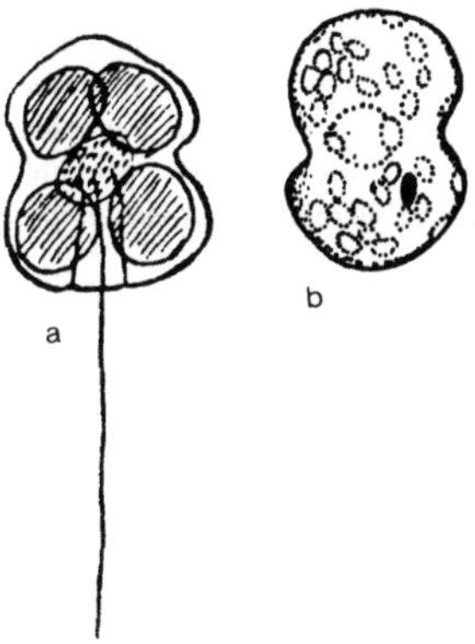

Fig. 105. *Gymnodinium simplex* (Lohmann) Kofoid and Swezy (*a* after Lebour, *b* after Bicudo and Skvortzov).

Dimensions: 25–35 μm × 12–18 μm.
Occurrence: lakes; USA.

4. *Gymnodinium marylandicum* Thompson 1947 (fig. 103)
Cells are egg-shaped and dorsiventrally flattened. The cingulum spirals to the left; the sulcus extends slightly into the epicone and in the hypocone deflects from the longitudinal axis. It does not reach the antapex. There are 4–6 parietal chloroplasts. Red corpuscles are often present in the cytoplasm. A stigma is lacking. Hypnospores are ovoid.
Dimensions: vegetative cells 17–22 μm × 13–19 μm, hypnospores 18–24 μm × 14–21 μm.
Occurrence: lakes; USA.

5. *Gymnodinium najadeum* Schiller 1928 (fig. 104)
Cells are ellipsoidal. The epi- and hypocones are rounded and equal in size. The wide, deep cingulum circumscribes the cell. The sulcus deflects to the right and enlarges near the antapex. The theca is covered with small papillae. Chloroplasts and stigma are absent.
Dimensions: 28–30 μm × 20–21 μm.
Occurrence: saline inland waters; Ukraine, Mediterrean Sea.

6. *Gymnodinium simplex* (Lohmann) Kofoid and Swezy 1921 (fig. 105)
Protodinium simplex Lohmann 1908
Cells are broadly elliptical, about 1.5 times longer than broad, and not flattened dorsiventrally. The cingulum circumscribes the cell. There is no sulcus. The discoid chloroplasts are numerous.
Dimensions: 35–45 μm × 23–31 μm.
Occurrence: European seas, in freshwater pool in Brazil.

The following invalid species, described by Schiller (1955), *G. viridaliut* and *G. huber-pestalozzii,* are probably *G. aeruginosum Stein 1883 or G. mitratum* Schiller 1933. *Gymnodinium baumeisteri* is probably *G. palustre* Schilling 1891. *Gymnodinium cyaneum* and *G. caerulescens* are probably *G. aeruginosum* Stein 1883. *Gymnodinium eufrigidum, G. submontanum, G. granii, G. devorans, G. achroum, G. legiconveniens, G. paradixiforme, G. posthiemale, G. absumens, G. peisonis, G. glacum, G. knollii, G. danubiense, G. schuettii,* and *G. amphiconicoides* are probably *G. lantzschii* Utermöhl 1925. *Gymnodinium stagnale* is probably related to *Amphidinium elenkinii* Skvorcov 1925.

Appendix

1. **Actiniscaceae** Kützing 1849

1. **Actiniscus** Ehrenberg 1843

Dichtyocha (Actiniscus) pentasterias Ehrenberg 1840 and the genus *Actiniscus* Ehrenberg 1843 were originally described as fossil. One species and variety are known recently from brackish and freshwaters.

1. *Actiniscus canadensis* Bursa 1969 (fig. 106)
Cells are spherical or subspherical. A delicately septate, narrow cingulum encircles the cell. The ends of the cingulum meet or are slightly displaced. The sulcus starts as a narrow dot and extends over the ventral surface as a simple furrow which bifurcates in the upper edge of the cingulum. The dense protoplasm is colorless, and contains spherical and amorphous lipids and small inclusions. Holophytic cells, in which patches of chloroplasts are found, contain spherical inclusions of reserve material. Haematochrome grains are most frequently observed in holozoic cells. Rudimentary pentasters occur only in a few individuals. Silica secreting vacuoles have been observed in some cells. Thin needleshaped trichochysts have been observed. Reproduction is by cellular fission starting at the antapex resulting in the formation of 2 even or uneven daughter cells.
Dimensions: 28–43 μm in diameter.
Occurrence: Great Bear Lake; Canada.

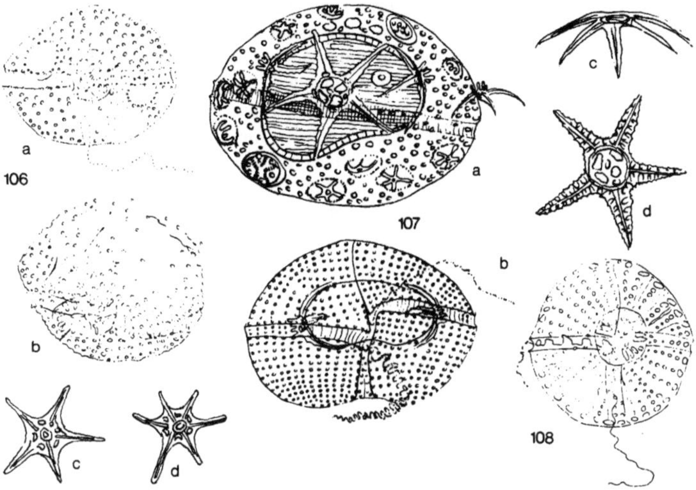

Fig. 106. *Actiniscus canadensis* Bursa, *a* a vegetative cell with 2 secretory vacuoles that produce pentasters, *b* a vegetative cell with ruptured cell wall, *c, d* pentasters (after Bursa).

Fig. 107. *Actiniscus pentasterias* var. *arcticus* Bursa, *a* a vegetative cell with 14 rudimentary peripheral pentasters and with 3 perinuclear pentasters, *b* a vegetative cell with 2 perinuclear pentasters, *c, d* pentasters (after Bursa).

Fig. 108. *Pseudoactiniscus apentasterias* Bursa, *a* ventral view (after Bursa).

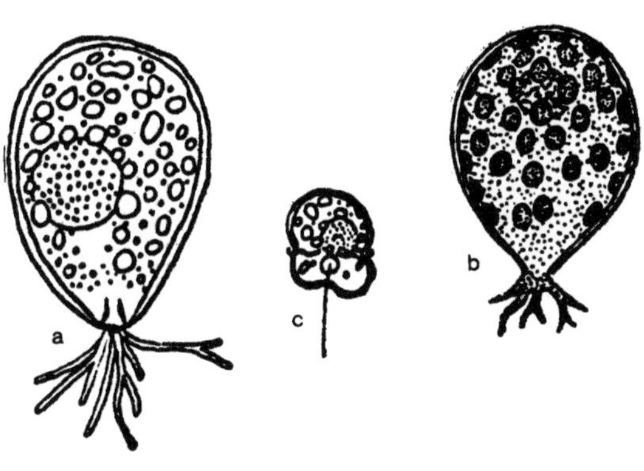

Fig. 109. *Oodinium limneticum* Jakobs, *a, b* vegetative parasitic stage, *c* zoospore (after Jakobs).

2. *Actiniscus pentasterias* var. *arcticus* Bursa 1969 (fig. 107)
Cells are spherical. A narrow cingulum with perpendicular trabeculae circumscribes the cell, its ends meeting or offset 2.5 × the cingulum width. The sulcus begins as an apical dot and extends as a shallow sigmoid or straight furrow; it is slightly forked close to the cingulum. At the longitudinal flagellar pole, the sulcus forms a deeper and wider groove, becoming widest on the dorsal surface of the antapex. The pellicle contains regularly arranged rows of small spherical inclusions. Cells contain 1–8 large and 14 peripheral rudimentary pentasters. The dense protoplasm contains spherical inclusions, oil droplets, reddishbrown haematochrome and diffuse chloroplast pigments. Nutrition is heterotrophic, although phagocytosis of cryptomonads, peridinians and chrysophytes has been recorded. Reproduction is by cell division resulting in two even or uneven-sized daughter cells.
Dimensions: 28–63 μm in diameter.
Occurrence: summer plankton from Ogak Lake on Baffin Island, Hudson Bay, Hudson Straight Arctic Ocean, and Canadian Arctic Archipelago.

2. **Pseudoactiniscus** Bursa 1969 (fig. 108)

Cells spherical or spherical with equatorial cingulum whose ends meet at the same level on the ventral side. Cells rarely dorsiventrally compressed. Holozoic colorless cells dominant: holophytic cells are rare and appear to contain chromatophores.

1. *Pseudoactiniscus apentasterias* Bursa 1969 (fig. 108)
Cells are spherical or subspherical; rarely slightly flattened dorsiventrally. A shallow cingulum with smooth slightly marked ridges circumscribes the cell whose ends either meet on the ventral side or are slightly displaced. The sulcus begins as a small apical dot and proceeds toward the cingular edge where it forks. The colorless, fragile pellicle may contain rows of fine spherical inclusions extending from apex to antapex. The protoplasm contains spherical inclusions, lipids, carbohydrates, haematochrome grains, and diffuse chloroplasts in holophytic cells. Rod-shaped trichocysts are present. Nutrition is holozoic and holophytic. Ingestion of cryptomonads and gymnodinioids has been recorded. Reproduction is by means of cell division producing two even or uneven zoospores.
Dimensions: 26–53 μm in diameter.
Occurrence: summer, Keyhole Lake; Arctic Canada.
Important references: Schiller 1937, Bursa 1969.

2. **Blastodiniaceae** Chatton 1920

1. **Oodinium** Chatton 1912 (fig. 109)

A parasitic genus that is predominantly marine; only one freshwater species is known: *Oodinium limneticum* Jakobs 1946 (fig. 109). Its colorless vegetative cells are non-motile and anchored into the substrate by pseudopodia or haustoria. They are either amoeboid and naked or egg-shaped with a definitive membrane. Numerous refractile, colorless bodies are present within the cell. Reproduction is by the formation of gymnodinium-like zoospores.
Dimensions: 30 μm × 22 μm.

Occurrence: parasitically on 9 species of freshwater fish; USA.
Important references: Chatton 1920, Jakobs 1946.

3. **Katodinium** Fott 1957 (fig. 46c)

Massartia Conrad 1926; non: de Wildemann 1897
Cells are spherical, oval, egg, or mushroom shaped. They are often dorsiventrally flattened. The cingulum is anteriorly placed and may spiral. Cells are usually colorless. The sulcus may be obscure and sometimes extends into the epicone. The cell wall is thin and structureless. A stigma has been recorded for some species. Motile cells divide obliquely into two daughter cells. Non-motile, *Cystodinedria*-like stages have been observed attached to planktonic diatoms. Sexual reproduction by isogametes has been recorded in *Katodinium fungiforme*. Cells are autotrophic, saprotrophic and holozoic.
Important references: Conrad 1926, Schiller 1933, Harris 1939/40, Huber-Pestalozzi 1950, Fott 1957, Loeblich III 1965, Loeblich and Loeblich III 1966, Matjenko and Litvinenko 1977, Popovský 1982.

Key to the species

1a. Sulcus extends into the epicone 2
1b. Sulcus does not extend into the epicone 6
2a. Sulcus is wide in the epicone . . . . **2. K. edax**
2b. Sulcus is narrow in the epicone 3
3a. Cells dorsiventrally flattened 4
3b. Cells not dorsiventrally flattened . . . . **8. K. schilleri**
4a. Chloroplasts present . . . . **7. K. planum**
4b. Chloroplasts absent 5
5a. Sulcus extends almost from the apex to antapex . . . . . . **13. K. woloszynskae**
5b. Sulcus extends into epicone slightly . . . . **10. K. stigmaticum**
6a. Chloroplasts present 7
6b. Chloroplasts absent, cyanells may be present 11
7a. Cells mushroom shaped 8
7b. Cells not mushroom shaped 9
8a. One chloroplast present, hypocone narrow at cingulum . . . . **4. K. mazuricum**
8b. Two to eight chloroplasts present, hypocone broadly rounded at base . . . . **1. K. bohemicum**
9a. Cells dorsiventrally flattened . . . . **5. K. monadicum**
9b. Cells not dorsiventrally flattened 10
10a. Hypocone obliquely cut . . . . **11. K. vierae**
10b. Hypocone rounded . . . . **12. K. vorticella**
11a. Mushroom shaped . . . . **3. K. fungiforme**
11b. Not mushroom shaped 12
12a. Cells dorsiventrally flattened 13
12b. Cells not dorsiventrally flattened . . . . **12. K. vorticella**
13a. Numerous cyanells present . . . . **6. K. montanum**
13b. Cyanells absent . . . . **9. K. spirodinioides**

1. Katodinium bohemicum (Fott) Litvinenko 1977

Gymnodinium bohemicum Fott 1938 (fig. 110)

Cells are ovoid with broadly rounded poles; they are elliptical in transverse section. The cingulum is relatively wide and deep; the sulcus is wide and not easily discernible. It does not reach the antapex. Chloroplasts are discoidal, lying freely in the center of the cell. They number from two to eight. The cell contains numerous highly refractive granules and red-colored droplets. A stigma is lacking. The presence of a "pouch shaped" pseudopodium has been reported in the area where the two furrows join. Nutrition is autotrophic or holozoic.

Dimensions: 10–16 μm × 5.8–11 μm.

Occurrence: peat bogs, oligotrophic lakes; Czechoslovakia, USA.

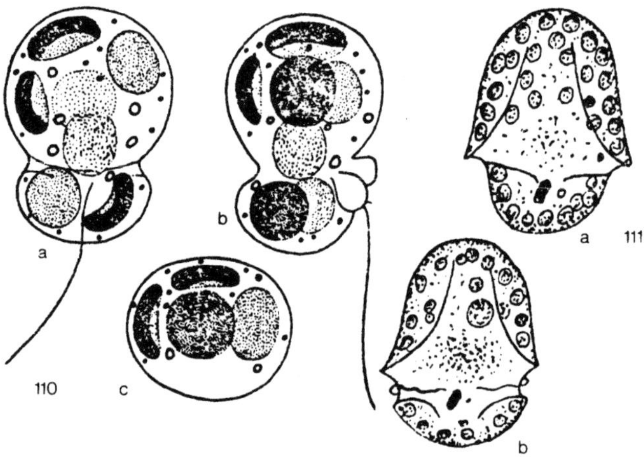

Fig. 110. *Katodinium bohemicum* (Fott) Litvinenko, *c* cross section of epicone (after Fott).

Fig. 111. *Katodinium edax* (Schiller) Loeblich III (after Schiller).

2. Katodinium edax (Schiller) Loeblich III 1965 (fig. 111)

Massartia edax Schiller 1954

Cells are ovoid in ventral view and elliptical in transverse section. The epicone is broadly rounded and helmet-shaped; the hypocone is hemispherical. The sulcus is very wide in the central part of the cell and narrows toward both poles. A stigma is present. Chloroplasts are lacking, but numerous cyanells are present in the colorless cytoplasm. Different algal taxa have been observed within the cell as well as red-colored droplets.

Dimensions: 17–20 μm × 16 μm.

Occurrence: in winter plankton, lakes; Austria.

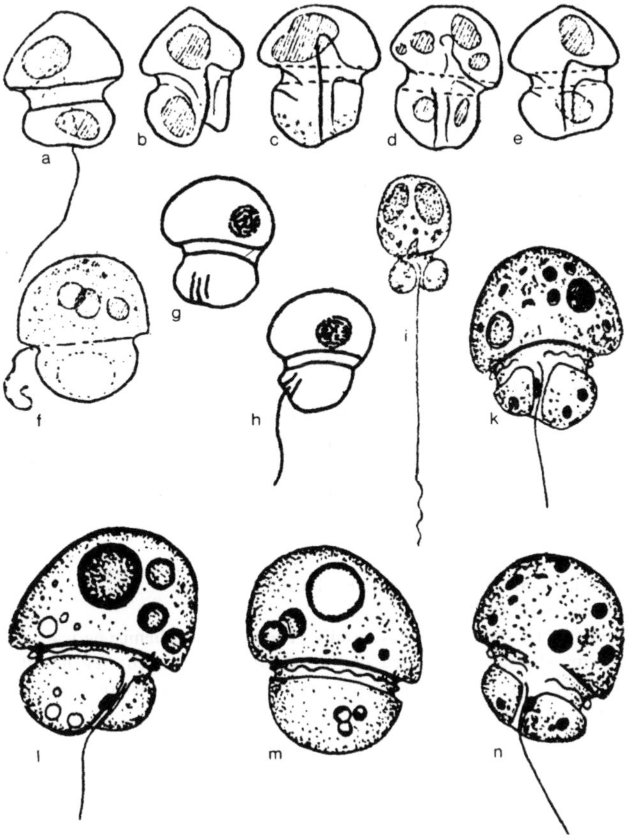

Fig. 112. *Katodinium fungiforme* (Anisimova) Loeblich III (*a–f* after Harris, *g, h* after Anisimova, *i–n* after Schiller).

3. Katodinium fungiforme (Anisimova) Loeblich III 1965 (figs. 52, 112)
Gymnodinium fungiforme Anisomova 1920, *Massartia crassifilum* Schiller 1954, *Katodinium crassifilum* (Schiller) Loeblich III 1965, *Massartia austriacum* Schiller 1955, *Katodinium austriacum* (Schiller) Loeblich III 1965, *Gymnodinium blax* Harris 1939/40

Cells are ovoid and flattened dorsiventrally. The epicone and hypocone are broadly rounded. The epicone is "mushroom" shaped. The sulcus reaches the antapex. Chloroplasts and stigma are absent. Numerous cyanells and several colorless droplets are sometimes present. Nutrition is holozoic.

Dimensions: 4–15 μm × 5–12 μm.
Occurrence: plankton of ponds and lakes, also in brackish waters; Europe, USA.

4. Katodinium mazuricum Javornický 1965 (fig. 113)
Cells are typically mushroom-shaped and are somewhat flattened dorsiventrally. The epicone is broadly rounded, the hypocone is narrowed at the cingulum. The girdle is only visible on the dorsal side in lateral view, the ventral side appears quite flat. The girdle usually forms an incomplete circle, rarely will spiral slightly. The single chloroplast has the shape of an irregularly lobed plate which almost fills the entire protoplast. A large bright-red stigma is present. Refractive droplets are present in the cytoplasm.
Dimensions: approx. 13 μm × 12 μm.
Occurrence: plankton, lakes; Poland.

5. Katodinium monadicum (Perty) Javornický 1970 (fig. 114)
Peridinium monadicum Perty 1852, *Gymnodinium monadicum* (Perty) Saville-Kent 1880, *Katodinium polyplastidum* Popovský 1968
Cells are elongated and ellipsoidal in shape with a bluntly conical epicone and a flatly rounded hypocone. They are flattened dorsiventrally. The furrows are wide; the sulcus does not reach the antapex. Several parietal chloroplasts are present which are oval to plate shaped. In some populations the stigma is absent.
Dimensions: 15–28 μm × 9–15.4 μm.
Occurrence: Central Europe.

6. Katodinium montanum (Schiller) Loeblich III 1965 (fig. 115)
Massartia montana Schiller 1955
Cells are ovoid and slightly flattened dorsiventrally. The epicone is hemispherical, the hypocone broadly rounded and excavated on the antapex. The sulcus reaches the antapex and extends very slightly onto the dorsal side of the hypocone. Chloroplasts are lacking. Numerous cyanells and a stigma are present. Digestion of green algae has been recorded to occur while they are in a non-motile state.
Dimensions: 16–20 μm × 13–17 μm.
Occurrence: concrete tank; Austria.

7. Katodinium planum (Fott) Loeblich III 1965 (fig. 116)
Massartia plana Fott 1938
Cells are broadly egg-shaped and flattened dorsiventrally. The epicone is broadly rounded and pentangular in longitudinal section. The flattened hypocone is broadly excavated on the antapex. The sulcus extends into 2/3 of the epicone. Two large chloroplast are present, the larger of which is in the epicone. A stigma is absent. Numerous refractive droplets are present in the cytoplasm.
Dimensions: 16–20 μm × 15–18 μm.
Occurrence: plankton, oligotrophic lake; Czechoslovakia.

8. Katodinium schilleri (Wołozyńska) Loeblich III 1965 (fig. 117)
Massartia schilleri Wołoszyńska 1936, *Massartia tetragonops* Harris 1939/40, *Katodinium tetragonops* (Harris) Loeblich III 1965, *Massartia ptyrtica* Harris 1939/40, *Katodinium ptyrticum* (Harris) Loeblich III 1965
Cells are ovoid; the epicone is conical to helmet-shaped, the hypocone varies from broadly conical to rounded in shaped. The cingulum is displaced. The sulcus is deep and

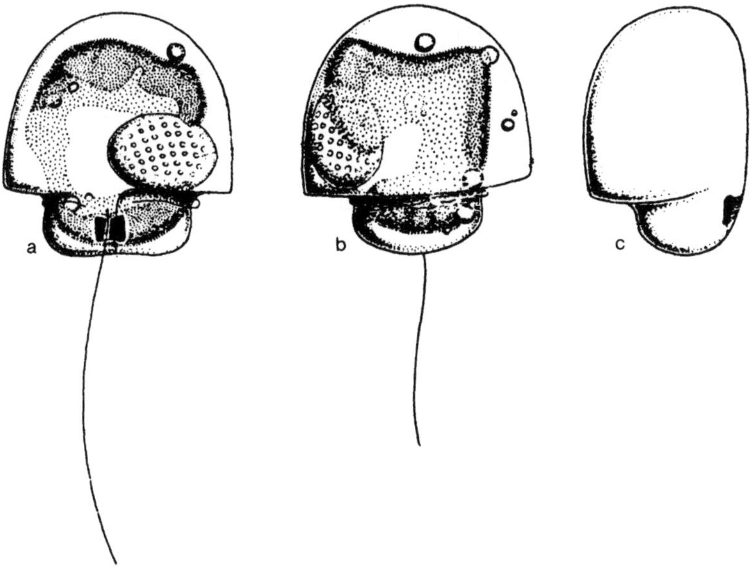

Fig. 113. *Katodinium mazuricum* Javornický (after Javornický).

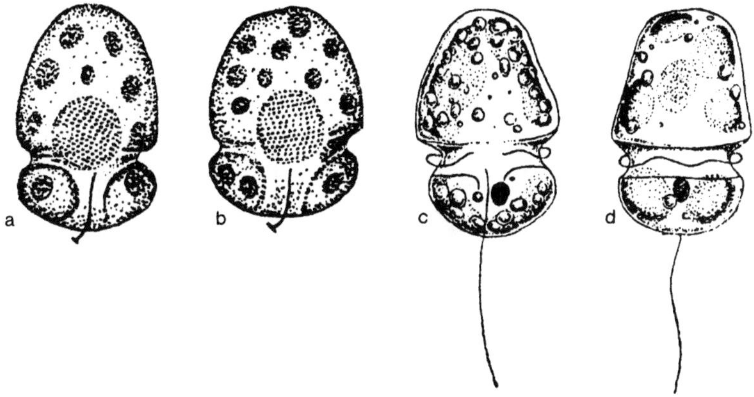

Fig. 114. *Katodinium monadicum* (Perty) Javornický (*a*,, *b* after Popovský, *c*, *d* after Javornický).

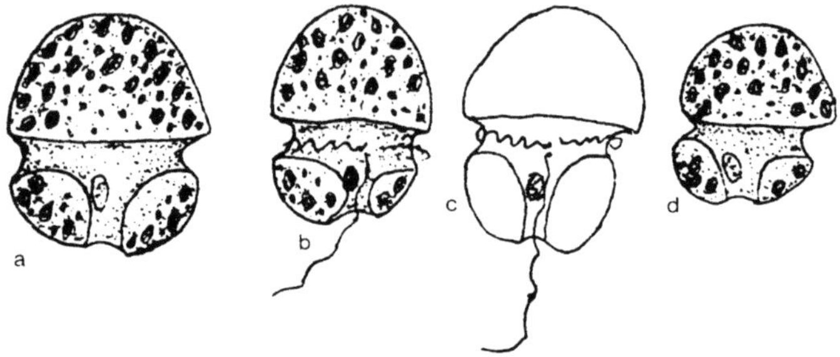

Fig. 115. *Katodinium montanum* (Schiller) Loeblich III (after Schiller).

Fig. 116. *Katodinium planum* (Fott) Loeblich III (after Fott).

116

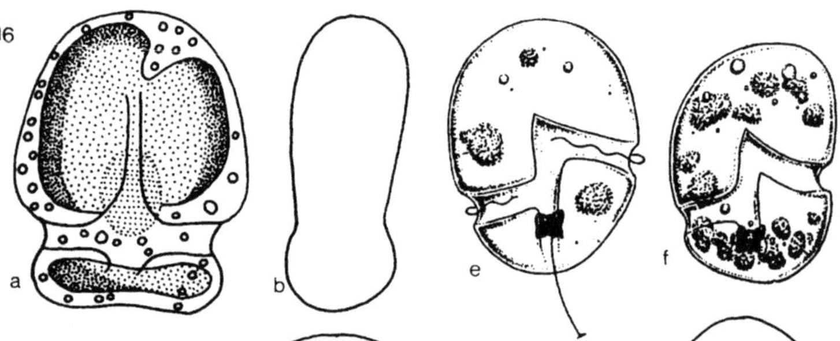

117

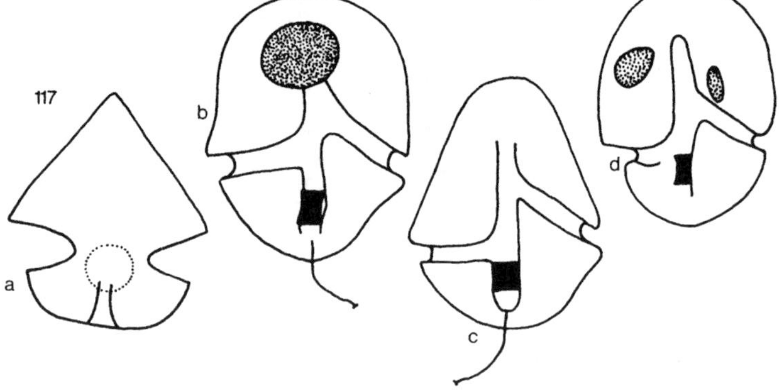

Fig. 117. *Katodinium schilleri* (Wołoszyńska) Loeblich III (*a* after Wołoszyńska, *b–d* after Harris).

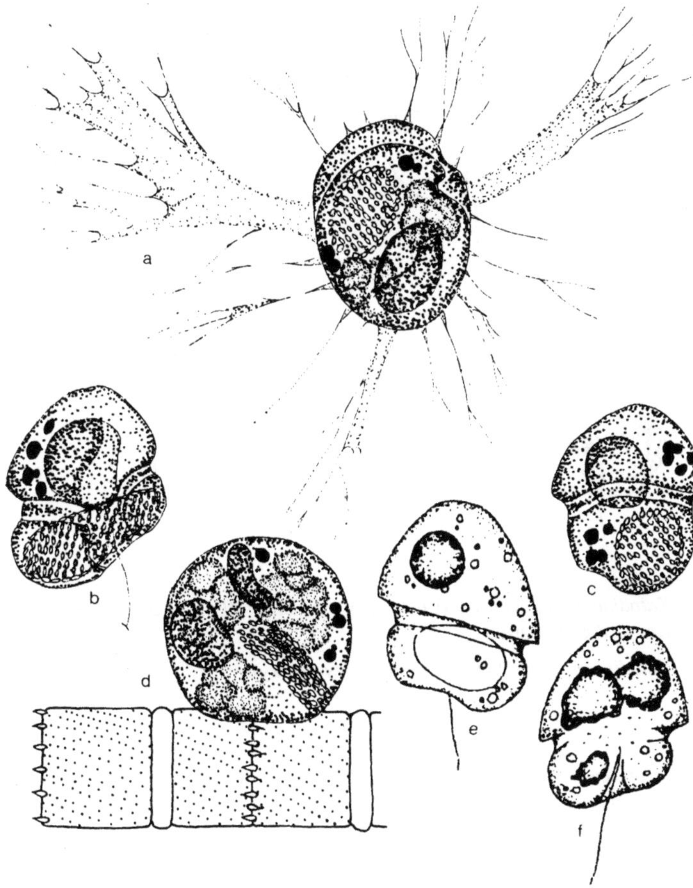

Fig. 118. *Katodinium spirodinioides* Christen, *a* cell with pseudopodia, *d* nonmotile cell attached to a *Melosira* sp. (*a–d* after Popovský, *e, f* after Christen).

forms a sharp sinus in the epicone. Chloroplasts are lacking. The stigma is usually large but may be small or even absent. Nutrition is holozoic.
Dimensions: 8–16.5 μm × 8–16 μm.
Occurrence: oligo- and eutrophic ponds and lakes.

9. Katodinium spirodinioides Christen 1961 (fig. 118)
Cells are ovoid and strongly flattened dorsiventrally. The epicone is broadly rounded; the rounded hypocone is cut obliquely. The cingulum spirals and is somewhat deeply incised on the dorsal side. The sulcus is wide and reaches the antapex. Chloroplasts and stigma are

119

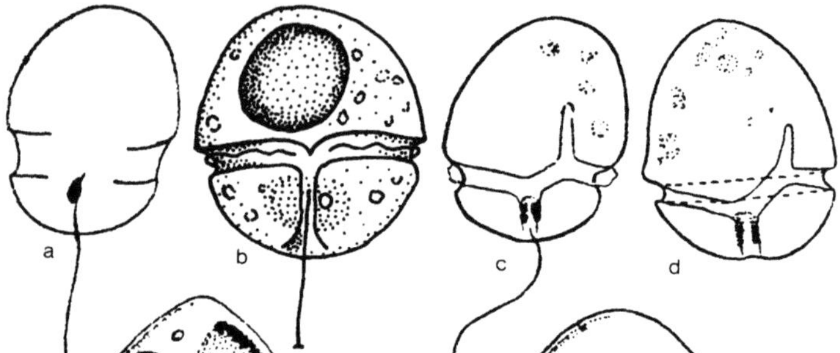

120

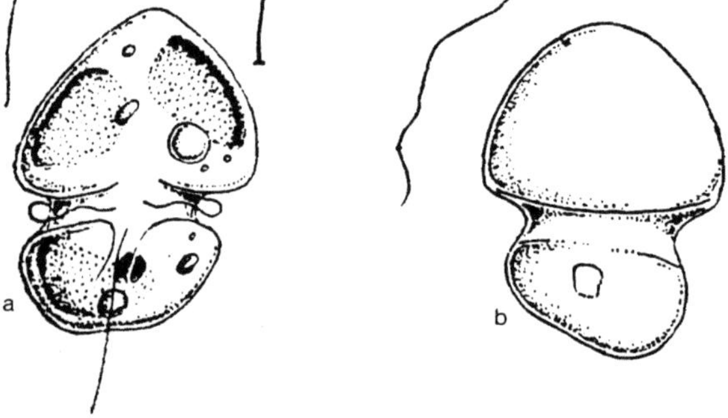

Fig. 119. *Katodinium stigmaticum* (Lindemann) Loeblich III (*a* after Lindemann, *b* after Christen, *c, d* after Harris).

Fig. 120. *Katodinium vierae* Javornický (after Javornický).

lacking. The presence of rhizopodia has been reported in free swimming cells which lacked flagella ("Cystodinedria"-shaped) and were attached to *Melosira*.
Dimensions: 10–15 µm × 8–10 µm.
Occurrence: peat bogs, lakes; Switzerland, Czechoslovakia.

10. Katodinium stigmaticum (Lindemann) Loeblich III 1965 (fig. 119)
Gymnodinium stigmaticum Lindemann 1928, *Massartia campylops* Harris 1939/40, *Katodinium campylops* (Harris) Loeblich III 1965, *Katodinium astigmatum* Christen 1959

Cells are ovoid and dorsiventrally flattened. The epicone is rounded or helmet-shaped: the hypocone broadly rounded. The cingulum is deep and somewhat spiral; the sulcus extends slightly into the epicone but does not reach the antapex on the hypocone. Chloroplasts are lacking. A stigma may be present. Nutrition is holozoic.
Dimensions: 12–25 µm × 10–21 µm.
Occurrence: eutrophic ponds and lakes; Switzerland, Germany, Great Britain.

11. Katodinium vierae Javornický 1970 (fig. 120)
Cells are ovoid or obovoid in shape, with a higher and wider, rounded or mitra-formed epicone. The hypocone is smaller, obliquely cut and slightly excavated. Cells are not

dorsiventrally flattened. The deep, wide cingulum spirals; the antapex is cut parallel to the cingulum. The sulcus is distinct only on the hypocone where it reaches the antapex. A stigma is present. Two or three chloroplasts shaped like irregular plates lay just beneath the periplast. Several spherical granules are scattered in the cytoplasm.
Dimensions: 13–15 μm × 9 μm.
Occurrence: peat bog; Sweden.

12. Katodinium vorticella (Stein) Loeblich III 1965 (fig. 121)
Gymnodinium vorticella Stein 1878, 1883, *Peridinium vorticella* Stein 1878, *Gymnodinium vorticella* (Stein) Lindemann 1928, *Massartia vorticella* (Stein) Schiller 1933, *Katodinium viride* Christen 1961, *Katodinium vernale* Christen 1961, *Massartia pratensis* Baumeister 1943, *Katodinium pratensis* (Baumeister) Loeblich III 1965

Cells are spherical or ovoid with a conical or dome-shaped epicone and a broadly rounded hypocone. The cingulum is deep and its ends are sometimes displaced one to one and one-half times its width. It rises on the left with the sulcus forming a sinus in the epicone. The sulcus is usually narrow and almost reaches the antapex. A stigma may be present. Chloroplasts are usually absent. If present there are 2 to 7 which are relatively small and greenish in color. Numerous refractive granules and (sometimes) reddish droplets are present in the cytoplasm. Nutrition is holozoic.
Dimensions: 13.5–33 μm × 10.5–30 μm.
Occurrence: peat bogs, ponds, pools, in melted snow; Europe, USA.

13. Katodinium woloszynskae (Schiller), Loeblich III 1965 (fig. 122)
Spirodinium vorticella Wołoszyňska 1917, *Massartica woloszynskaae* Schiller 1933, *Gymnodinium musaei* Danysz in Pouchet 1887, *Massartia musaei* (Danysz) Schiller 1933, *Katodinium musaei* (Danysz) Loeblich III 1965, *Katodinium piscinale* Fott 1957, *Massartia woloszynskae* var. *notata* Skuja 1956, *Katodinium woloszynskae* var. *notata* (Skuja) Loeblich III 1965, *Katodinium notatum* (Skuja) Christen 1961

Cells are ovoid and flattened dorsiventrally. The epicone is helmet-shaped, and is greater and wider than the nearly rounded hypocone. The cingulum is spiralled, sometimes severely so. The sulcus is deep, extending nearly from the apex to the antapex. Chloroplasts are lacking. A stigma present. Numerous refractive droplets and algal remains are present in the protoplasm.
Dimensions: 20–50 μm × 16–48 μm.
Occurrence: ponds, pools; France, Czechoslovakia, Poland.

Appendix

1. *Katodinium auratum* Bursa 1970 (fig. 123)

Cells elliptical and dorsiventrally flattened. The epicone is broadly rounded on the apex. The cingulum encircles the cell. The sulcus is very short and ends abruptly, it does not reach the antapex. Chloroplasts are numerous and parietal.
Dimensions: 23.8–26.5 μm × 15–16.5 μm.
Occurrence: highmountain brooks; Colorado, USA.

2. *Katodinium hyperxanthum* (Harris) Loeblich III 1965 (fig. 124)

Massartia hyperxantha Harris 1939/40, *Katodinium hyperxanthoides* (Harris) Loeblich III 1965, *Massartia hyperxanthoides* Harris 1939/40, *Katodinium molopicum* (Harris)

121

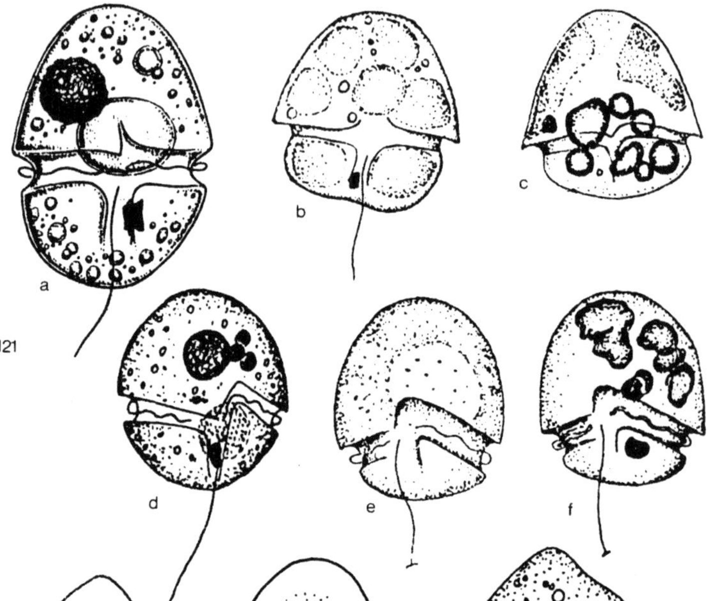

122

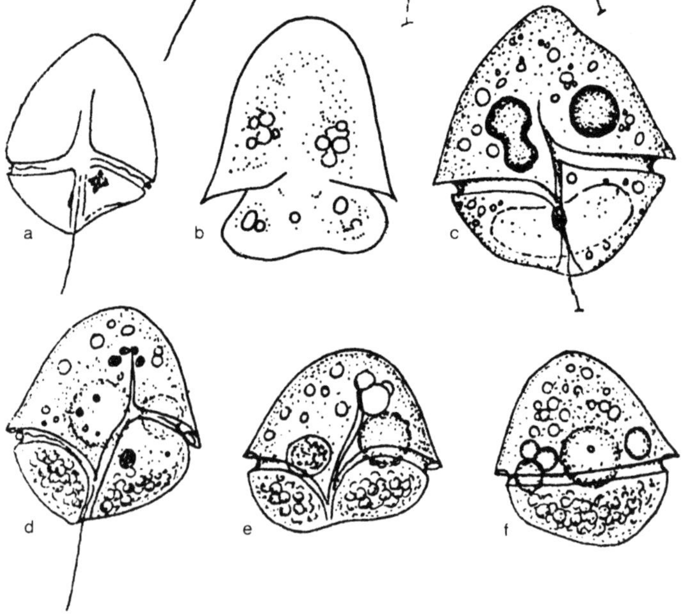

Fig. 121. *Katodinium vorticella* (Stein) Loeblich III (*a* after Javornický, *b, c* after Christen, *d* after Thompson, *e, f* after Baumeister).

Loeblich III 1965, *Massartia molopica* Harris 1939/40. These species are evidently in fact dividing monads from the genera *Amphidinium* or *Gymnodinium*.

3. *Katodinium fontinalis* Bicudo and Skvortzov 1970 (fig. 125)
Cells are relatively asymmetrical, somewhat oval in front view and dorsiventrally flattened. The epicone is subovoidal, the hypocone subglobose. The cingulum spirals slightly to the left; the sulcus is poorly defined. Chloroplasts are numerous, discoid, and located in the epicone.
Dimensions: 8–9 μm × 4 μm.
Occurrence: temporary pool with *Tribonema;* Brasil.

4. *Katodinium sanctipaulense* Bicudo and Skvortzov 1970 (fig. 126)
Cells are approximately elliptical to subcylindrical and dorsiventrally flattened. The epicone is semi-elliptical; the hypocone subhemispherical. The cingulum is distinct and oblique while the sulcus is poorly defined. Chloroplasts are numerous, granular, and located in the epicone.
Dimensions: 13–16 μm × 6–8 μm.
Occurrence: permanent pool with *Spirogyra;* Brasil.

4. **Gyrodinium** Kofoid and Swezy 1921 (fig. 46d)

Gymnodinium Schütt 1896 partim, *Spirodinium* Schütt 1896; non: *Spirodinium* Fiorentini 1890, *Glenodinium* Seligo 1885

The cingulum spirals to the left as it descends and is diplaced more than 0.2 of the total body length. The sulcus does not extend into the epivalve. The cell membrane or pellicle is smooth, without known structure or ornamentation. Chloroplasts may be present. Corpuscles are present in the cytoplasm. Hypnospores are spherical and enveloped by a thick membrane.
Important references: Kofoid and Swezy 1921, Popovský 1868, von Stosch 1973.

Key to the species

1a. Chloroplasts present . . . . **2. G. pusillum**
1b. Chloroplasts absent . . . . **1. G. hyalinum**

1. Gyrodinium hyalinum (Schilling) Kofoid and Swezy 1921 (fig. 128)
Gymnodinium hyalinum Schilling 1891, *Spirodinium hyalinum* (Schilling) Lemmermann 1900

Cells are ovoid and asymmetric. The broad, slanted epicone is helmet-shaped and much smaller than the hypocone. The hypocone is broadly rounded and may be obliquely cut at the bottom. The deeply incised cingulum spirals severely to the left as it descends. The deeply incised sulcus reaches the antapex. Chloroplasts are lacking; a stigma is present. Several refractile bodies and occasionally green algae are present in the cytoplasm. Nutrition is holozoic and holophytic. Hypnospores are spherical with a firm wall. They have been observed dividing.

Fig. 122. *Katodinium woloszynskae* (Schiller) Loeblich III (*a* after Fott, *b* after Danysz, *c* after Christen, *d–f* after Skuja).

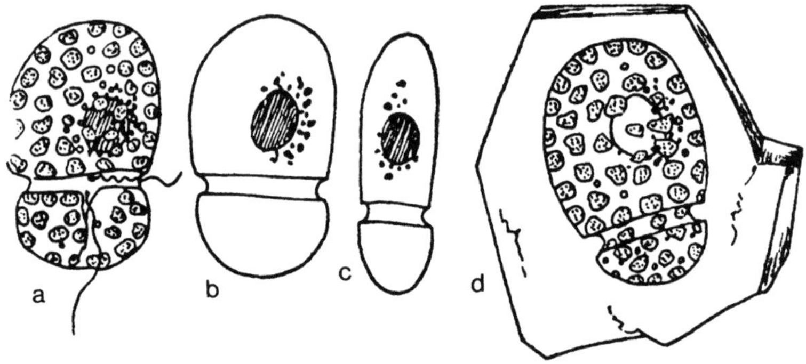

Fig. 123. *Katodinium auratum* Bursa, *d* cell attached by its ventral side to pebble (after Bursa).

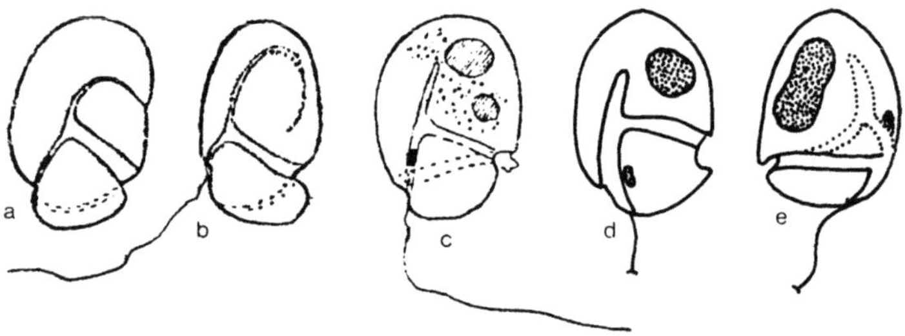

Fig. 124. *Katodinium hyperxanthum* (Harris) Loeblich III (after Harris).

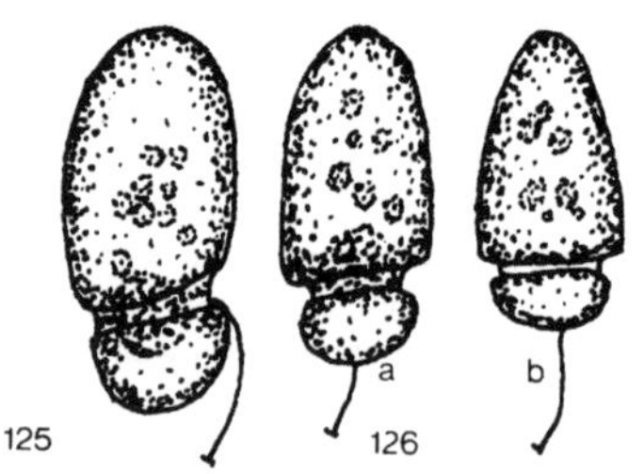

Fig. 125. *Katodinium fontinalis* Bicudo and Skvortzov (after Bicudo and Skvortzov).

Fig. 126. *Katodinium sanctipaulense* Bicudo and Skvortzov (after Bicudo and Skvortzov).

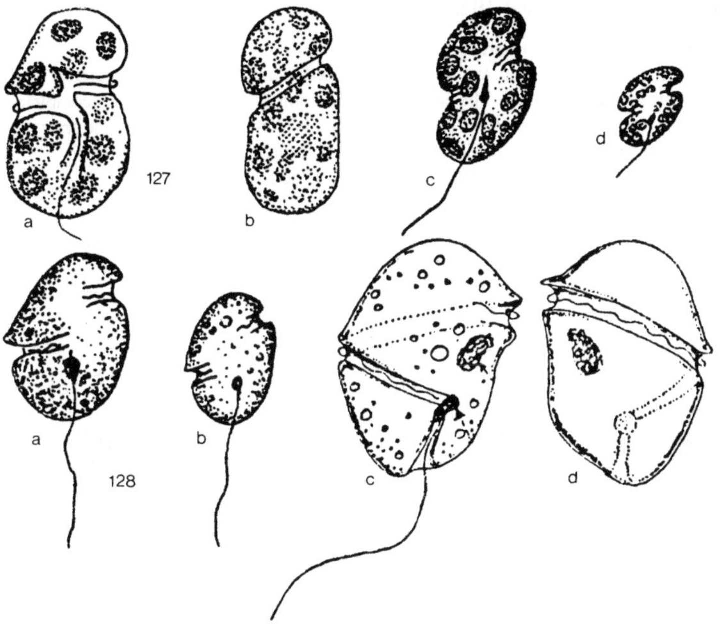

Fig. 127. *Gyrodinium pusillum* (Schilling) Kofoid and Swezy (*a* after Thompson, *b* after Popovský, *c, d* after Schilling).

Fig. 128. *Gyrodinium hyalinum* (Schilling) Kofoid and Swezy (*a* after Schilling, *b* after Nygaard, *c, d* after Javornický).

Dimensions: 23–31 μm × 20–23 μm.
Occurrence: pools, ponds, eutrophic waters; Europe.

2. Gyrodinium pusillum (Schilling) Kofoid and Swezy 1921 (fig. 127)
Gymnodinium pusillum Schilling 1891, *Spirodinium pusillum* (Schilling) Lemmermann 1900

Cells are strongly ellipsoidal and asymmetric. The epicone is slanted and rounded; the hypocone elongated and rounded at the antapex. The epicone is smaller than the hypocone. The deeply incised cingulum severly spirals to the left as it descends. The sulcus does not reach the antapex. Several parietal, oval chloroplasts are present. A stigma is usually present. Hypnospores are sphèrical with a firm wall.
Dimensions: 16–23 μm × 9–18 μm.
Occurrence: peat bogs, pools, ponds; Europe.
Note: *Gyrodinium elongatum* Schiller 1955 (invalidly described) is probably *Gyrodinium pusillum* (Schilling) Kofoid and Swezy 1921.

5. **Bernardinium** Chodat 1923 (fig. 46e)

Hemidinium Stein 1883 partim

Naked ellipsoidal cells, that are slightly flattened dorsiventrally. The epicone and hypocone are broadly rounded. The epicone may be conical. The cingulum is incomplete, curving downward to the left. A narrow sulcus extends from the cingulum to the antapex. A stigma and numerous, small, diffuse, parietal chloroplasts may be present.

Important references: Chodat 1923, Javornický 1962.

1. **Bernardinium bernardinense** Chodat 1923 (fig. 129)
   Hemidinium bernardinense (Chodat) Huber-Pestalozzi 1950, *H. soligenum* Skuja 1964

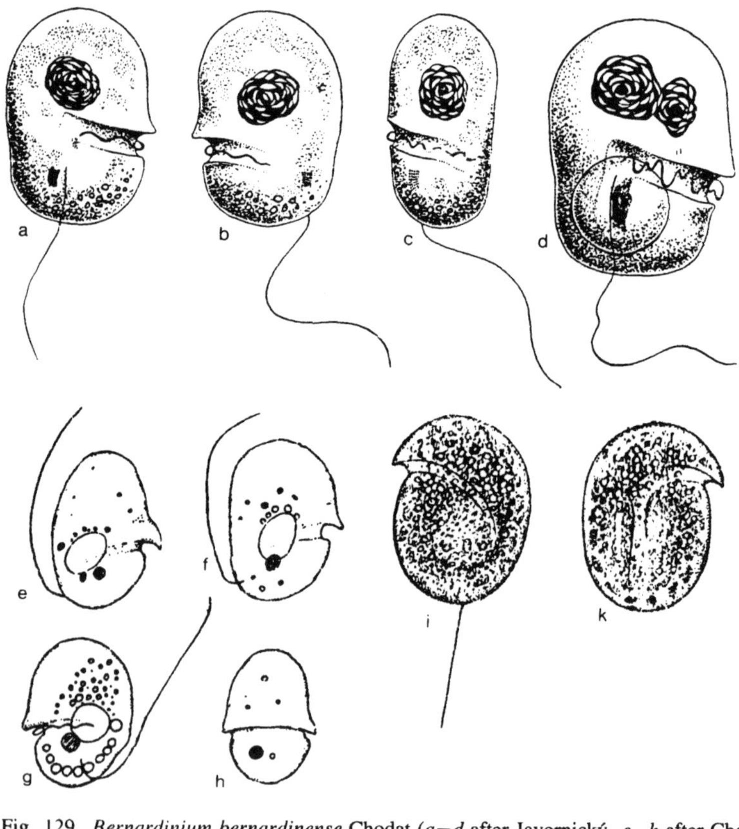

Fig. 129. *Bernardinium bernardinense* Chodat (*a–d* after Javornický, *e–h* after Chodat, *i, k* after Skuja).

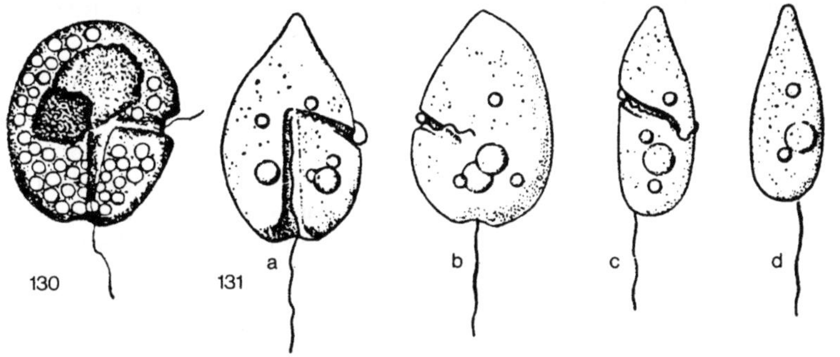

Fig. 130. *Bernardinium salinum* (Anisimova) Javornický (after Anisimova).

Fig. 131. *Bernardinium thiophyllum* (Conrad) Javornický (after Conrad).

The epicone and hypocone are broadly rounded. Cells are colorless; a stigma is present. Nutrition is holophytic.
Dimensions: 15–25 μm × 10–20 μm.
Occurrence: pools, rivers, ponds, puddles; Switzerland, USA, Czechoslovakia.

Appendix

Two species have been reported from inland saline waters.

1. *Bernardinium salinum* (Anisimova) Javornický 1962 (fig. 130)
   Hemidinium salinum Anisimova 1926

This is a poorly described species, which differs from *B. bernardinense* in that it lacks a stigma and differs slightly in sulcal morphology.
Dimensions: 15.5–18.5 μm × 24 μm.
Occurrence: halophile, Lake Srednoe; USSR.

2. *Bernardinium thiophilum* (Conrad) Javornický 1962 (fig. 131)
   Hemidinium thiophilum Conrad 1939

Cells are pear shaped, dorsiventrally flattened, and with a slightly metabolic cell covering. The epicone is tapered and smaller than the hypocone which is excavated at the bottom. The incomplete cingulum is located above the middle of the cell. The sulcus clearly reaches the antapex where it forms a sinus. Several refractile corpuscles are present in the cytoplasm.
Dimensions: 24–30 μm × 15–19 μm.
Occurrence: halophile, salt marshes (H_2S); Belgium.

2. Family Woloszynskiaceae Bourrelly 1970

Cells are covered with a delicate theca composed of small, numerous platelets that are usually hexangular in shape. A suture (carina) is usually present on the ventral side of the epitheca. The theca splits along this zone during reproduction.

1. **Woloszynskia** Thompson 1950 (fig. 46a)

Gymnodinium Stein 1878/83 partim

A slightly spiralled cingulum divides vegetative cells into an epicone and a hypocone. The sulcus may extend to the antapex. The cell wall is composed of very fine, numerous platelets which are usually hexangular and visible only on empty cells. Cells contain a dinokaryon, stigma, and numerous refractive droplets (storage products, remnants etc). Chloroplasts are present in some taxa. Motile and non-motile cells may divide producing two zoospores. The protoplasts may also divide within the parent cell forming as many as 8 zoospores or gametes. Gametic fusion has been reported. Resting stages (hypnospores) are usually angular having one or two corners or ends and are triangular or square in transverse section. Many species have been described but it is only possible to determine 9 species unambiguously of which 7 are reported from Europe. Nutrition is autotrophic, holozoic or holophytic.

Occurrence: plankton of oligo-, eu-, and dystrophic pools and lakes.

Important references: Huber-Pestalozzi 1950, Thompson 1950, Bourrelly 1970, Bibby and Dodge 1972, v. Stosch 1973.

Key to the species

1a. Vegetative cell oval; one or both apices extended forming long cones or spines . . . 2
1b. Vegetative cell oval; apices not extended . . . 4
2a. Cells strongly flattened dorsiventrally . . . **7. W. tenuissima**
2b. Cells slightly flattened dorsiventrally . . . 3
3a. Both apices extended forming long conical processes or spines . . . **1. W. apiculata**
3b. Apex only extended into a long conical process . . . **8. W. tylota**
4a. Epi- and hypocone equal in size . . . **2. W. coronata**
4b. Epi- and hypocone not of equal size . . . 5
5a. 2 to 4 chloroplasts are present . . . **4. W. ordinata**
5b. Chloroplasts numerous, radially arranged . . . 6
6a. Epicone is larger than the hypocone . . . 7
6b. Epicone is smaller than the hypocone . . . **5. W. pascheri**
7a. Cingulum indistinct and does not reach the antapex . . . **3. W. neglecta**
7b. Cingulum is distinct and reaches the antapex . . . **6. W. pseudopalustris**

1. Woloszynskia apiculata von Stosch 1973 (fig. 132)

Cells are pyriform or broadly pyriform in ventral view. The epicone is ovate-conical but slightly apiculate near the carina. The hypocone is conical or campanulate-conical and somewhat sharply pointed at the antapex. The cingulum spirals to the left with its ends displaced by approximately their own width. The chloroplasts are numerous, narrowly conical and radially arranged. A stigma is present. Platelets are polygonal reaching 7 μm in length, but are extremely variable in shape and size. A carina is present on the epicone. The resting stage (hypnozygote) is rhomboidal in ventral view with a blunt or minutely biapiculate apex and a somewhat sharply pointed and sometimes (obliquely) protracted

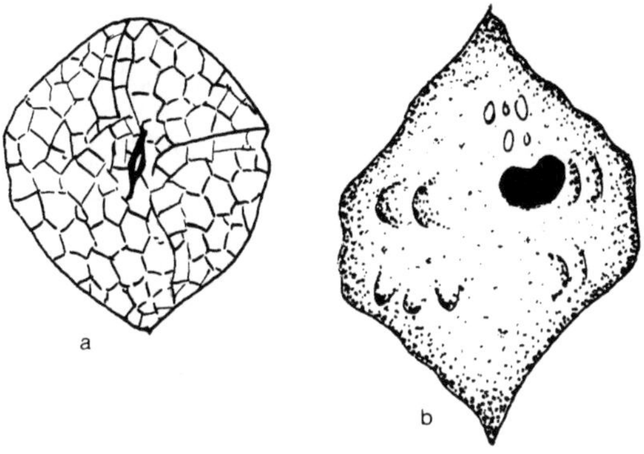

Fig. 132. *Woloszynskia apiculata* von Stosch (after von Stosch).

antapex. They are moderately compressed dorsiventrally and slightly concave at their ventral sides. The cingulum is slightly constricted but its ventral and lateral margins are swollen and ornamented by a simple or double row of tubercles. Reproduction vegetative by the formation of two to 8 zoospores per cell; sexual by fusion of gametes.
Dimensions: vegetative cells 25–46 µm × 18–36 µm, gametes 21–25 µm × 14–17 µm, planozygotes 47 µm × 34 µm.
Occurrence: plankton of a pond; Federal Republic of Germany (Marburg/L.).

2. **Woloszynskia coronata** (Wołoszyńska) Thompson 1950 (fig. 133)
Gymnodinium coronatum Wołoszyńska 1917, *G. nygaardii* Christen 1950, *Woloszynskia nygaardii* (Christen) Loeblich III 1969, *W. nigaardii* (Christen) Litvinenko 1977, *Gymnodinium coronatum* var. *glabrum* Wołoszyńska 1917, *Woloszynskia coronata* var. *glabra* (Wołoszyńska) Thompson 1950, *Glenodinium sanguineum* Marchesoni 1941, *Gymnodinium pulvisculus* var. *oculatum* Largajolli 1907

Cells are nearly spherical and slightly flattened dorsiventrally. The epicone is somewhat conical with an approximately median girdle and a wide sulcus on the hypocone. The hexangular platelets are usually irregularly arranged on the epicone and ornamented with visible nipple-like projections on the hypocone. There is a short carina on the epicone. Numerous chloroplasts and a stigma are present. Reproduction is by the formation of 2 to 8 zoospores. Hypnospores, triangular in median section are ornamented with horn-like processes.
Dimensions: vegetative cells 19–35 µm × 14–32 µm, zoospores 10 µm × 8 µm.
Occurrence: in plankton of ponds and lakes over the temperate zone of Europe and N. America.

3. **Woloszynskia neglecta** (Schilling) Thompson 1950 (fig. 134)
Glenodinium neglectum Schilling 1891, *Gymnodinium neglectum* (Schilling) Lindemann 1929, *G. neglectum* var. *astigmatica* Christen 1959

Cells are oval in ventral view and severely flattened dorsiventrally. Both cones are rounded – the epicone somewhat more so. The cingulum spirals slightly. The sulcus does

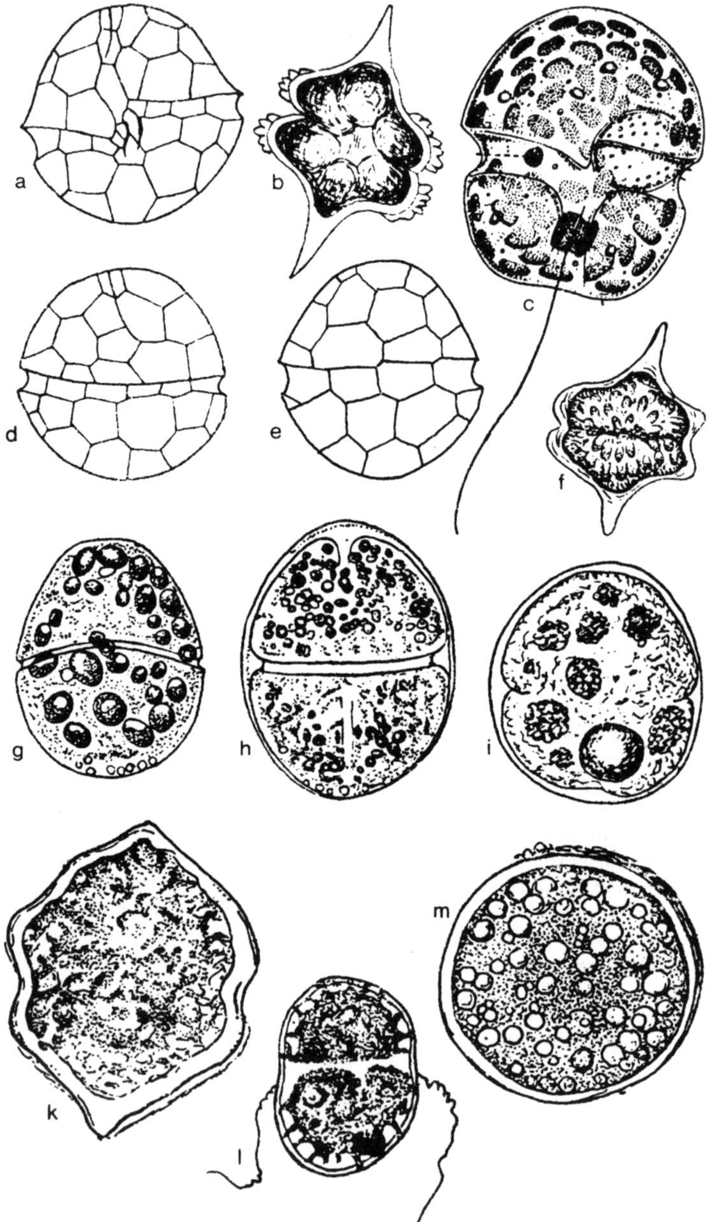
a
b
c
d
e
f
g
h
i
k
l
m

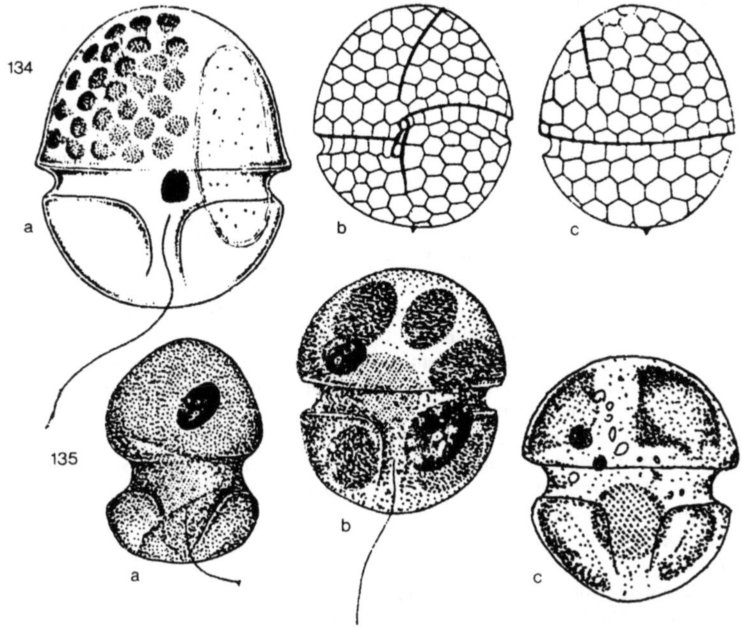

Fig. 134. *Woloszynskia neglecta* (Schilling) Thompson (*a* after Javornický, *b, c* after Wołoszyńska).

Fig. 135. *Woloszynskia ordinata* (Skuja) Thompson (*a, b* after Popovský, *c* after Skuja).

not reach the antapex. The theca is covered by regular hexagonal platelets. The numerous chloroplasts are parietal. A stigma is present. Reproduction is by zoospore formation. Dimensions: vegetative cells 32–35 µm × 28–30 µm, zoospores 22 µm × 18 µm, planozygotes 40–45 µm × 36–40 µm.
Occurrence: oligo-eutrophic pools and ponds; Europe.
Note: Schiller 1955 invalidly described *Gyrodinium pallidum, Glenodinium ampliconicum* and *Glenodinium gessneri,* which are related to *Woloszynskia pascheri* (Suchlandt) von Stosch 1973. *Glenodinium vindobonense* is probably *Woloszynskia neglecta* (Schilling) Thompson 1950.

Fig. 133. *Woloszynskia coronata* (Wołoszyńska) Thompson, *b, f, k* hypnozygotes, *m* resting stages (*a, b, d* after Wołoszyńska, *c* after Javornický, *e, f* after Nygaard, *g–l* after Baldi, *m* after Huber-Pestalozzi).

4. Woloszynskia ordinata (Skuja) Thompson 1950 (fig. 135)

Gymnodinium ordinatum Skuja 1939, *G. ordinatum* var. *sparsum* Popovský 1971

Cells are spherical or oval and flattened dorsiventrally with both cones rounded. The cingulum is broad and deep; the sulcus almost reaches the antapex. Chloroplasts are oval and from 3 to 6 in number. Refractive and red droplets are present in the cytoplasm. A stigma is lacking. Nutrition is autotrophic, holozoic, and holophytic.
Dimensions: vegetative cells 10–16 µm × 8–11 µm, zoospores 5–9 µm × 4–7 µm.
Occurrence: plankton of pools, lakes; Europe.

5. Woloszynskia pascheri (Suchlandt), v. Stosch 1973 (fig. 136)

Glenodinium pascheri Suchlandt 1916, *G. polonicum* Wołoszyńska 1916, *Gymnodinium polonicum* Wołoszyńska 1917, *G. hiemale* Wołoszyńska 1917; non *G. hiemale* (Schiller) Popovský 1984; *G. carinatum* var. *hiemalis* Wołoszyńska 1917, *G. tatricum* Wołoszyńska 1919, *G. mirum* Utermöhl 1923, 1925, *G. veris* Lindemann 1925, *Gyrodinium silvaticum* Lindemann 1925, *Gymnodinium undulatum* Wołoszyńska 1925, *G. hiemale* Skvorcov 1927, *Gyrodinium nivale* Lindemann 1929, *Gyr. pascheri* (Suchlandt) Schiller 1933, *Gymnodinium skvortzovii* Schiller 1933, *G. inversum* Nygaard 1949, *G. inversum* var. *elongatum* Nygaard 1949, *Woloszynskia polonica* (Wołoszyńska) Thompson 1950, *W. vera* (Lindemann) Thompson 1950, *W. hiemale* (Wołoszyńska) Thompson 1950, *W. mira* (Utermöhl) Kiselev 1954, *W. tatrica* (Wołoszyńska) Kiselev 1954, *Gymnodinium pascheri* (Suchlandt) Schiller 1954, *Dinastridium verrucosum* Baumeister 1957, *Glenodinium muriaticum* Skuja 1964

Cells vary in shape from spherical to oval to ovoid. The epicone is rounded, or conical or helmet-shaped. The hypocone is rounded or conical and sometimes excavated at the antapex. The epicone is usually smaller than the hypocone. Small undulations have rarely been observed on the hypocone. The cingulum is wide and spirals distinctly to the left. The sulcus is wide and almost reaches the antapex. Regular hexagonal platelets cover the cell. The chloroplasts are ribbon-shaped, and parietal. A stigma is absent. Hypnospores are cube-shaped with one blunt conical wall and bladder-shaped swellings at the corners ("*Dinastridium*"). Reproduction is by means of zoospores.
Dimensions: vegetative cells 18–50 µm, hypnospores 40–50 µm.
Occurrence: oligo- to eutrophic pools and ponds, cryophilic; Europe.

6. Woloszynskia pseudopalustris (Wołoszyńska) Kiselev 1954 (fig. 137)

Gymnodinium palustre fo. Wołoszyńska 1917, *G. excavatum* Nygaard 1954, *G. excavatum* var. *dextrorsum* Nygaard 1954, *G. pseudopalustre* (Wołoszyńska) Schiller 1933

Cells are spherical or slightly ovoid and somewhat flattened dorsiventrally in ventral view. The epicone is larger than the hypocone. Both cones are rounded: the hypocone is excavated. The cingulum spirals slightly to the left. The sulcus reaches the antapex. The morphology and arrangement of platelets (1–2 µm in diameter) are not yet well documented. Chloroplasts are numerous and parietal. A stigma is sometimes present. Light and red refractive droplets are often present in the cytoplasm. Reproduction is by division of motile cells into two daughter cells. Thick walled hypnospores are spherical and covered by short, dense bristles.
Dimensions: up to 50 µm × 50 µm, zoospores 26–42 µm × 21–34 µm.
Occurrence: ponds, lakes; Europe.

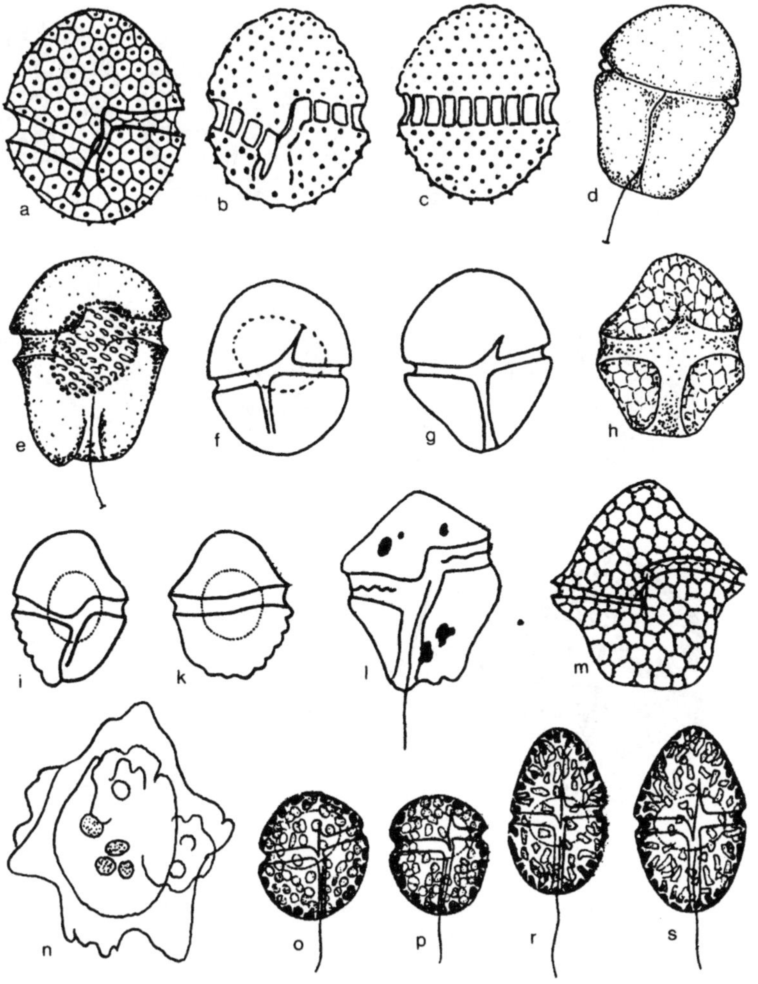

Fig. 136. *Woloszynskia pascheri* (Suchlandt) von Stosch, *n* hypnozygote (*a*, *b*, *c*, *i*, *k*, *m* after Wołoszyńska, *d* after Suchlandt, *e* after Javornický, *f*, *g* after Utermöhl, *h* after Schiller, *l* after Lindemann, *n* after Baumeister, *o–s* after Nygaard).

137

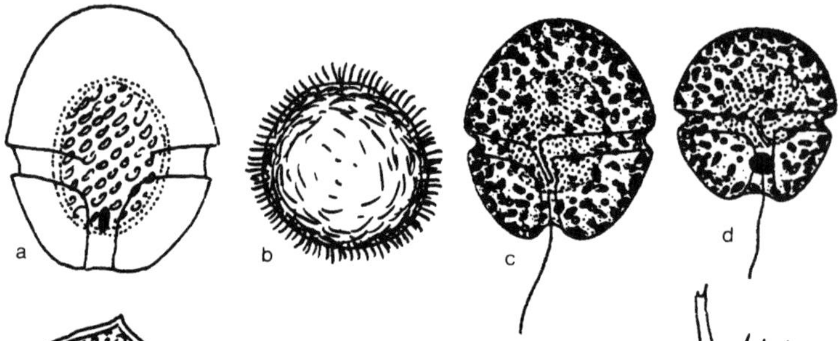

138

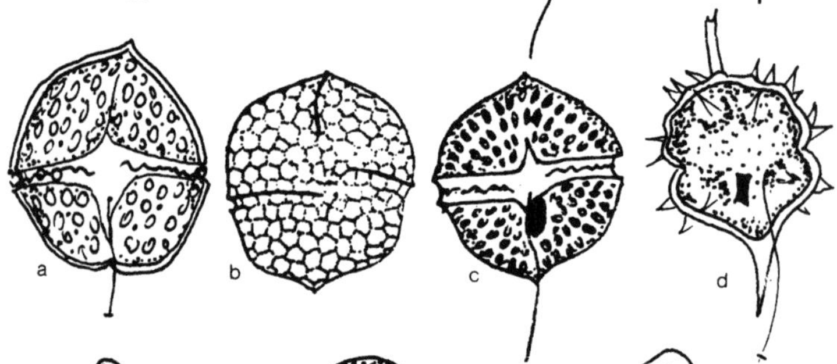

139

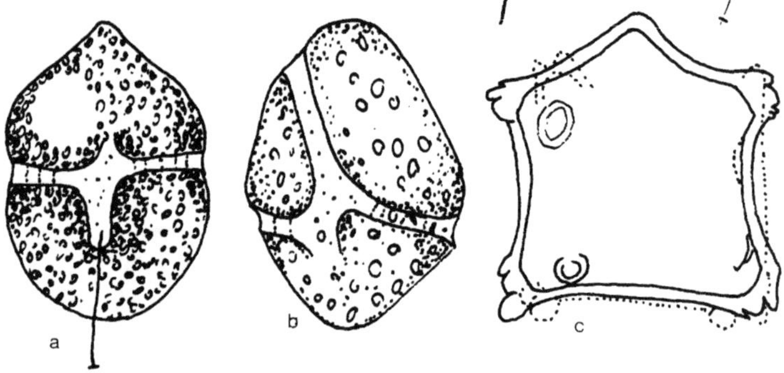

Fig. 137. *Woloszynskia pseudopalustris* (Wołoszyńska) Kiselev, *b* hypnozygote (*a, b* after Wołoszyńska, *c, d* after Nygaard).

Fig. 138. *Woloszynskia tenuissima* (Lauterborn) Thompson, *d* hypnozygote (*a* after Schiller, *b, d* after Wołoszyńska, *c* after Lindemann).

Fig. 139. *Woloszynskia tylota* (Mapletoft et al.) Bibby and Dodge, *c* hypnozygote (after Mapletoft et al.).

7. Woloszynskia tenuissima (Lauterborn) Thompson 1950 (fig. 138)
Gymnodinium tenuissimum Lauterborn 1894, *G. leopoliense* Wołoszyńska 1917, *G. lens* Fott 1957, *Woloszynskia leopoliensis* Thompson 1950

Cells are oval and flattened dorsiventrally. The apex is apiculate, the antapex may be so. The epicone is helmet-shaped, the hypocone rounded or helmet shaped. The cingulum is composed of a series of plates. The sulcus is narrow and does not reach the antapex. Chloroplasts are oval or irregular in shape and numerous. A stigma is present in zoospores. Hypnospores are square and usually have a long process at each end and shorter, thick spine-like processes over the cell surface.
Dimensions: vegetative cell 40 µm × 35 µm, planozygotes 66–80 µm × 60–72 µm, zoospores 35 µm × 30 µm.
Occurrence: plankton of ponds and lakes; Europe.

8. Woloszynskia tylota (Mapletoft et al. 1966) Bibby and Dodge 1972 (fig. 139)
Gymnodinium tylotum Mapletoft et al. 1966

Oval cells with a bluntly conical epicone and a rounded hypocone. The cingulum spirals to the left; the sulcus extends slightly into the epicone and on the hypocone reaches to the antapex. Numerous discoidal chloroplasts are present. A stigma is lacking. The square hypnospore resembles *Tetradinium*. Reproduction is by zoospore formation.
Dimensions: vegetative cells 30–35 µm × 25–30 µm, planozygotes 35 bis 45 µm × 30–40 µm, hypnospores 40–50 µm in length.
Occurrence: oligotrophic pools; Great Britain.

Appendix

1. *Woloszynskia cestocoetes* (Thompson) Thompson 1950 (fig. 140)
Gymnodinium cestocoetes Thompson 1947

Cells are ovoid in ventral view and not flattened dorsiventrally. The epivalve is large, rounded and shorter than the hypovalve. The hypovalve is conical with a sinus on the antapex. The median cingulum is deep. The sulcus extends slightly into the epitheca and widens to the antapex on the hypotheca. Chloroplasts are numerous and parietal. A stigma is absent. Reproduction by means of two zoospores. Hypnospores are oval in shape and spinescent.
Dimensions: vegetative cells 23–25 µm × 19–21 µm.
Occurrence: pool; USA.

2. *Woloszynskia limnetica* Bursa 1958 (fig. 141)

Cells are subspherical with a conical or dome-shaped epicone which varies in height. The hypocone may be slightly longer (up to one-third) than the epicone and is broad at the end. The epicone comes to a tapered point fusing with the sulcus. The sulcus is shallow in the epicone and deeper in the hypocone. The cingulum is straight in some cells and displaced about four widths. One or both lateral edges of the epicone overlap the hypocone. In adult cells, the larger membraneous plates and wider sulcus are composed of up to two rows of square plates. A finer reticulation of small hexagonal plates occurs in young motile individuals. Chloroplasts are spherical or subspherical and 5 to 9 per cell. In some cells, however, they may be elongated, and very numerous. The pusule is spherical with an exit-canal close to the centre of the epicone. A stigma is present. The cytoplasm is filled with starch grains. Spherical and subspherical cysts with fine hexagonal membrane reticulations or hyaline processes are formed.

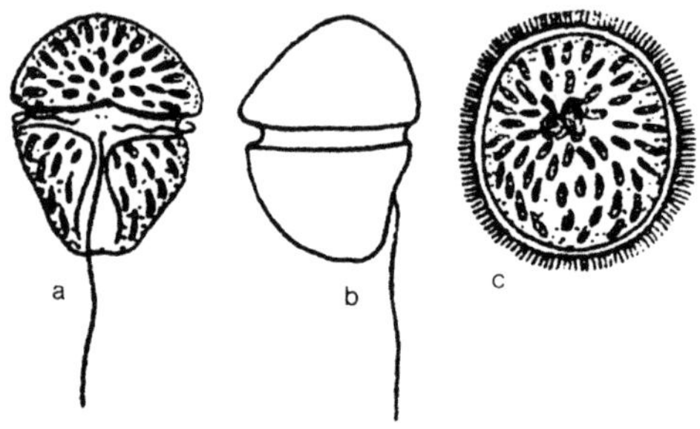

Fig. 140. *Woloszynskia cestocoetes* Thompson, *c* hypnozygote (after Thompson).

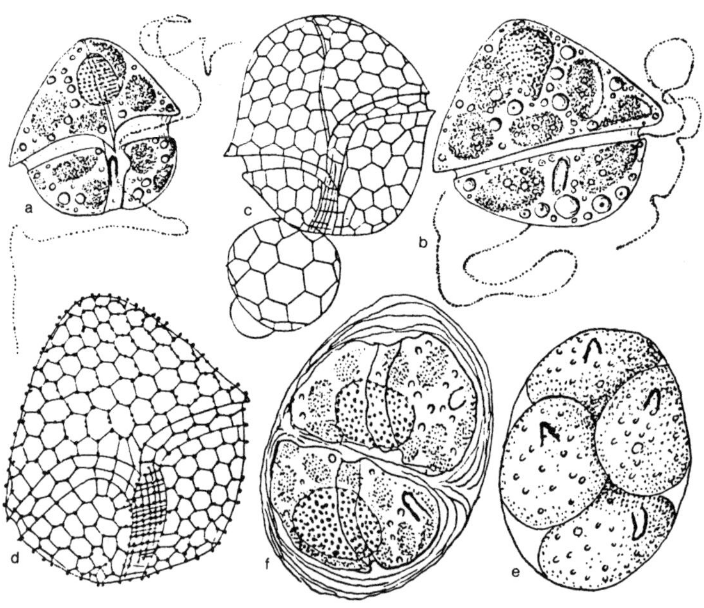

Fig. 141. *Woloszynskia limnetica* Bursa (after Bursa).

Dimensions: vegetative cells 28.4–63 μm × 31.5–60 μm, cysts 35–68 μm in diameter. Occurrence: pond; England.

3. *Woloszynskia reticulata* Thompson 1950 (fig. 142)

Cells are rhomboid-ovoid in ventral view and dorsiventrally flattened. The epivalve is delicate, smooth and bell-shaped with a carina over the apex from the ventral to the dorsal

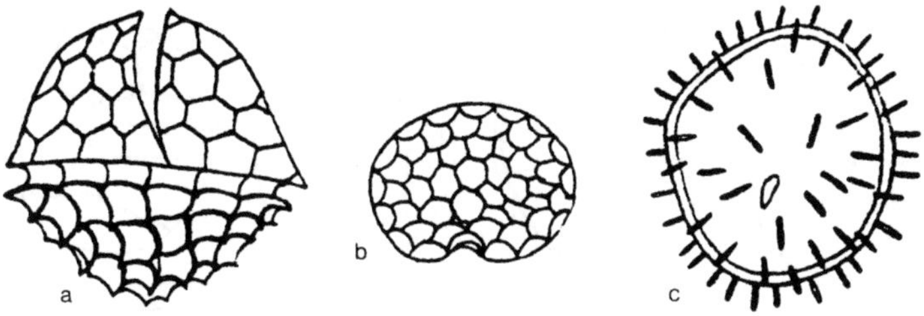

Fig. 142. *Woloszynskia reticulata* Thompson, *c* hypnozygote (after Thompson).

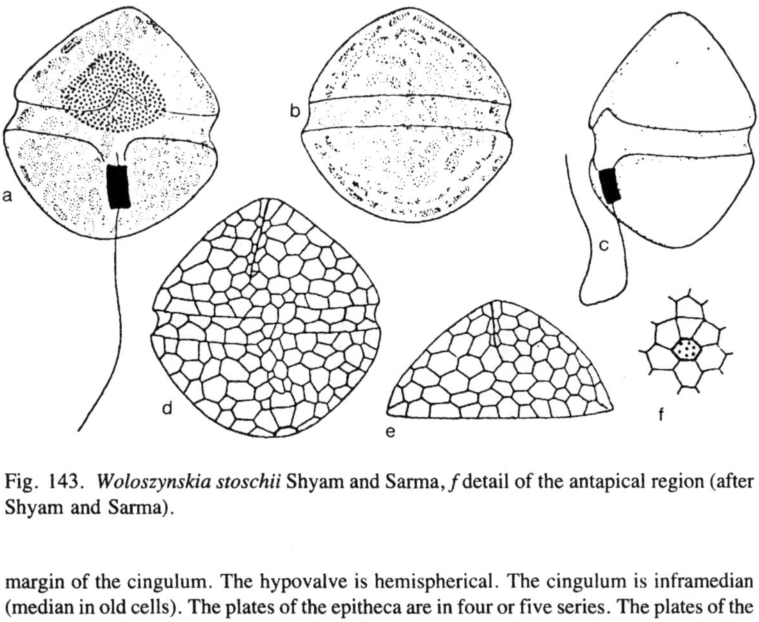

Fig. 143. *Woloszynskia stoschii* Shyam and Sarma, *f* detail of the antapical region (after Shyam and Sarma).

margin of the cingulum. The hypovalve is hemispherical. The cingulum is inframedian (median in old cells). The plates of the epitheca are in four or five series. The plates of the hypovalve are in four series of smooth, thick, concave, platelets. There is some variability in this grouping. The platelets forming the sulcus are curved inward more deeply than the surrounding ones, thus producing a continuous channel. Numerous chloroplasts and one stigma are present. Reproduction is by formation of two or four zoospores. The resting stage is spherical or slightly rhomboidal in shape and spinescent.
Dimensions: vegetative cells 25–52 μm × 21–46 μm.
Occurrence: lakes, ponds, river oxbow; USA.

4. *Woloszynskia stoschii* Shyam and Sarma 1975 (fig. 143)
Cells somewhat ovoid or conical in shape and slightly flattened dorsiventrally. The epicone is conical and slightly shorter than the widely rounded hypocone. The cingulum is

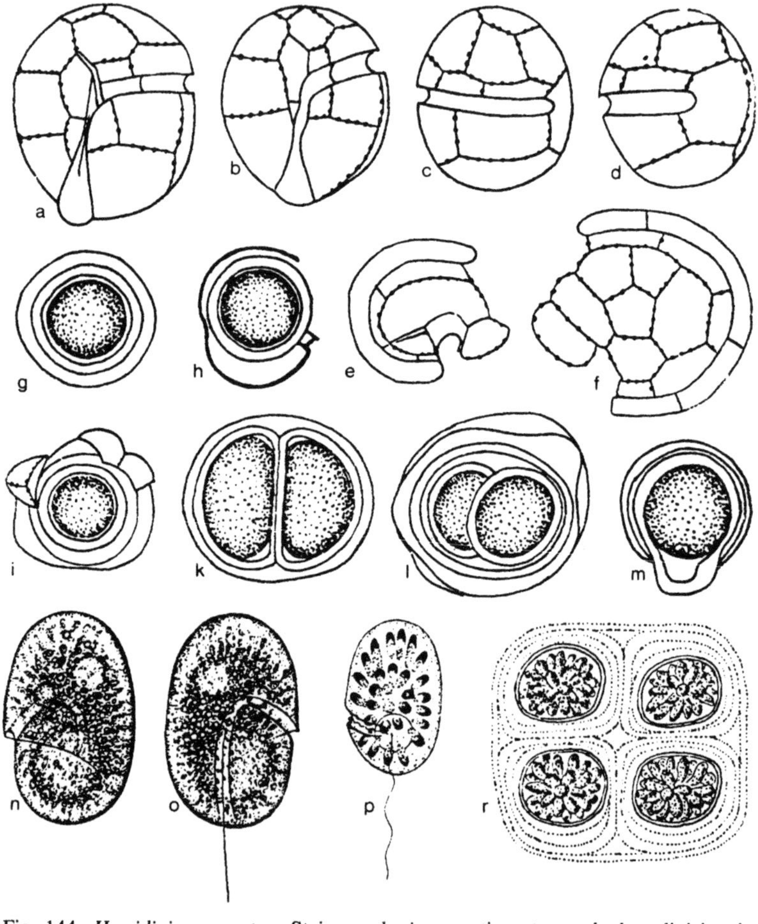

Fig. 144. *Hemidinium nasutum* Stein, *g, h, i, m* resting stages, *k, l, r* division in nonmotile stage *("Gloeodinium")* (*a–m* after Wołoszyńska, *n, o* after Skuja, *p* after Bicudo and Skvortzov, *r* after Thompson).

wide and slightly spiral. The sulcus extends slightly into the epicone. It is shallow and extends into half of the hypocone. The theca is covered with numerous polygonal platelets of different shapes and sizes. The platelets lack papillae. The conspicuous carina consists of double rows of narrow platelets, extending from the apex to 3/4 of the ventral and 1/4 of the dorsal side of the epitheca. An antapical platelet is present. Chloroplasts are numerous, parietal and elongated to ellipsoidal in shape. The tetragonal stigma lies in the

sulcal area. Reproduction is by the formation of 2 to 8 zoospores, in a non-motile cell. Fusion of two isogametes which produce a planozygote has been reported.
Dimensions: 29.7–37.2 µm × 28.4–34.6 µm.
Occurrence: pond (vegetative coloring), Varanasi; India.

3. Family **Hemidiniaceae** Bourrelly 1970

Cells with a delicate theca and incomplete cingulum. Chloroplasts are present.
Plate formula: 6′, 0a, 6″; 5‴, lp, 1⁗.

1. **Hemidinium** Stein 1879 (fig. 46e)

Nonsymmetrical cells that are elliptical or slightly ovoid and dorsiventrally flattened with the antapex broadly rounded and the anterior pole less so being cone-shaped. The incomplete cingulum curves down to the left; a narrow sulcus extends from the cingulum to the posterior pole. Numerous elongated chloroplasts are radially arranged. The cell is covered by a very thin theca with indistinctly marked aerolate plates. Cells are spherical and enveloped by layers of mucilage in the nonmotile stage (= *Gloeodinium*). No plates, furrows, or flagella are present. The cell content is difficult to discern in this stage. The chloroplasts are radially arranged and red droplets are usually present. The shape of the colonies and mucilage can vary greatly. Reproductive methods are not completely known.
Important references: Klebs 1912, Kilian 1924, Wołoszyńska 1930, Baumeister 1938, 1943, Bourrelly 1970, Popovský 1971b.

1. **Hemidinium nasutum** Stein 1878 (figs. 28, 51, 144, 145, 146, 147)
Gloeodinium montanum Klebs 1912 partim. *G. cracoviense* Starmach 1963 ad inter., *Hemidinium brasiliense* Bicudo and Skvortzov 1970
Dimensions: motile vegetative cells 22–36 µm × 15–30 µm, zoospores (gametes?) 15–28 µm × 10–17 µm. solitary cells of nonmotile stages: 40–49 µm × 16–39 µm. colonies: 60–93 µm × 52–73 µm.

1.1. var. *tatricum* Wołoszyńska 1930 (fig. 148)
Gloeodinium montanum Klebs 1912 partim, *Hemidinium nobile* Skuja 1964
Cells elongated and kidney-shaped; very flattened dorsiventrally with both poles rounded. The sulcus reaches the antapex. There is only one large plate on the right side of the cingulum.
Dimensions: motile vegetative cells 30–50 µm × 18–30 µm, zoospores 15 bis 25 µm × 8–15 µm.
Occurrence: peat bogs and dystrophic lakes of temperate and subarctic zones; Europe, N. America, New Zealand.

Appendix

The following species, originally described as *Hemidinium* appear to be species belonging to the genus *Bernardinium* since a cell wall has not been described.

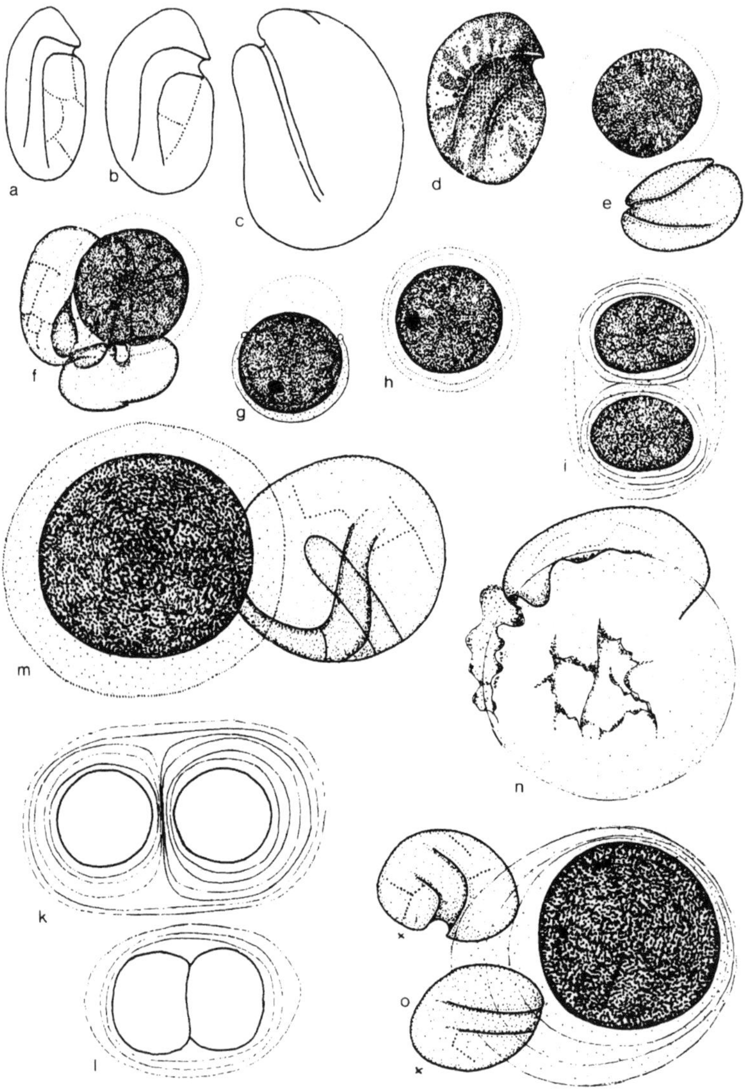

Fig. 145. *Hemidinium nasutum* Stein, *a–d* zoospores or gametes, *e* release of zoospore or gamete protoplast embedded in gelatinous material, *f, o* copulation? aplanozygote?, *g, h* resting stages embedded in gelatinous material, *i, k, l* division in the nonmotile stage *("Gloeodinium")*; different shapes of layered mucilage, *m* release of a vegetative cell protoplast surrounded by a fine gelatinous material, *n* empty cell walls following perhaps the release of motile spores (after Popovský).

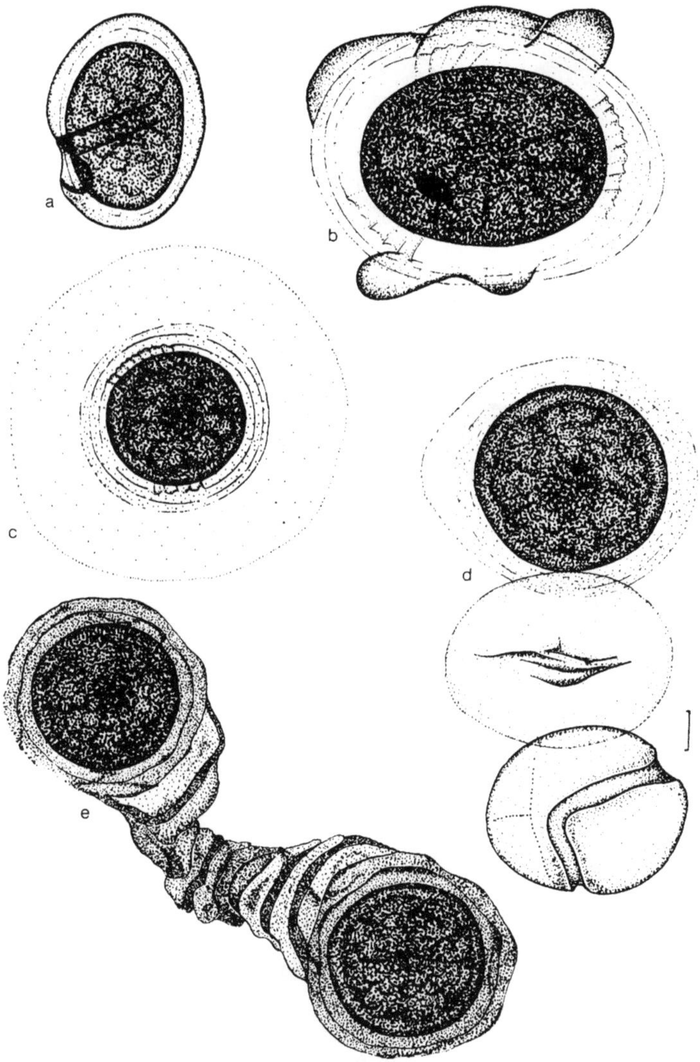

Fig. 146. *Hemidinium nasutum* Stein, *a, b* encystement within the theca, *c* resting stage in doubled gelatinous material composed of concentric rings, *d* empty theca showing single-layered mucilage produced by protoplast which later moved out of it and formed a concentric layers of mucilage around itself, *e* dendroid colonial form produced by two nonmotile cells (after Popovský).

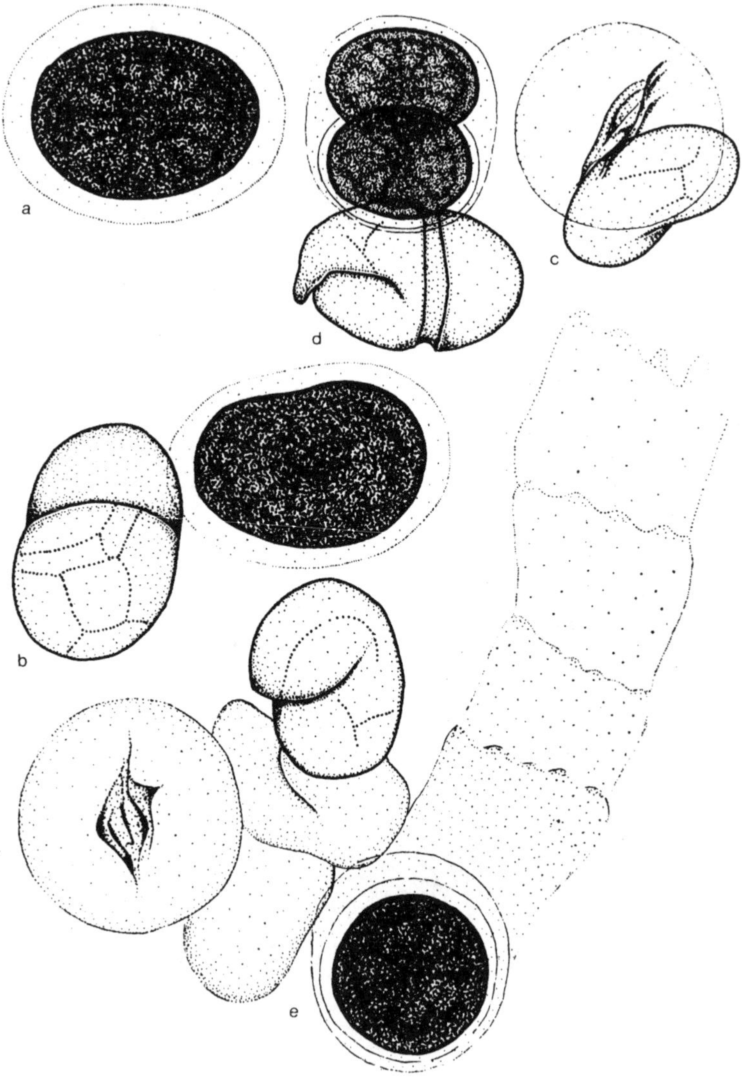

Fig. 147. *Hemidinium nasutum* Stein, *a* resting stage, *b* encystment outside the theca, *c* empty membranes, *d* encystment outside the theca following by division of the cyst, *e* ecdysis followed by repeated production of single-layered mucilage and migration out; final stage shows cell producing very fine mucilagenous sleeve.

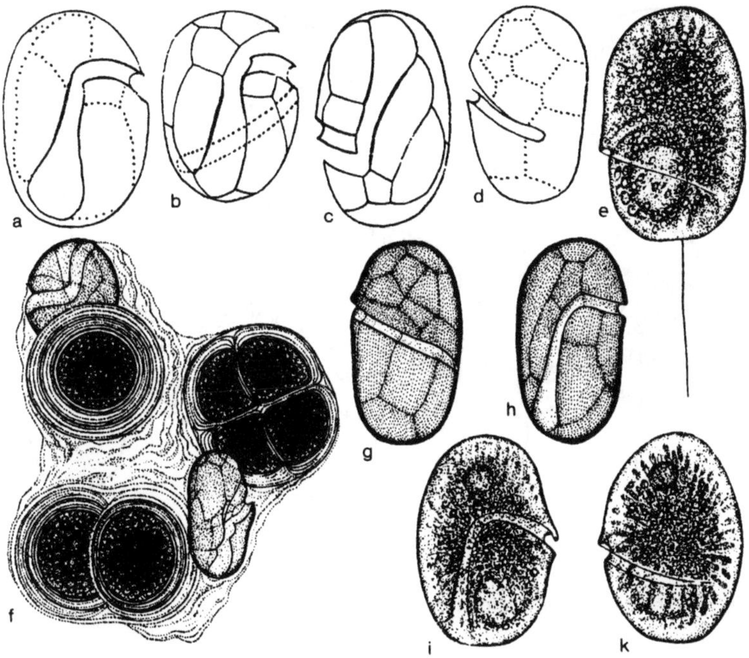

Fig. 148. *Hemidinium nasutum* var. *tatricum* Wołoszyńska, *f* dividing resting stages (*a–d* after Wołoszyńska, *e–k* after Skuja).

1. *Hemidinium ochraceum* Levander 1900 (fig. 149)
   H. pseudoochraceum Bicudo and Skvortzov 1970, *H. montanum* Bicudo and Skvortzov 1970

Cells are somewhat dorsiventrally flattened. The cingulum spirals as it descends to the left. A single, massive, laminated chloroplast is present in only the epicone or it may fill the cell. Some cells have a few smaller, discoid chloroplasts present.
Dimensions: 11–33 μm × 8–26 μm.
Occurrence: saline puddles; Finland. Freshwater temporary pools; Brazil.

2. *Hemidinium ucrainicum* Proškina-Lavrenko 1945 (fig. 150)

Cells are large, ovoid and dorsiventrally flattened. The cingulum spirals as it descends to the left. The sulcus almost reaches the antapex. The numerous chloroplasts are large. Refractive bodies are present in the cytoplasm.
Dimensions: 15–18 μm × 10–14 μm.
Occurrence: saline inland lakes; Ukraine.

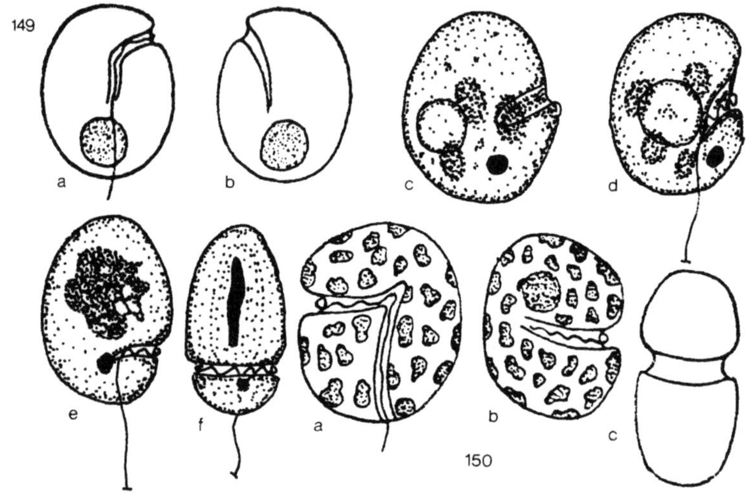

Fig. 149. *Hemidinium ochraceum* Levander (*a, b* after Levander, *c–f* after Bicudo and Skvortzov).

Fig. 150. *Hemidinium ucrainicum* Proškina-Lavrenko (after Proškina-Lavrenko).

4. Family **Glenodiniopsidaceae** Schiller 1937

Glenodiniaceae Lindemann 1928b partim, *Pyrophagaceae* Lindemann 1928b
Cells with six to seven postcingular plates on the hypotheca. Chloroplasts present. Plate formula: 3–4′, (3a–4a), (7–8)″; (6–7)‴, 2⁗.
Important reference: Schiller 1937.

Key to the genera

1a. Apical pore present; plate pattern symmetric . . . . **2. Sphaerodinium** (p. 157)
2b. Apical pore absent; plate pattern asymmetric . . . . **1. Glenodiniopsis** (p. 156)

1. **Glenodiniopsis** Wołoszyńska 1916

Glenodinium partim, *G. uliginosum* Schilling 1891
Cells are slightly oval and somewhat flattened dorsiventrally. No apical pore present. Two different plate patterns have been reported.
Important references: Klebs 1912, Wołoszyńska 1916, Schiller 1937, Huber-Pestalozzi 1950, Popovský 1971 a.

Key to the species

1a. Plate formula is 4′, 4a, 8″; (6–8)‴, 2⁗ . . . . **1. G. steinii**
1b. Plate formula is 3′, 3a, 8″; 7‴, 2⁗ or 2′, 4a, 8″; 7‴, 2⁗ . . . **2. G. pretiosa**

1. Glenodiniopsis steinii (Lemmermann) Wołoszyńska 1916 (fig. 151)
Glenodinium cinctum Stein 1878, *G. uliginosum* Schilling 1891, *G. steinii* Lemmermann 1900, 1910, *Gloeodinium montanum* Klebs 1912 partim

Cells are somewhat oval and flattened dorsiventrally. The epicone is rounded and larger than the hypocone which is rounded at the antapex. The cingulum is relatively deep and spirals to the left. The sulcus extends from half the length of the epicone where it is narrow and reaches almost to the antapex. Both furrows are composed of plates. The numerous chloroplasts are parietally or radially arranged. There is no stigma. Numerous corpuscles are present within the cytoplasm. The resting stages are spherical with a multilayered mucilage ("Gloeodinium montanum"). Reproduction is only known to occur in the resting stage by cell division forming pseudocolonies.
Dimensions: vegetative cells 27–50 μm × 21–39 μm, non-motile stages 25 bis 48 μm × μm in diameter.
Occurrence: dystrophic lakes, peat-bogs; Europe.

2. Glenodiniopsis pretiosa Lindemann 1931 (fig. 152)
Cells are approximately oval and flattened dorsiventrally. The epicone is rounded and larger than the hypocone which is also rounded. The cingulum is deep and spirals to the left. The sulcus does not extend into the epicone; it does extend to the antapex. Chloroplasts are numerous and parietal. A stigma is lacking.
Dimensions: 30–40 μm × 26–39 μm.
Occurrence: peat bogs; Poland and Sumatra.

2. Sphaerodinium Wołoszyńska 1916

Cells spherical or oval; epicone and hypocone equal in size. The cingulum spirals to the left. Plate formula: 4′, 4a, 7″; 6‴, 2⁗. An apical pore is present. Reproduction occurs by division of motile cells into two daughter cells (gametes?).
Important references: Wołoszyńska 1916, 1917, Huber-Pestalozzi 1950.

1. Sphaerodinium cinctum (Ehrenberg) Wołoszyńska 1917 (fig. 153)
Sphaerodinium cracoviense Wołoszyńska 1916, *Glenodinium cinctum* (O. F. Müller) Ehrenberg 1830, *Sphaerodinium polonicum* var. *tatricum* Wołoszyńska 1916, *Sphaerodinium cinctum* var. *polonicum* (Wołoszyńska) Huber-Pestalozzi 1950, *Sphaerodinium limneticum* Wołoszyńska 1916, *Sphaerodinium cinctum* var. *limneticum* (Wołoszyńska) Huber-Pestalozzi 1950, *Sphaerodinium javanicum* Wołoszyńska 1930

Cells are spherical or oval and slightly flattened dorsiventrally. The cingulum spirals to the left; the sulcus extends into 1/3 of the epitheca. It narrows along it length reaching almost to the antapex. The theca is smooth and ornamented with fine dots. Plate 3′ is hexangular. Oval chloroplasts and a stigma are present. Planozygotes have been reported. Hypnozygotes are spherical with a thick membrane.
Dimensions: 25–50 μm × 25–48 μm.
Occurrence :pools, ponds, swamps; Europe.

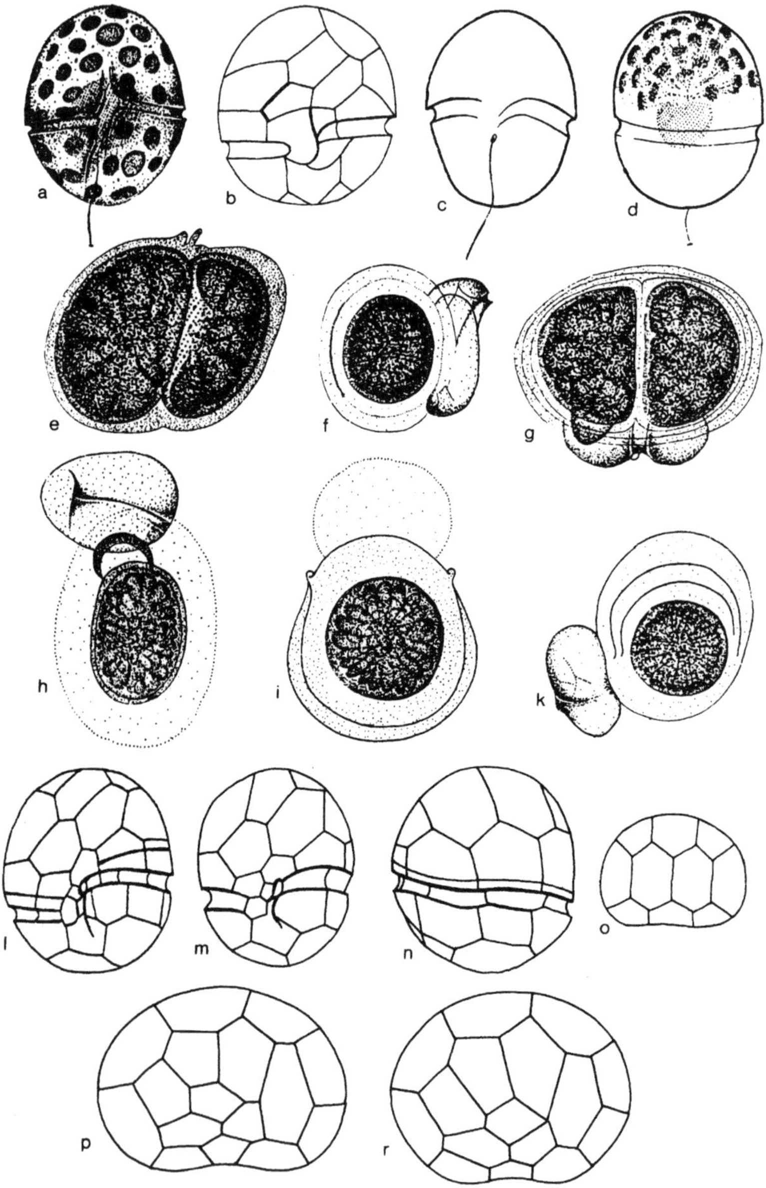
a
b
c
d
e
f
g
h
i
k
l
m
n
o
p
r

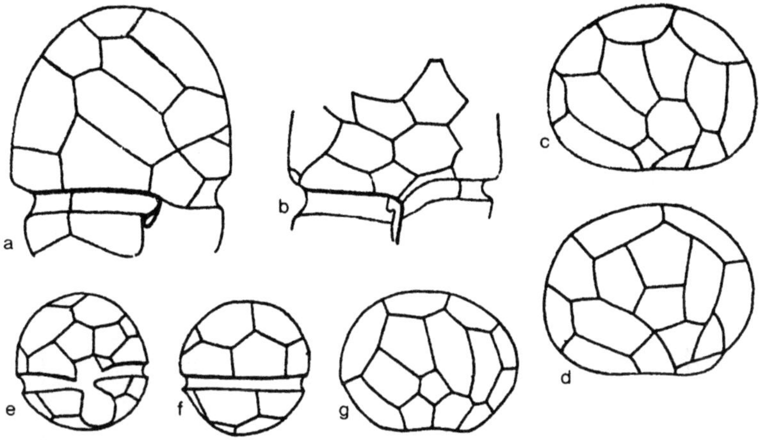

Fig. 152. *Glenodiniopsis pretiosa* Lindemann (*a–d* after Popovský, *e–g* after Lindemann).

Appendix

1. *Sphaerodinium fimbriatum* Thompson 1950 (fig. 154)

This taxon has been reported from the USA. Cells are oval or ellipsoid. The cingulum spirals to the left; the wide sulcus reaches both apices. The plates are ornamented with toothed wing-like processes on both the epi- and hypothecae. Chloroplasts and a stigma are present.

Dimensions: 42–53 µm × 42–46 µm.

Occurrence: lakes; USA.

Appendix

1. *Glenodinium* Ehrenberg 1837 partim

The validity of this genus is in question. Since its taxa are thecate they must probably belong to genera such as *Peridinium, Peridiniopsis, Glenodiniopsis, Sphaerodinium,* or *Woloszynskia.* The taxa listed below are included for completeness.

1. *Glenodinium alpestre* Chodat 1921 (fig. 155)
   Dimensions: 14–16 µm × 12 µm. Probably *Peridinium umbonatum.*

Fig. 151. *Glenodiniopsis steinii* (Lemmermann) Wołoszyńska, *e* resting stage *("Gloeodinium"), f, g, h, k* ecdysis followed by encystment and mucilage production, *i* germinated cyst (*a–k, o, p* after Popovský, *l–n, r* after Wołoszyńska).

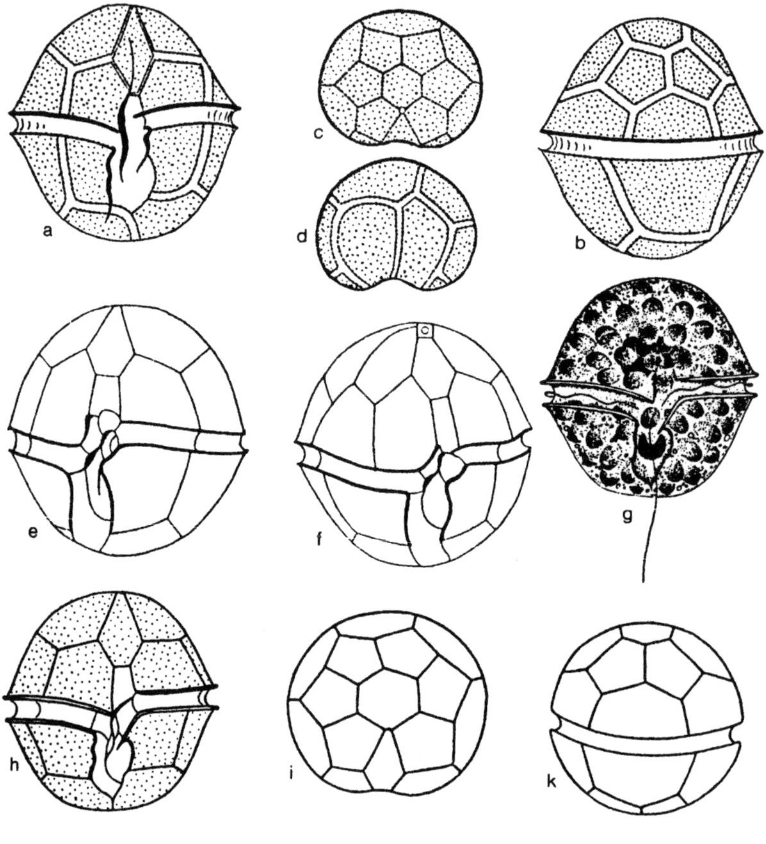

Fig. 153. *Sphaerodinium cinctum* Wołoszyńska (*a–f* after Wołoszyńska, *g–k* after Thompson).

2. *Glenodinium apiculatum* Zacharias 1901 (fig. 156)
   Colorless. Dimensions: 44 µm × 24 µm. Probably *Gymnodinium helveticum.*
3. *Glenodinium armatum* Levander 1900 (fig. 157)
   Dimensions: 18–25 µm × 16–20 µm. Probably a variety of *Peridinium umbonatum.*
4. *Glenodinium bernardinense* Chodat and Zender 1923 (fig. 158)
   Dimensions:24–26 µm × 21–29 µm.
5 *Glenodinium berghii* Lemmermann 1900 (fig. 159)
   syn. *Glenodinium cinctum* Bergh 1881
   Dimensions: 31–39 µm × 28–34 µm.

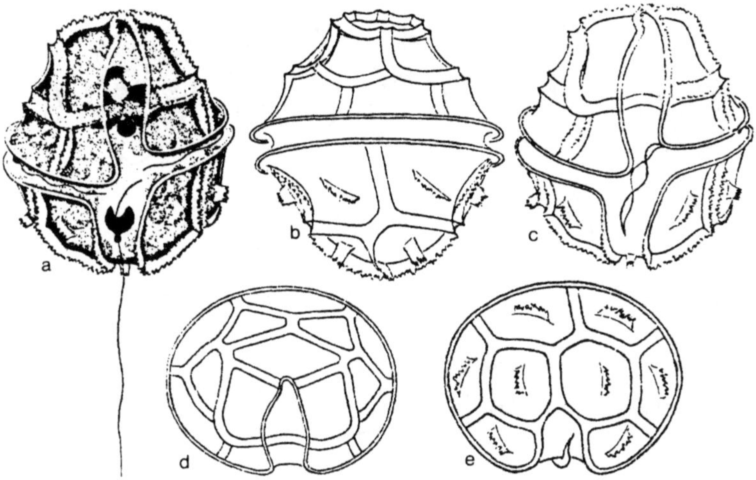

Fig. 154. *Sphaerodinium fimbriatum* Thompson (after Thompson).

6. *Glenodinium dangeardii* (Dangeard 1883) Lemmermann 1910 (fig. 160)
syn. *Glenodinium cinctum* Dangeard 1883
Dimensions: 33–42 µm × 25–32 µm.
7. *Glenodinium helicozoster* Harris 1939/40 (fig. 162)
Dimensions: 19–24 µm × 8.5–10 µm.
8. *Glenodinium inaequale* Chodat 1921 (fig. 161)
Dimensions: 20 µm × 14–16 µm. Probably *Peridinium umbonatum*.
9. *Glenodinium lemmermannii* Zacharias 1901, 1903 (fig. 163)
Dimensions: 60 µm × 40 µm.
10. *Glenodinium leptodermum* Harris 1939/40 (fig. 164)
Colorless. Dimensions: 14–24 µm × 12–16 µm.
11. *Glenodinium limos* Harris 1939/40 (fig. 165)
Dimensions: 46 µm × 40 µm
12. *Glenodinium pseudostigmosum* Skuja 1956 (fig. 166)
Dimensions: 30–38 µm × 25–33 µm. Probably a *Peridinium* or *Peridiniopsis* species.
13. *Glenodinium punctulatum* Harris 1939/40 (fig. 167)
Colorless. Dimensions: 15–24 µm × 12–18 µm. Probably a species of *Woloszynskia*.
14. *Glenodinium pulvisculus* (Ehrenberg) Stein 1883 (fig. 168)
Dimensions: 23–35 µm × 13–29 µm. Probably *Sphaerodinium*, *Glenodiniopsis* or *Dinosphaera palustris*.

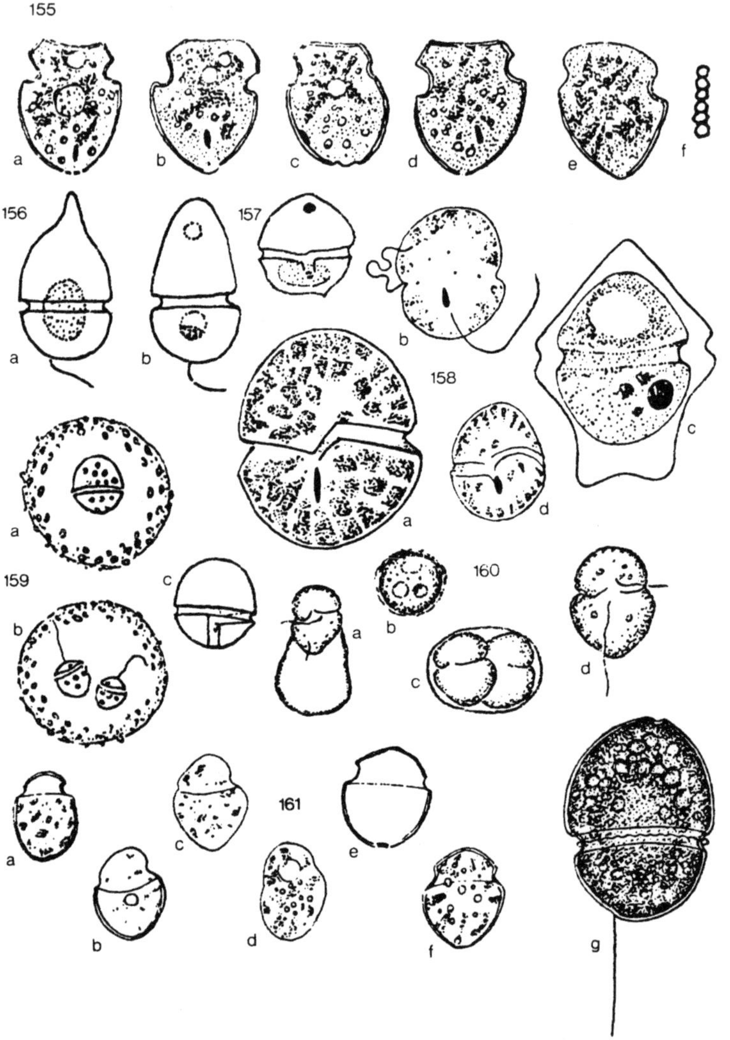

Fig. 155. *Glenodinium alpestre* Chodat, *f* thecal surface structure (after Chodat).
Fig. 156. *Glenodinium apiculatum* Zacharias (after Zacharias).
Fig. 157. *Glenodinium armatum* Levander (after Levander).
Fig. 158. *Glenodinium bernardinense* Chodat and Zender (after Chodat and Zender).
Fig. 159. *Glenodinium berghii* Lemmermann (after Lemmermann).
Fig. 160. *Glenodinium dangeardii* Lemmermann (after Lemmermann).
Fig. 161. *Glenodinium inaequale* Chodat (after Chodat, *g* after Skuja).

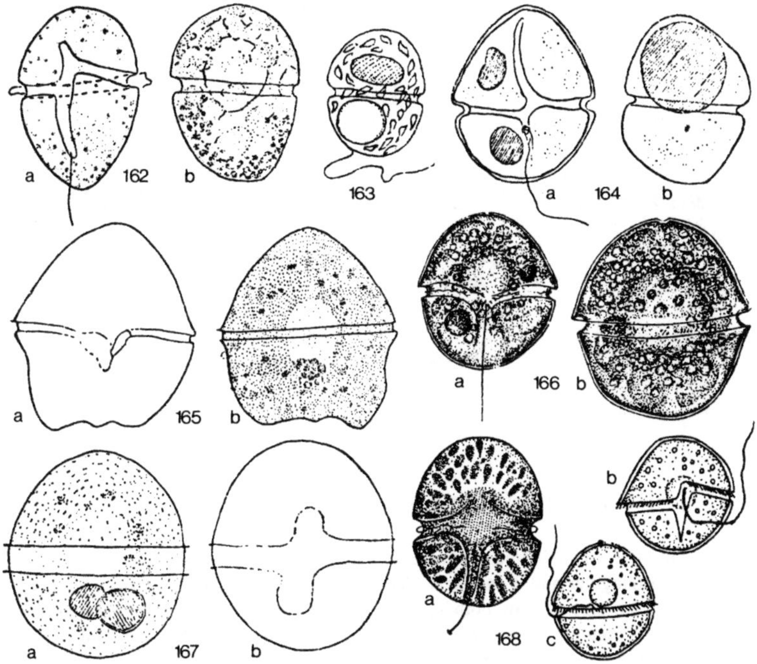

Fig. 162. *Glenodinium helicozoster* Harris (after Harris).

Fig. 163. *Glenodinium lemmermannii* Zacharias (after Zacharias).

Fig. 164. *Glenodinium leptodermum* Harris (after Harris).

Fig. 165. *Glenodinium limos* Harris (after Harris).

Fig. 166. *Glenodinium pseudostigmosum* Skuja (after Skuja).

Fig. 167. *Glenodinium punctulatum* Harris (after Harris).

Fig. 168. *Glenodinium pulvisculus* (Ehrenberg) Stein (*a* after Popovský, *b, c* after Stein).

15. *Glenodinium sanguineum* Marchesoni (1941) (fig. 133)
 syn. *Glenodinium pulvisculus* var. *oculatum* Largajolli 1907
 Dimensions: 12–23 µm × 11–30 µm. Synonym of *Woloszynskia coronata*.
16. *Glenodinium smreczyniense* Wołoszyńska 1952, nomen nudum (figs. 45, 169)
 No description was given. Probably a *Sphaerodinium* species.

Important references: Lefèvre 1932, Harris 1939/40, Huber-Pestalozzi 1950, Wołoszyńska 1952, Skuja 1956, Prager 1963, Starmach 1974.

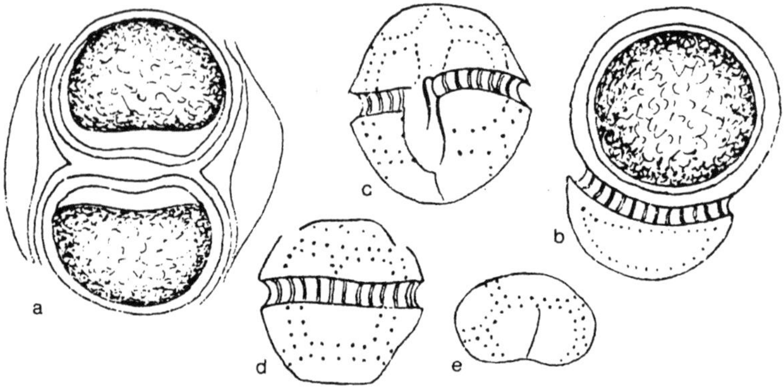

Fig. 169. *Glenodinium smreczyniense* Wołoszyńska (after Wołoszyńska).

5. Family **Peridiniaceae** Ehrenberg 1838

Cells are motile, with a definitive number of thecal plates. The hypotheca has a total of 7 plates (5 postcingular and 2 antapical places, respectively). Tremendous variability in position and arrangement of both epi- and hypothecal plates and sutures has been reported. The cingulum is complete, helicoidal and composed of plates. An apical pore, trichocysts, pseudopodia, chloroplasts, stigma and pyrenoid may be absent. Vegetative reproduction is by the production of two zoospores. Sexual reproduction is by isogamy. Thick-walled hypnospores may be embeded with thick mucilage. Division of the nonmotile form in mucilage ("*Gloeodinium montanum*"-like) has been reported. There are two freshwater genera: *Peridinium* and *Peridiniopsis,* with the plate formulas 4′, (2a–3a), 7″; 5‴, 2⁗ and (3–5)′, (0a–1a), (6–7)″; 5‴, 2⁗, respectively.
Important references: Schiller 1937, Huber-Pestalozzi 1950, Bourelly 1970, Popovský 1983.

Key to the genera

1a. Two to three intercalary plates . . . . **1. Peridinium** (p. 164)
1b. Zero to one intercalary plates . . . . **2. Peridiniopsis** (p. 186)

1. **Peridinium** Ehrenberg 1832

Vorticella O. F. Müller 1786, *Glenodinium* Ehrenberg 1838, *Ceratophorus* Diesing 1850, *Ceratium* Claparède et Lachmann 1858/61, *Protoperidinium* Bergh 1882, *Nephrodinium* Meunier 1910, *Chalubinskia* Wołoszyńska 1916, *Protoperidinium* Meunier 1919, *Minuscula* Lebour 1925.
Cells are ellipsoidal or spherical. The hypotheca is equal or smaller than the epitheca. The theca may be ornamented with spines, teeth and/or wings. Plate formula: 4′, (2a–3a), 7″; 5‴, 2⁗.

Important references: Lefèvre 1932, Kiselev and Cvetkov 1935, Schiller 1937, Huber-Pestalozzi 1950, Bourrelly 1968a, 1970, Starmach 1970, Matvienko et Litvinenko 1977.

Key to the subgenera

1a. Apical pore present . . . . **2. Poroperidinium** (p. 171)
1b. Apical pore not present . . . . **1. Cleistoperidinium** (p. 165)

1. Subgenus **Cleistoperidinium** Lefèvre 1932
Epithecal plate formula 4′, 3a, 7″
section **willei** (p. 165); plate pattern symmetric, plates 1′ and 3′ do not touch (fig. 4)
section **striolatum** (p. 165); plate pattern symmetric, plates 1′ touches plates 2′, 3′, 4′ (fig. 5)
section **cinctum** (p. 166); plate pattern asymmetric along dorsiventral axis (fig. 6)
Epithecal plate formula 4′, 2a, 7″
section **palatinum** (p. 170); plate pattern asymmetric along dorsiventral axis. (fig. 7)

Section **willei** (fig. 4)

1. Peridinium willei Huitfeld-Kaas 1900 (figs. 31a, 32, 50, 170)
P. alatum Garbini 1902, *P. guestrowiense* Lindemann 1918, *P. kincaidi* Wailes 1933, *P. volzii* Lemmermann 1905, *P. volzii* var. *cinctiforme* Lefèvre 1932, *P. tabulatum* Playfair 1912, *P. willei* var. *botanicum* Playfair 1919, *P. volzii* var. *botanicum* (Playfair) Lindemann 1931, *P. tabulatum* var. *westii* Playfair 1912, *P. tabulatum* var. *hieroglyphicum* Playfair 1912, *P. australe* Playfair 1919, *P. hieroglyphicum* Playfair 1919, *P. volzii* var. *meandricum* Lauterborn 1910, *P. volzii* fo. *compressum* and *P. volzii* fo. *sinuatum* (Lindemann) Lefèvre 1932, *P. striolatum* Wailes 1928, *P. vancouverense* Wailes 1931, *P. volzii* fo. *vancouverense* (Wailes) Lefèvre 1932, *P. volzii* var. *maximum* Bernard 1936, *P. volzii* var. *australe* G. S. West 1916; non: *Glenodinium tabulatum* Ehrenberg 1938 partim, *Peridinium tabulatum* (Ehrenberg) Claparède and Lachmann 1858/61 partim, *P. striolatum* Playfair 1919

Cells are spherical, often wider than longer and dorsiventrally flattened. The hypotheca is smaller than the epitheca. The cingulum spirals to the left and is flanged, the sulcus extends from the epitheca to the hypotheca and may reach the antapex. Apical and antapical plates are often flanged with wings up to 2 μm in height. Plates are distinctly reticulated. One or two teeth-like structures may be present on the sulcal margins. Chloroplasts are numerous and parietal. Reproduction is by a division of the motile vegetative stage or by fusion of isogametes ultimately resulting in a hypnozygote.
Dimensions: 38–78 μm × 36–80 μm.
Occurrence: very common world wide in pools, lakes and peat bogs.

Section **striolatum** (fig. 5)

2. Peridinium striolatum Playfair 1919 (fig. 171)
non. *P. striolatum* Wailes 1928

Cells elongated, egg-shaped and dorsiventrally flattened. The cingulum spirals to the left. The sulcus extends into half of the epitheca; it enlarges on the hypotheca where it reaches the antapex. The plates are concave or flat with rows of fine points. Plate 1′ is relatively small, and hexangular or oval.
Dimensions: 41–57 μm × 34–53 μm.
Occurrence: Australia.

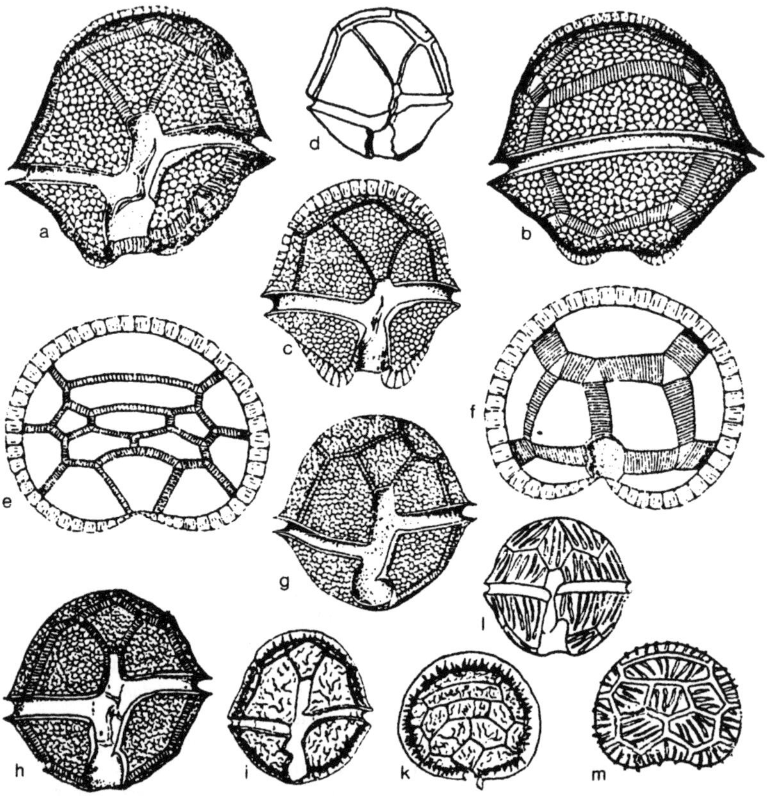

Fig. 170. *Peridinium willei* Huitfeld-Kaas (*a–g, l, m* after Lefèvre, *h* after Huber-Pestalozzi, i, k after Lindemann).

Section **cinctum** (fig. 6)

1a. Cells are strongly flattened dorsiventrally, plate 2a is squared . . . . **5. P. raciborskii**

1b. Cells are somewhat flattened dorsiventrally, plate 2a is pentangular **2**

2a. The epitheca is larger than the hypotheca; the cingulum extends further into the epivalve. . . . . **3. P. cinctum**

2b. The epi- and hypothecae are approximately equal in size; the cingulum barely extends into the epivalve . . . . **4. P. gatunense**

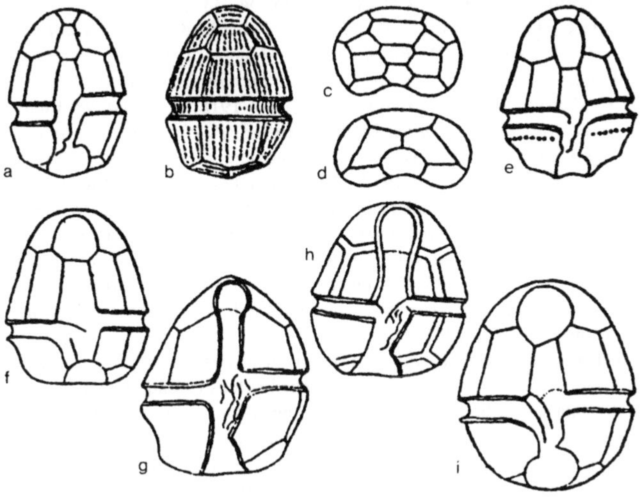

Fig. 171. *Peridinium striolatum* Playfair (after Playfair).

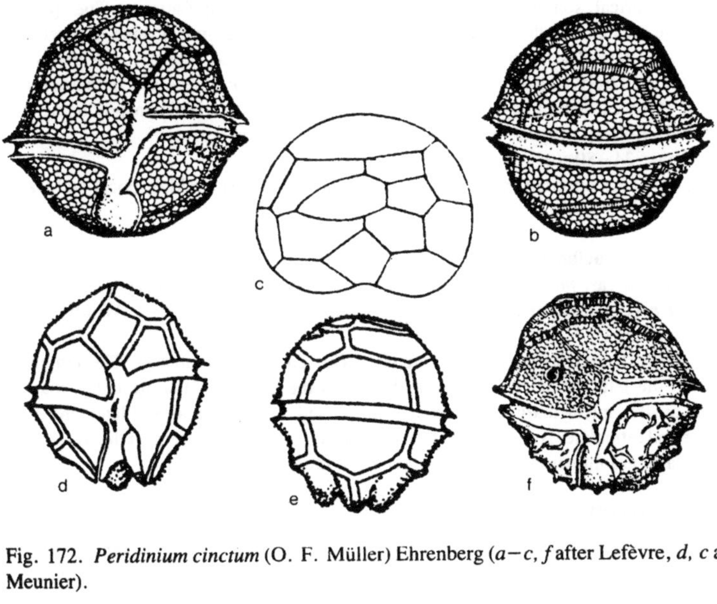

Fig. 172. *Peridinium cinctum* (O. F. Müller) Ehrenberg (*a–c, f* after Lefèvre, *d, c* after Meunier).

3. **Peridinium cinctum** Ehrenberg 1838 (figs. 31b, 33, 34, 38e, 172)

Vorticella cincta O. F. Müller 1786, *P. tabulatum* Penard 1891, *P. cinctum* var. *lemmermannii* G. S. West 1909, *P. tuberosum* Meunier 1919, *P. westii* var. *aureolatum* Lemmermann 1908, *P. cinctum* var. *laesum* Lindemann 1918, *P. germanicum* Lindemann 1918, *P. eximium* Lindemann 1918, *P. rhenanum* Lindemann 1919, *P. cinctum* var. *regulatum* Lindemann1918, *P. cinctum* fo. *regulatum* (Lindemann) Lefèvre 1932, *P. cinctum* var. *irregulatum* Lindemann 1918, *P. cinctum* var. *angulatum* Lindemann 1918, *P. cinctum* f. *angulatum* (Lindemann) Lefèvre 1932, *P. cinctum* f. *ovoplanum* Lindemann 1918, *P. westii* Lemmermann in W. et G. S. West 1906, *P. tabulatum* var. *meandrica* Lauterborn 1910, *P. meandricum* Brehm 1907, *P. cinctum* f. *meandricum* Lefèvre 1932, *P. cinctum* f. *westii* (Lemmermann) Lefèvre 1932, *P. cinctum* var. *tuberosum* (Meunier) Lindemann 1928, *P. cinctum* var. *carinatum* Steinecke and Lindemann 1923, *P. cinctum* f. *tuberosum* (Meunier) Lefèvre 1932.

Cells are spherical or elongated and oval, and dorsiventrally flattened. The cingulum spirals to the left and is sometimes flanged. The flanged sulcus extends into 1/3 of the epitheca and to the antapex in the hypotheca. The epitheca with its irregularly arranged plates is larger than the hypotheca. The right antapical plate (2'''') is larger than the left one. Plates are thick and usually reticulated, rarely are they ornamented with irregularly arranged ribs (rugose). Chloroplasts are parietal and numerous.

Dimensions: 40–78 μm × 40–73 μm.

Occurrence: wide spread in pools and lakes.

4. **Peridinium gatunense** Nygaard 1925 (fig. 173)

P. cinctum var. *gibbosum* Lefèvre 1927, *P. zonatum* Playfair 1919

Cells are spherical, somewhat wider than long and dorsiventrally flattened. The epi- and hypothecae are equal in the size and bluntly conical. The cingulum is wide, flanged and spirals to the left. The sulcus widens slightly and reaches the antapex. The plates are reticulated or rarely rugose and irregularly arranged. Chloroplasts are numerous and parietal. Bundles of trichocysts are present rarely. Vegetative division producing two naked cells follows ecdysis. Sexual reproduction is by isogamy.

Dimensions: 34–80 μm × 29–75 μm.

Occurrence: cosmopolitan, rare, scarcely in Europe (Czechoslovakia), pools and lakes; USA.

5. **Peridinium raciborskii** Wołoszyńska 1912

P. cinctum var. *palustre* Lindemann 1919, *P. chalubinskii* Wołoszyńska 1919, *P. palustre* (Lindemann) Lefèvre 1932, *P. palustre* var. *raciborskii* (Wołoszyńska) Lefèvre 1932

Cells spherical, and dorsiventrally flattened. The 2a plate is squared. The cingulum spirals to the left; the sulcus is wide and reaches the antapex. Hyaline wings may be present on the sulcal margin and on the apical and antapical plates. Plates are often concave and reticulate. Numerous teeth-like projections may be present on the hypotheca. Chloroplasts are numerous, and often radially arranged.

Key to the varieties

1a. Cells are larger than 80 μm × 70 μm . . . . 5.1. var. *raciborskii*
1b. Cells are smaller than 80 μm × 70 μm . . . . 5.2. var. *palustre*

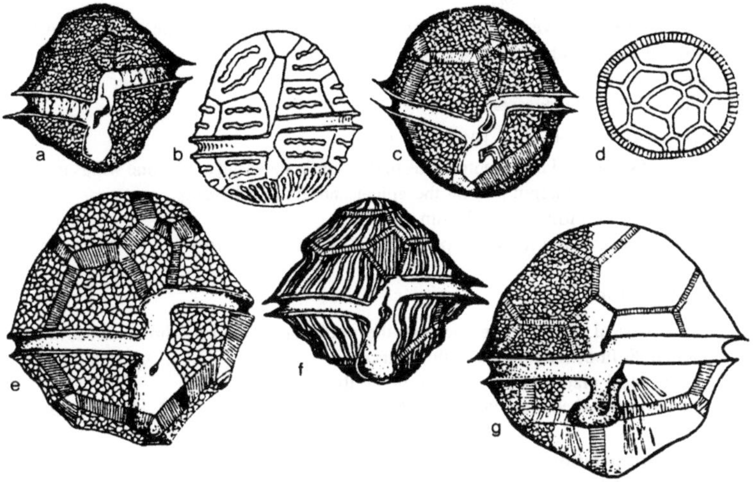

Fig. 173. *Peridinium gatunense* Nygaard (*a, c, e, f* after Lefèvre, *b* after Playfair, *d* after Nygaard, *g* after Popovský).

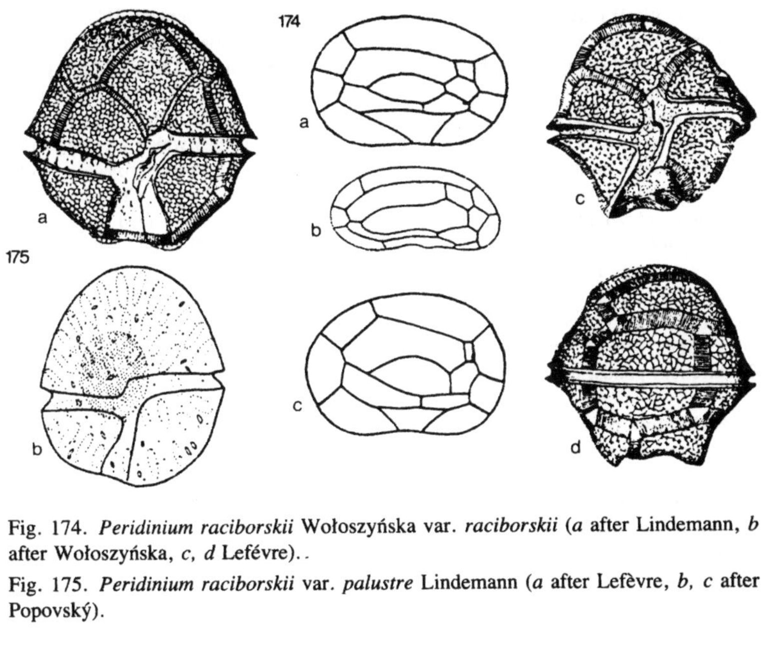

Fig. 174. *Peridinium raciborskii* Wołoszyńska var. *raciborskii* (*a* after Lindemann, *b* after Wołoszyńska, *c, d* Lefévre)..

Fig. 175. *Peridinium raciborskii* var. *palustre* Lindemann (*a* after Lefèvre, *b, c* after Popovský).

5.1. var. *raciborskii* (fig. 174)
Cells are flattened dorsiventrally. Numerous teeth-like projektions are present, most of which are on the hypotheca.
Dimensions: 80–100 μm × 70–90 μm.
Occurrence: Java, Madagascar.

5.2. var. *palustre* Lindemann 1928 (fig. 175)
Cells are not as flattened dorsiventrally as in the nominate variety. Hyaline wings may be present on the sulcal margin and on the apical and antapical plates.
Dimensions: 45–80 μm × 55–70 μm.
Occurrence: sporadically in pools, ponds and small lakes; Europe.

Section **palatinum** (fig. 7).

1a. Asymmetrical plate arrangement in the epitheca plates are almost concave and pointed. Sutures are concave without striation . . . . **6. P. palatinum**
1b. Symmetrical plate arrangement in the epitheca, plates are almost convex. Sutures with striae . . . . **7. P. pseudolaeve**

6. Peridinium palatinum Lauterborn 1896 (fig. 31 c, 176)
P. marssonii Lemmermann 1900, *P. laeve* Huitfeld-Kaas 1900, *P. laeve* subsp. *marssonii* Lindemann 1919, *P. anglicum* G. S. West 1909, *P. laeve* subsp. *marssonii* var. *contactum* Lindemann 1919, *P. laeve* var. *minor* Denis 1925, *P. palatinum* var. *minor* (Denis) Lefèvre 1932, *P. palatinum* fo. *cristatum* Lindemann 1919, *P. apiculatum* var. *cristatum* (Lindemann) Lindemann 1928, *P. palatinum* f. *anglicum* (G. S. West) Lefèvre 1932, *P. apiculatum* (G. S. West) Lindemann 1928.
Cells are ovoid, and severely flattened dorsiventrally. A "pseudoapex" (lacking a pore) is visible. The epitheca is bell shaped and larger than the hypotheca. The cingulum spirals slightly to the left. The sulcus widens on the hypotheca and does not reach the antapex. The first apical plate is relatively small and does not reach the apex. Plates on the epitheca are almost concave and ornamented with points or small spines which are irregularly arranged. Plates on the hypotheca have clusters of spines or teeth-like projections. The sutures are not striated and are concave. Suture margins may be flanged. Chloroplasts are numerous and radially arranged. A stigma may be present. Smooth-walled resting cysts (hypnozygotes) are formed within the theca. Reproduction by division producing two naked zoospores follows ecdysis. The daughter cells are motionless until they have formed a new theca. Plates of this new theca are deeply concave with sutures raised as prominent sharp flanges.
Dimensions: 30–55 μm × 25–48 μm.
Occurrence: ponds, pools, mainly in cool waters; cosmopolitan.

7. Peridinium pseudolaeve Lefèvre 1925 (fig. 177)
P. laeve Lindemann 1919; non: *P. laeve* Huitfeld-Kaas 1900
Cells are ellipsoidal and slightly flattened dorsiventrally. Both epi- and hypothecae are equal in size. The cingulum spirals to the left, the sulcus widens on the hypothea and does not reach the antapex. Plate arrangement is symmetric. Plates are convex with fine reticulations.
Dimensions: 35–48 μm × 32–35 μm.
Occurrence: pools, ponds, lakes, gravel pits; Germany, France.

176

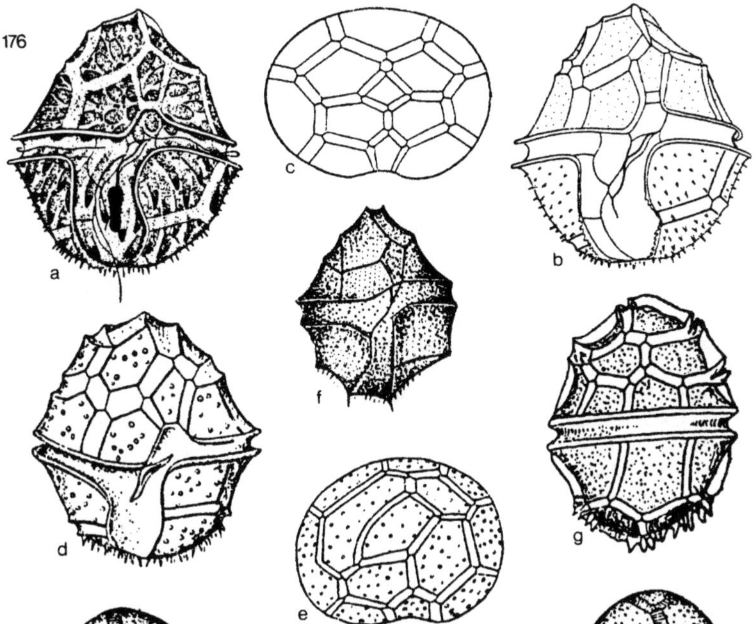

177

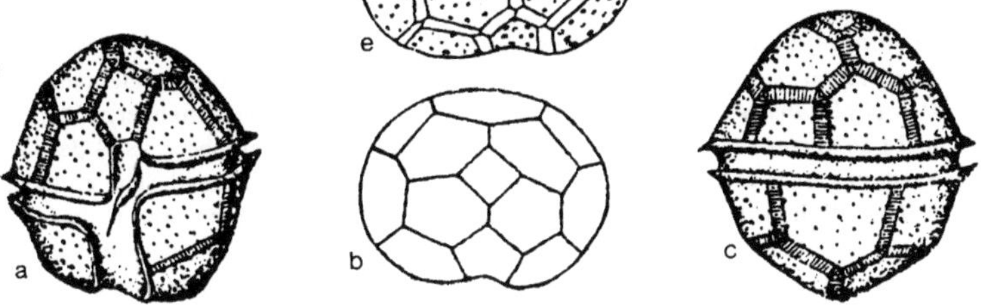

Fig. 176. *Peridinium palatinum* Lauterborn (*a–c* after Thompson, *e* after Lefèvre, *f* after Popovský, *g* after Lindemann).

Fig. 177. *Peridinium pseudolaeve* Lefèvre (*a–c* after Lefèvre).

2. Subgenus **Poroperidinium** Lefèvre 1932
Plate formula 4′, 3 a, 7″
section **bipes** (p. 172); plate pattern symmetric, plate 1 a does not touch plate 4″ (fig. 8)
section **gutwinskii** (p. 175); plate pattern may be asymmetric, plate 1 a touches plate 4″ (fig. 9)
section **lomnickii** (p. 175); plate pattern asymmetric, plate 1 a is pentagonal (fig. 11)
section **allorgei** (p. 179); plate pattern asymmetric, plate 1 a is squared (fig. 12)
Plate formula 4′, 2 a, 7″
section **umbonatum** (p. 180); plate pattern symmetric (fig. 10)

Section **bipes** (fig. 8)

1a. Chloroplasts are present, cell shape spherical ellipsoidal or pentangular **2**
1b. Chloroplasts are absent, cell shape is rhomboidal . . . . **8. P. achromaticum**
2a. Cell shape is spherical or ovoid **3**
2b. Cell shape is pentangular with two small horn-like processes on the hypotheca . . . . **10. P. limbatum**
3a. Cells are spherical to ellipsoidal, two well-developed triangular wings are present on the hypotheca . . . . **9. P. bipes**
3b. Cells are spherical, no triangular wings present on the hypotheca . . . . **11. P. playfairii**

8. Peridinium achromaticum Levander 1902 (fig. 178)
Cells are rhomboidal, and slightly flattened dorsiventrally in ventral view. The epitheca is conical, the hypotheca conical and excavated at the antapex. The cingulum encircles the cell; the sulcus is broad and reaches the antapex. Two teeth-like projections are present on the left side of hypotheca. The right side of the sulcus is flanged. Plates bear fine reticulations. Chloroplasts are absent. Hypnospores are spherical with a thick cell wall. Nutrition saprotrophic.
Dimensions: 28–48 μm × 24–40 μm.
Occurrence: pools, ponds with plants, also saline and marine species; Europe and Asia.

9. Peridinium bipes Stein 1883 (figs. 30, 179)
P. bipes var. *exisum* Lemmermann 1900, *P. bipes var. ocultatum* Lindemann 1918, *P. bipes* f. *tabulatum* (Ehrenberg) Lefèvre 1932, *P. tabulatum* (Ehrenberg) Claparède and Lachmann 1858/61 partim, *Glenodinium tabulatum* Ehrenberg 1838 partim.
Cells are egg or pear shaped to spherical and dorsiventrally flattened. The epitheca is conical or bell shaped and is larger than the hypotheca. The cingulum spirals to the left and is flanged, the sulcus extends from 1/3 of epitheca to the antapex where it widens. A hyaline wing is present on the apical plates and on the sulcal sutures. Both antapical plates have two, triangular flanges (wings). Plates are reticulated and almost concave. Chloroplasts are numerous and parietal. Protoplast may become spherical, surrounded by a thick wall and released from the theca. This then acts as resting stages.
Dimensions: 40–95 μm × 38–90 μm.
Occurrence: very common in ponds and lakes; Europe.

10. Peridinium limbatum (Stokes) Lemmermann 1900 (fig. 180)
Protoperidinium limbatum Stokes 1888
Cells are pentangular in ventral view and deeply excavated on the antapex which is bifurcate. Cells are severely flattened dorsiventrally and deltoid shaped in side view. The epitheca is conical and larger than the hypotheca. The relatively narrow cingulum spirals to the left; the sulcus widens on the hypotheca where it reaches the antapex. Plates are concave and reticulated. The wings extend from the apical pore to both horns in adult specimens. Chloroplasts are numerous and parietal.
Dimensions: 80–93 μm × 60–82 μm.
Occurrence: pools and ponds with *Sphagnum* and mosses; USA, Ireland.

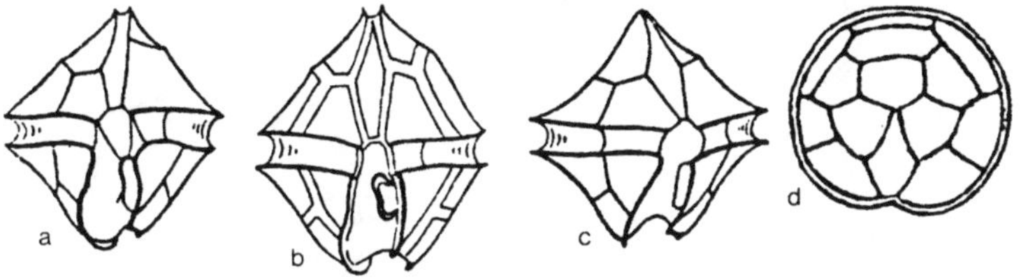

Fig. 178. *Peridinium achromaticum* Levander (*a–c* after Wołoszyńska, *d* after Lebour).

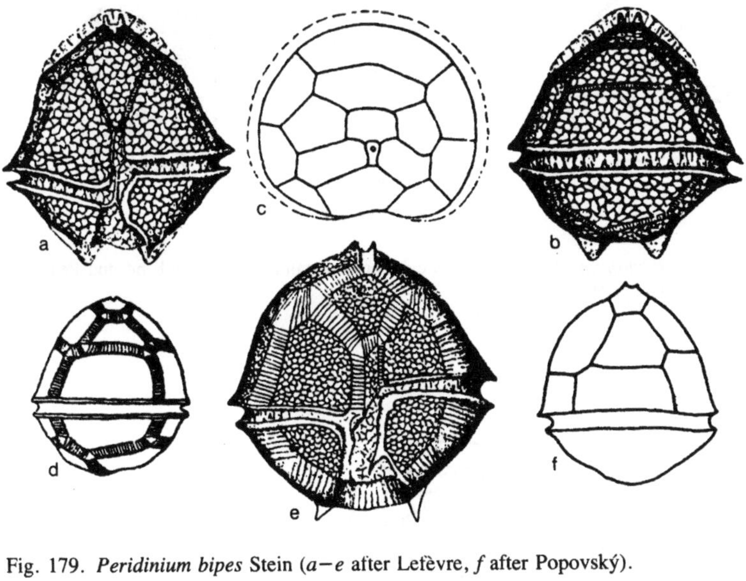

Fig. 179. *Peridinium bipes* Stein (*a–e* after Lefèvre, *f* after Popovský).

11. Peridinium playfairii Lindemann 1931 (fig. 181)
Cells are spherical or egg-shaped, and slightly flattened dorsiventrally. Both the epi- and hypothecae are equal in size. The cingulum spirals to the left; the sulcus reaches the antapex. The plates are finely reticulated.
Dimensions: 40–50 µm × 40–50 µm.
Occurrence: Malaysia.
Note: Playfair (1919) poorly and vaguely described *Peridinium australe*. Several specimens are depicted on his figures. One belongs to the section willei (subgenus

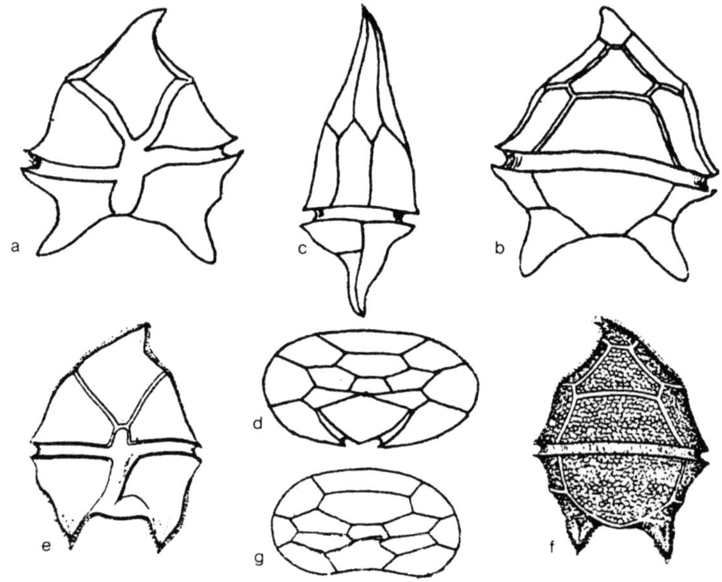

Fig. 180. *Peridinium limbatum* (Stokes) Lemmermann (*a–d* after Lind and Pearsall, *e–g* after Huber-Pestalozzi).

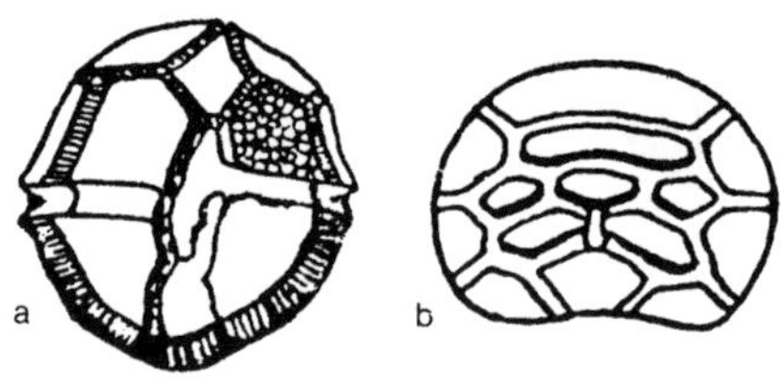

Fig. 181. *Peridinium playfairii* Lindemann (after Lindemann).

Cleistoperidinium) and the second to the section bipes (subgenus Poroperidinium). Consequently Thomasson (1974) proposed a new name, *Peridinium sydneyense* for specimens which possessed an apical pore, – section bipes. *Peridinium sydneyense* is very similar to *Peridinium playfairii*.

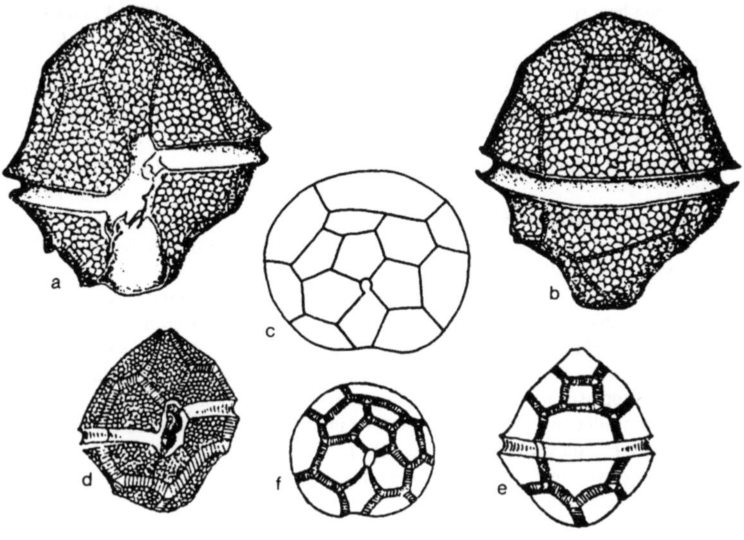

Fig. 182. *Peridinium gutwinskii* Wołoszyńska (*a–c* after Lefèvre, *d–e* after Wołoszyńska).

Section **gutwinskii** (fig. 9)

12. Peridinium gutwinskii Wołoszyńska 1912 (fig. 182)

Cells are irregularly spherical, ovoid or egg-shaped, and not dorsiventrally flattened. The epi- and hypothecae are equal in size. The deep cingulum spirals to the left. The flanged sulcus widens along the hypotheca and does not reach the antapex. The antapical plates are strongly concave. Plates bear distinct reticulations. Chloroplasts are numerous.
Dimensions: 40–60 μm × 35–60 μm.
Occurrence: in plankton of ponds and pools; Java, Sumatra, Madagascar.

Section **lomnickii** (fig. 11)

1a. Cells with spines on the hypotheca . . . . **13. P. aciculiferum**
1b. Cells without spines on hypotheca **2**
2a. Epitheca is slightly spherical. **3**
2b. Epitheca is pentangular or conical **4**
3a. Plate pattern is approximately symmetrical . . . . **15. P. lomnickii**
3b. Plate pattern is asymmetrical **6**
4a. Hypothecal plates bear fine reticulations . . . . **13.1. P. aciculiferum** f. *inerme*
4b. Hypothecal plates bear small projections, points or spicules **5**

5a. All plates bear spicules . . . . **15.2. P. lomnickii** var. *splendida*
5b. Epithecal plates bear fine reticulations, hypothecal plates bear spicules . . . . **14. P. godlewskyi**
6a. Cells spherical, plates with fine nipple-like projections . . . . **15.3. P. lomnickii** var. *wierzejskii*
6b. Cells deltoid, plates with spicules . . . . **15.2. P. lomnickii** var. *splendida*

13. Peridinium aciculiferum Lemmermann 1900 (fig. 183)

P. umbonatum var. *aciculiferum* Lemmermann 1908, *P. stagnale* Meunier 1919, *Glenodinium aciculiferum* (Lemmermann) Lindemann 1982.

Cells are egg-shaped and slightly flattened dorsiventrally. The epitheca is somewhat spherical with a prominent apex; the hypotheca is also somewhat spherical. The cingulum spirals to the left. The sulcus widens along the hypotheca and reaches the antapex. Three or four spine-like projections are present on the hypotheca. Plates bear reticulations. The numerous chloroplasts are disc-shaped. Vegetative reproduction takes place in a non-motile stage embeded by layered mucilage. The thick-walled hypnospores are spherical.

13.1. fo. *inerme* Wołoszyńska 1928 (fig. 184)

Cells without or with one spine on the hypotheca.

Dimensions: 35–51 μm × 29–42 μm.

Occurrence: lakes and ponds, also marine; Europe.

14. Peridinium godlewskii Wołoszyńska 1916 (fig.185)

Cells are egg-shaped and slightly flattened dorsiventrally. The epitheca is pentangular, the hypotheca broadly rounded. Both, epi- and hypotheca are equal in size. The cingulum spirals to the left; the sulcus extends slightly into the epitheca, it widens along the hypotheca and does not reach the antapex. Plates are concave. Epithecal plates bear delicate reticulations while those on the hypotheca are ornamented with numerous spicules. Chloroplasts are small and numerous.

Dimensions: 25–30 μm × 30 μm.

Occurrence: ponds; Poland.

15. Peridinium lomnickii Wołoszyńska 1916

P. lomnickii var. *punctulatum* Lindemann 1924, *Glenodinium lomnickii* (Wołoszyńska) Lindemann 1928, *Chalubinskia tatrica* Wołoszyńska 1916, *Peridinium tatrae* Schiller 1937.

Cells are egg-shaped, and slightly flattened dorsiventrally. The epitheca is bell-shaped and larger than the hypotheca. The hypotheca is rounded and somewhat excavated on the antapex. The cingulum is wide and spirals to the left. The sulcus does not reach the antapex. Epithecal plates are ornamented with points or aerolae and on the hypotheca with spicules. Chloroplasts, if are present are small and numerous. The thick-walled hypnospores are egg-shaped.

Dimensions: 20–50 μm × 22–50 μm.

Occurrence: plankton of pools and lakes, widespread, winterform.

15.1. var. *lomnickii* (fig. 186)

15.2. var. *splendida* Wołoszyńska 1916 (fig. 187)

Glenodinium lomnickii var. *splendida* (Wołoszyńska) Lindemann 1928

The plane of symmetry of epithecal plates is shifted to the right. Cells are pentangular and strongly flattened dorsiventrally.

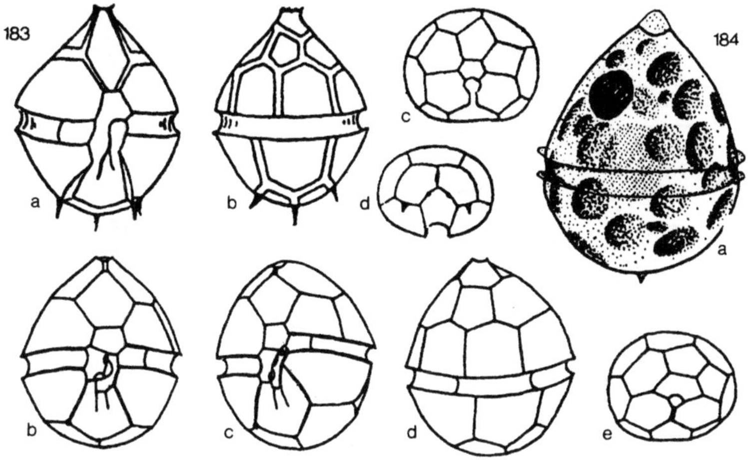

Fig. 183. *Peridinium aciculiferum* Lemmermann (*a–d* after Wołoszyńska).

Fig. 184. *Peridinium aciculiferum* fo. inerme Wołoszyńska (*a* after Popovský, *b–e* after Wołoszyńska).

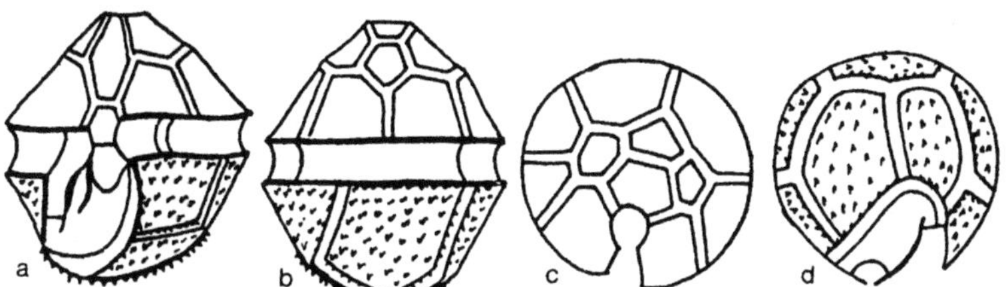

Fig. 185. *Peridinium godlewskii* Wołoszyńska (*a, b, d* after Wołoszyńska, *c* after Huber-Pestalozzi).

15.3. var. *wierzejskii* (Wołoszyńska) Lindemann 1928 (fig. 188)

P. wierzejskii Wołoszyńska 1916, *P. wierzejskii* var. *minor* Wołoszyńska 1916.

The plane of symmetry of epithecal plates is shifted to the right. The spicules on hypotheca are absent. Often without chloroplasts.

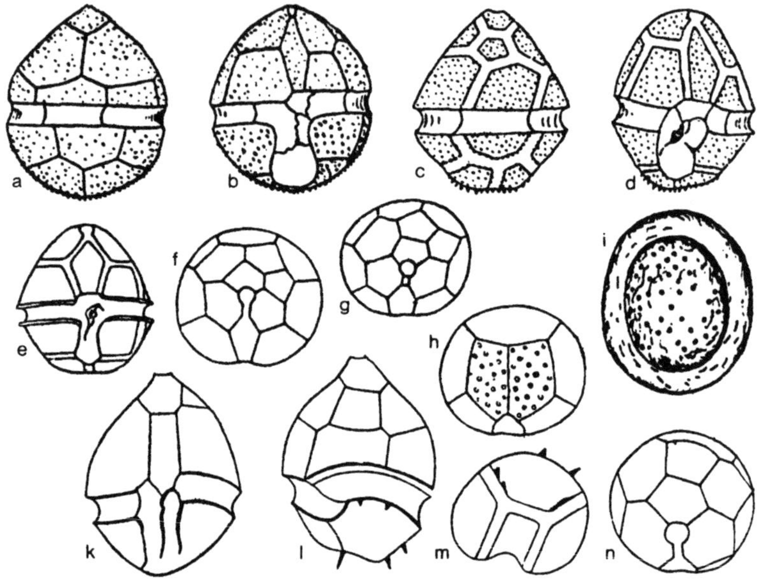

Fig. 186. *Peridinium lomnickii* Wołoszyńska var. *lomnickii* (*a, b, i* after Lindemann, *c, d, g, k–n* after Wołoszyńska, *e, f, h* after Lefèvre).

Appendix

1. *Peridinium baliense* Lindemann 1931 (fig. 189)

Cells are spherical, rarely egg-shaped, and with a large conical epitheca. The cingulum encircles the cell, the sulcus extends slightly into the epitheca and widens along the hypotheca reaching the antapex. Plates are almost concave and finely reticulated. Wing-like structures, located between three spines are present on the hypotheca.
Dimensions: 45–55 μm × 44–55 μm.
Occurrence: Indo-malaysia, Bali.

2. *Peridinium keyense* Nygaard 1926 (fig. 190)

Cells are spherical and slightly flattened dorsiventrally. The epi- and hypothecae are equal in size. The cingulum spirals to the left. The sulcus barely extends into the epitheca; it widens greatly along the hypotheca and reaches the antapex. Plates are concave and reticulate.
Dimensions: 45–57 μm × 47–54 μm.
Occurrence: lake; Malaysia.

187

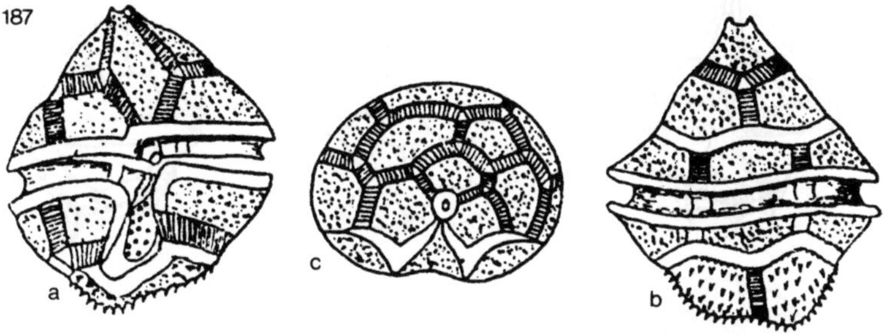

188

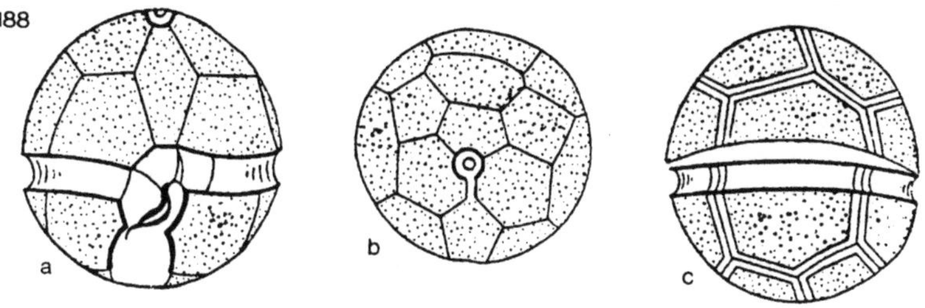

Fig. 187. *Peridinium lomnickii* var. *splendida* Wołoszyńska (after Wołoszyńska).

Fig. 188. *Peridinium lomnickii var. wierzejskii* (Wołoszyńska) Lindemann (after Wołoszyńska).

3. *Peridinium subsalsum* Ostenfeld 1908 (fig. 191)
P. tabulatum fo. *brasiliana* Möbius 1892.

Cells are broadly egg-shaped; the epitheca is conical, the hypotheca rounded. The cingulum spirals to the left. The sulcus is narrow and does not reach the antapex. Three spine-like projections are present on the hypotheca. Plates are finely reticulated or have fine dots in rows. Some plates are often flanged. Chloroplasts are radially arranged. Plate pattern: (3–4)′, 4a, 7″; 4‴, (1–2)⁗.
Dimensions: 22–60 µm × 20–56 µm.
Occurrence: pool in Cuba; saline waters – Aral Lake, USSR; Brasil.

Section **allorgei** (fig. 12)

1a. Cells are rounded . . . . **16. P. allorgei**
1b. Cells are spindle-shaped . . . . **17. P. wisconsinense**

16. Peridinium allorgei Lefèvre 1927, 1932 (fig. 192)

Cells vary from spherical to lens-shaped and are slightly flattened dorsiventrally. The narrow cingulum spirals to the left. The sulcus barely extends into the epitheca; it widens greatly along the hypotheca and reaches the antapex. Plates are convex and ornamented with rows of small papillae.
Dimensions: 25–43 µm × 28–43 µm.
Occurrence: ponds; West Europe.

189

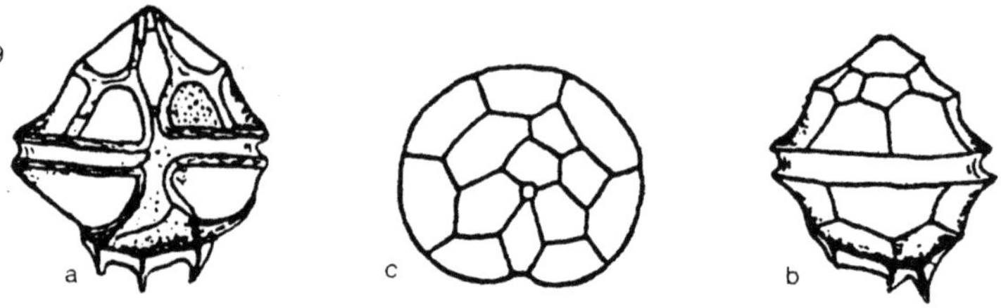

190

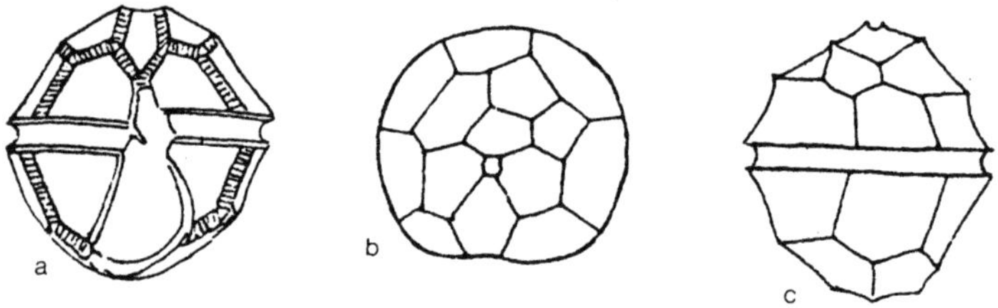

Fig. 189. *Peridinium baliense* Lindemann (after Lindemann).

Fig. 190. *Peridinium keyense* Nygaard (after Nygaard).

17. Peridinium wisconsinense Eddy 1930 (fig. 193)

Cell are spindle-shaped and slightly flattened dorsiventrally. The epi- and hypothecae form a prominent cone which extends into a single, stout sharply-pointed horn. The broad cingulum spirals to the left. The sulcus extends only half-way down the hypotheca. Plates are reticulate and mostly concave. Plate 1'''' forms the posterior horn. Resting stages are broadly ovoid with one pole sharply pointed; the other bluntly so; the wall is thick and lamellate at the poles.

Dimensions: 55–64 μm × 48–56 μm.

Occurrence: lakes; North America.

Section **umbonatum** (fig. 10)

1a. Rectilinear sutures **2**
1b. Curved sutures . . . . **19. P. morzinense**
2a. Cingulum spirals to the left . . . . **20. P. umbonatum**
2b. Cingulum is equatorial . . . . **18. P. africanum**

18. Peridinium africanum Lemmermann 1907 partim (fig. 194)

P. marchicum var. *javanicum* Wołoszyńska 1912, *P. ornamentosum* Lindemann 1931, *P. tabulatum* var. *intermedium* Playfair 1912, *P. africanum* var. *intermedium* (Playfair) Lefèvre 1932, *P. africanum* var *javanicum* (Wołoszyńska) Lefèvre 1932, *P. intermedium* Playfair 1919; non *P. intermedium* Thompson 1950

Cells are pentangular, and not dorsiventrally flattened in ventral view. The epicone is broadly conical and slightly larger than the hypocone. The cingulum encircles the cell. The sulcus extends slightly into the epicone, it widens along the hypocone and reaches the

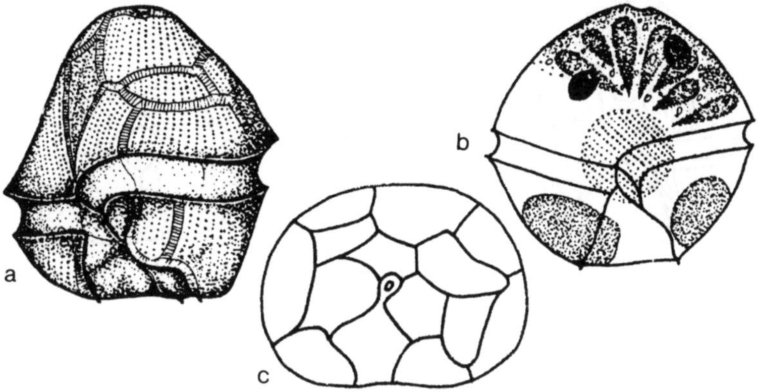

Fig. 191. *Peridinium subsalsum* Ostenfeld (after Popovský).

192

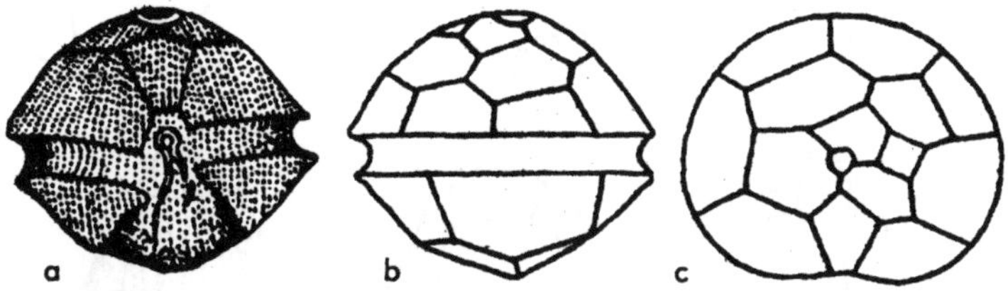

193

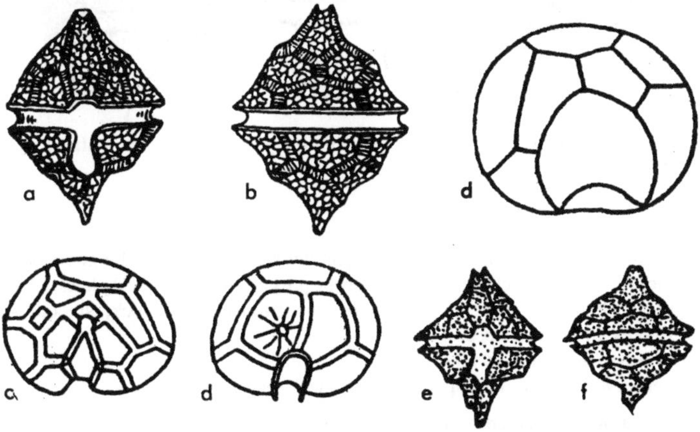

Fig. 192. *Peridinium allorgei* Lefèvre (after Lefèvre).

Fig. 193. *Peridinium wisconsinense* Eddy (after Eddy).

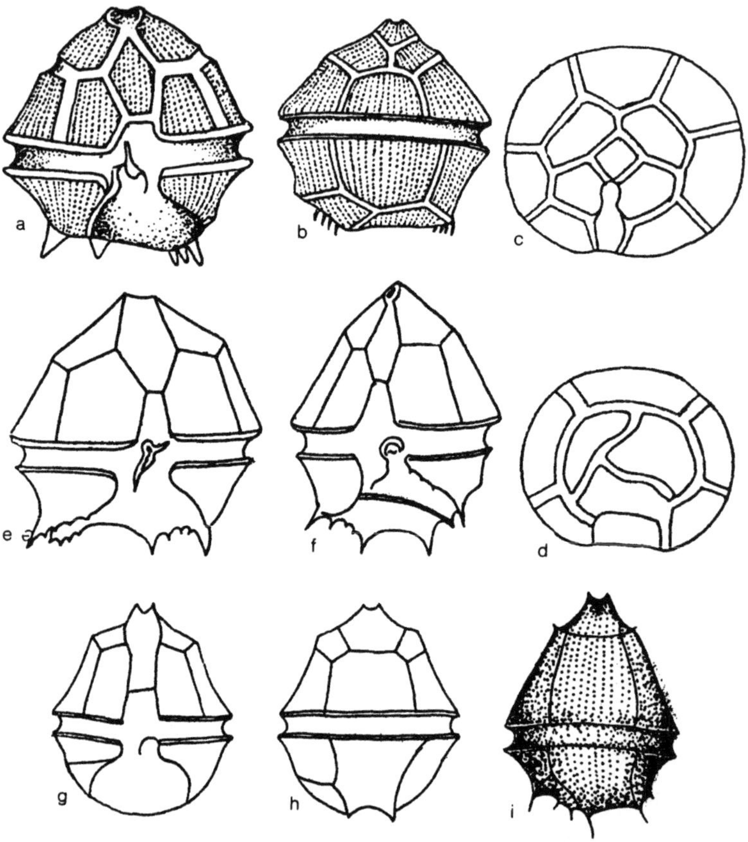

Fig. 194. *Peridinium africanum* Lemmermann (*a–f* after Lefèvre, *g, h* after Playfair, *i* after Popovský).

antapex. It bears several small teeth-like projections along its margins. Large single spine-like processes or smaller clusters there of may be present on the antapical plates. Plates are straight or slightly concave with rows of fine papillae. The thick, smooth-walled hypnospores are heart shaped.
Dimensions: 20–40 µm × 20–35 µm.
Occurrence:widespread in temperate and tropical zones; lakes, ponds.

19. Peridinium morzinense Lefèvre 1928, 1932 (fig. 195)

P. elegans Lefèvre 1925, *P. lubieniense* fo. *elegans* (Lefèvre) Lindemann 1928
Cells are oval or egg-shaped, not flattened dorsiventrally. The epitheca is bell-shaped and larger than the hypotheca which is broadly rounded. The cingulum spirals slightly

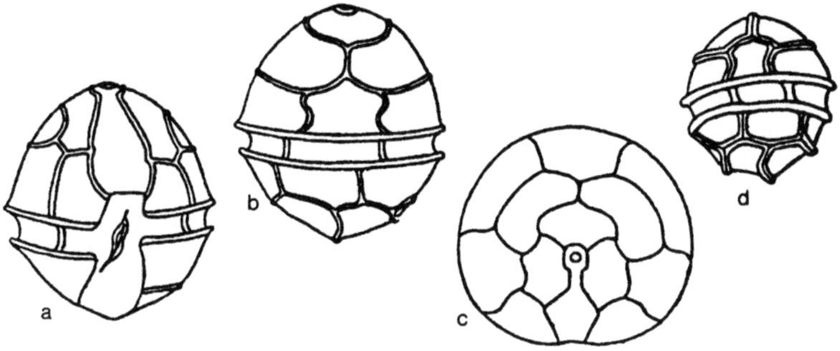

Fig. 195. *Peridinium morzinense* Lefèvre (after Lefèvre).

to the left. The sulcus extends slightly into the epitheca and widens in the hypotheca where it extends to the antapex. The plates lack ornamentation and are connected by sinuous sutures. The sutures are not striated in the planozygotes but rather form a triangular ridge between the plates. Chloroplasts are sometimes lacking.

Dimensions: 30–41 μm × 20–35 μm.

Occurrence: peat bogs; Poland, France.

Note: *P. morzinense* exhibits morphological thecal features, similar to *Peridinium avellana* (Meunier) Lebour 1925 found in brackish waters.

20. Peridinium umbonatum Stein 1883

P. umbonatum var. *inaequale* Lemmermann 1910, *P. umbonatum* var. *papilliferum* Lemmermann 1908, *P. linzium* Lindemann 1918 partim, *P. umbonatum* fo. *spiniferum* (Lindemann) Levèvre 1932, *P. ambiguum* Lindemann 1931, *P. inconspicuum* Lemmermann 1899, *P. minimum* Wołoszyńska 1916, *P. munusculum* Lindemann 1918, *P. tabulatum* var. *caudatum* Playfair 1912, *P. caudatum* Playfair 1919, *P. geminum* var. *excavatum* Playfair 1919, *P. geminum* var. *elegans* Playfair 1919, *Glenodinium lefevrei* Lindemann 1928, *Peridinium orrei* Huitfeld-Kaas 1906, *Glenodinium guildfordense* (Playfair) Lindemann 1931, *Peridinium steinmanni* Wołoszyńska 1930, *P. parvulum* Wołoszyńska 1930, *P. javanicum* Bernard 1908, *P. inconspicuum* var. *excavatum* (Playfair) Lefèvre 1932, *P. geminum* var. *angulosum* Playfair 1919, *P. inconspicuum* var. *armatum* Lemmermann 1910, *P. inconspicuum* fo. *armatum* (Lemmermann) Lefèvre 1932, *P. marchicum* Lemmermann 1910, *P. munusculum* fo. *spiniferum* Lindemann 1918, *P. inconspicuum* fo. *spiniferum* (Lindemann) Lefèvre 1932, *P. marchicum* var. *keyense* Nygaard 1926, *P. inconspicuum* var. *balatonicum* Entz 1937, *P. tatricum* Wołoszyńska 1916, *P. tatricum var. spinulosa* Wołoszyńska 1916, *P. africanum* fo. *tatricum* (Wołoszyńska) Lefèvre 1932, *P. africanum* var. *spinulosum* (Wołoszyńska) Lefèvre 1932, *Glenodinium geminum* (Playfair)

Lindemann 1931, *Glenodinium pusillum* Pénard 1891, *Peridinium pusillum* (Pénard) Lemmermann 1901, *Gymnodinium oligoplacatum* Skuja 1956.

Cells oval, egg-shaped to pentangular in ventral view, and dorsiventrally flattened. The epicone is bell-shaped or conical and larger than the hypocone which is rounded. The wide cingulum spirals to the left. The sulcus extends slightly into the epicone and widens along the hypocone where it reaches the antapex. The theca is relatively thin, the dorsal side may be ornamented with fine teeth-like or large spine-like processes. The plates are convex or concave with rows of papillae. Chloroplasts are mainly disc-shaped, parietal, rarely club-shaped and radially arranged. Chloroplasts may be lacking. Nutrition is autotrophic or saprotrophic. Vegetative reproduction is by a division of armored vegetative cells. Fusion of two thecate isogametes occurs with zygote formation completed outside the fusing thecae. Repeated ecdysis of the growing spherical zygote occurs several times such that it becomes peanut-shaped.
Dimensions: 15–45 µm × 12–32 µm.
Occurrence: small to big standing waters, from oligo to eutrophic; widespread.

Key to the varieties

1a. Cells are spherical . . . . 20.1. var. *centenniale*
1b. Cells are not spherical 2
2a. Plates with rows of dots 4
2b. Plates irregularly sculptured with dots 3
3a. Chloroplasts are lacking, one spine on hypotheca . . . . 20.3. var. *goslaviense*
3b. Chloroplasts are present, without spines on hypotheca . . . 20.4. var. *lubieniense*
4a. Epitheca conical two large spines present on hypotheca . . . 20.2. var. *deflandrei*
4b. Epitheca bell shaped, with or without numerous teeth-like processes or spiculae . . . . 20.5. var. *umbonatum*

20.1. var. *centenniale* Playfair 1919 (fig. 196)
Peridinium umbonatum var. *ovale* Playfair 1919, *P. umbonatum* var. *globosum* Lefèvre 1927, *P. centenniale* (Playfair) Lefèvre 1932

Cells spherical and slightly flattened dorsiventrally. The cingulum is wide, relatively deep and spirals to the left. The sulcus extends slightly into the epicone, it widens slightly along the hypocone where it reaches the antapex. The apical pore is shifted to the left; the plate pattern is asymmetric. Plates are convex with irregularly distributed papillae.
Dimensions: 28–43 µm × 25–40 µm.

20.2. var. *deflandrei* (Lefèvre) Popovský and Pfiester 1986 (fig. 197)
Peridinium deflandrei Lefèvre 1927

Cells doubly conical and slightly flattened dorsiventrally. The epicone is larger than the bluntly conical hypocone, which bears two large projections at the antapex one of which ends in a large spine. The cingulum spirals to the left; the sulcus extends slightly into the epicone and widens along the hypocone where it reaches the antapex. The convex plates are ornamented by rows of delicate points. Resting stages are egg-shaped.
Dimensions: 28–35 µm × 26–32 µm.

196

197

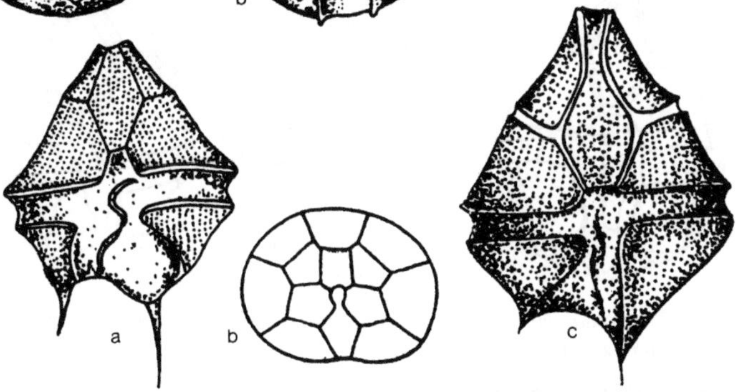

Fig. 196. *Peridinium umbonatum* var. *centenniale* Playfair (*a, b* after Lefèvre, *c* after Playfair).

Fig. 197. *Peridinium umbonatum* var. *deflandrei* (Lefèvre) Popovský and Pfiester (*a, b* after Lefèvre, *c* after Popovský).

20.3. var. *goslaviense* (Wołoszyńska) Popovský and Pfiester 1986 (fig. 198)
Peridinium goslaviense Wołoszyńska 1916

Cells are egg or pear-shaped, and slightly flattened dorsiventrally. The epitheca is bell-shaped with the hypotheca broadly rounded. The cingulum is wide and spirals to the left. The sulcus slightly extends into the epitheca. It widens along the hypotheca and nearly reaches the antapex. The plates are convex and ornamented with fine, irregularly placed papillae. One large spine is present on the antapex. Chloroplasts are lacking.
Dimensions: 20–40 µm × 16–35 µm.

20.4. var. *lubieniense* (Wołoszyńska) Popovský and Pfiester 1986 (fig. 199)
Peridinium lubieniense Wołoszyńska 1916, *P. dzieduszyckii* Wołoszyńska 1916, *P. lubieniense* fo. *dzieduszyckii* (Wołoszyńska) Lefèvre 1932, *P. linzium* Lindemann 1918 partim

Cells oval shaped and slightly flattened dorsiventrally. The epicone is bell-shaped and a little larger than the hypocone. The cingulum is wide and spirals to the left. The obviously small papillae are present on the concave antapical plates.
Dimensions: 35–45 µm × 30–32 µm.

198

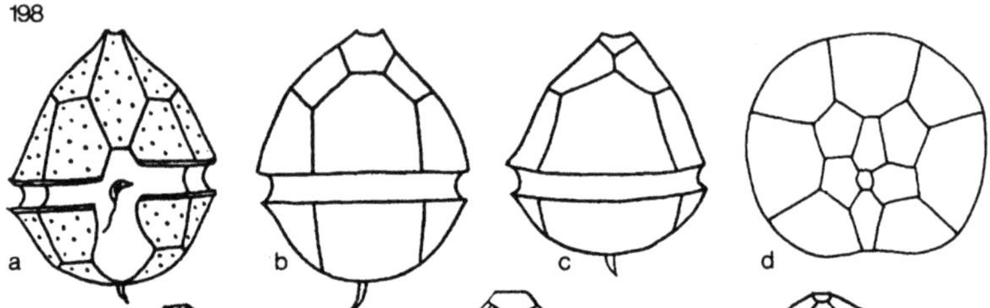

199

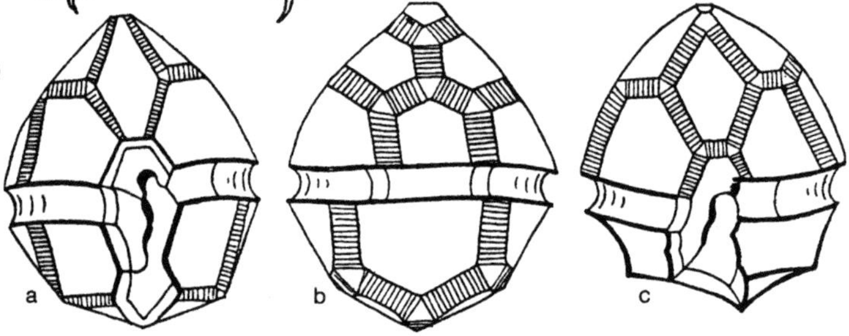

Fig. 198. *Peridinium umbonatum* var. *goslaviense* (Wołoszyńska) Popovský and Pfiester (*a, b* after Lefèvre, *c* after Wołoszyńska, *d* after Lindemann).

Fig. 199. *Peridinium umbonatum* var. *lubieniense* (Wołoszyńska) Popovský and Pfiester (after Wołoszyńska).

20.5 var. umbonatum (figs. 35, 36, 37, 48, 200)
Cells egg-shaped and dorsiventrally flattened. The epicone is conical or bell-shaped and larger than the hypocone which is rounded and sometimes sinuate, often minute. The sulcus reaches the antapex.
Dimensions: 15–40 μm × 12–30 μm.
Note: the invalid species described by Schiller: *Glenodinium denticulatum, Gl. kampneri,* and *Gl. sciculiferum* are probably *Peridinium umbonatum* Stein 1883.

2. **Peridiniopsis** Lemmermann 1904

Glenodinium auct., *Diplopelta* Stein in Pavillard 1913, *Staszicella* Wołoszyńska 1916, *Stylodinium* Klebs 1912 partim

Cells are ellipsoidal, or spherical. The hypotheca is equal to or smaller than the epitheca. The theca may be ornamented with spines, teeth-like projections or wings.
Plate formula: (3–5)′, (0a–1a), (6–8)″; 5‴, 2⁗
plate formula with 7″ plates:
section **lindemanni** (p. 188); 3′, 1a, 7″, (fig. 14)
section **elpatiewskyi** (p. 189); 4′, 0a, 7″, (fig. 15)
section **polonicum** (p. 192); 4′, 1a, 7″, (fig. 13)
section **thompsonii** (p. 195); 5′, 0a, 7″, (fig. 21)

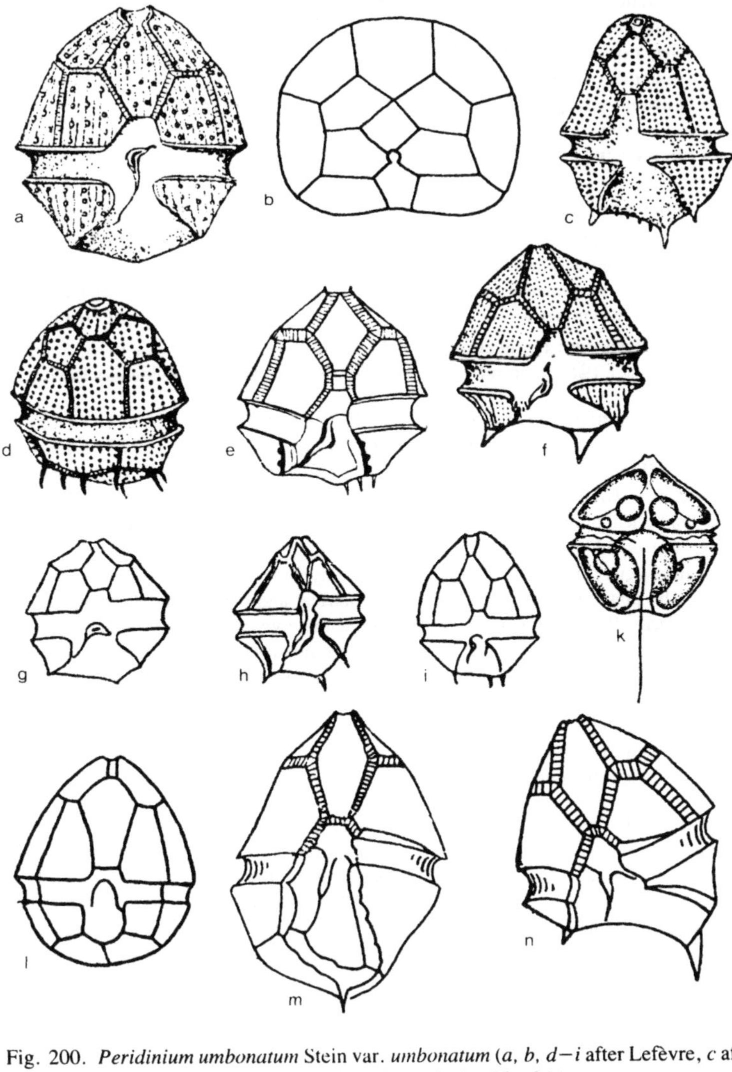

Fig. 200. *Peridinium umbonatum* Stein var. *umbonatum* (*a, b, d–i* after Lefèvre, *c* after Stein, *e, m, n* after Wołoszyńska, *k* after Skuja, *l* after Playfair).

section **quadridens** (p. 195); 5′, 1a, 7″, (fig. 19)
plate formula with 6″ plates:
section **borgei** (p. 195); 3′, 1a, 6″, (fig. 20)
section **penardii** (p. 198); 4′, 0a, 6″, (fig. 17)
section **cunningtonii** (p. 202); 5′, 0a, 6″ or 4′, 1a, 6″ (fig. 16)
plate formula with (6−8)″ plates:
section **dinobryonis** (p. 204); (4−5)′, 1a, (6−8)″, (fig. 25)

section **lindemanni** (fig. 14)

1a. Cells are pentangular . . . . **2. P. lindemanni**
1b. Cells are ovoid or spherical **2**
2a. Plate pattern asymmetric . . . . **1. P. balticum**
2b. Plate pattern symmetric . . . . **3. P. oculatum**

Important references: Lemmermann 1904, Klebs 1912, Wołoszyńska 1916, Schiller 1937, Baumeister 1943, Bourrelly 1968b, 1970, Matvienko and Litvinenko 1977.

1. Peridiniopsis balticum (Levander) Bourrelly 1968 (fig. 201)
Glenodinium cinctum Levander 1892, *G. balticum* Levander 1894, *Peridinium balticum* (Levander) Lemmermann 1910, *P. dybowskii* Wołoszyńska 1916, *Glenodinium dybowskii* (Wołoszyńska) Lindemann 1925/26

Cells are spherical or oval shaped, and slightly flattened dorsiventrally. The epitheca and hypotheca are equal in size, or the epitheca may be slightly larger. The cingulum spirals to the left, the sulcus is wide, and found only on the hypotheca where it does not reach the antapex. The plate arrangement is very asymmetrical and the rhomboidal plate (1′) is extremely large. Plates of younger cells are fine, without visible structure, but adult plates are ornamented with rows of dots. The sutures are not striated in the planozygotes but rather form a ridge between the plates. Chloroplasts are disc-shaped and numerous.
Dimensions: 22−40 µm × 20−35 µm.
Occurrence: ponds, Central Europe; in brackish water, Baltic Sea.
Note: Carty and Cox (1986) proposed the new generic name *Durinskia* for *Peridiniopsis balticum* (Levander) Bourrelly 1968.

2. Peridiniopsis lindemanni (Lefèvre) Bourrelly 1968 (fig. 202)
Peridinium lindemanni Lefèvre 1927, *Glenodinium lindemanni* Lefèvre 1932

Cells are pentangular, sligthly flattened dorsiventrally. The epitheca is conical, with a rounded top, the hypotheca is trapezoid in shape. The wide cingulum spirals slightly to the left. The sulcus widens greatly in the hypotheca and reaches the antapex. The plates are ornamented with very fine ridges. Several (usually 3−4) spines are present on the hypotheca.
Dimensions: 10−12 µm × 9−11 µm.
Occurrence: Spain, Madagascar.

3. Peridiniopsis oculatum (Stein) Bourrelly 1968 (fig. 203)
Glenodinium oculatum Stein 1883, *Peridiniopsis aculeatum* (Stein) Bourrelly 1968 in Starmach 1974

Cells are ovoid or spherical and slightly flattened dorsiventrally. The epi- and hypothecae are rounded. The cingulum circumscribes the cell or spirals slightly to the left. The sulcus is wide and does not reach the antapex. The plates bear very minute punctae. Chloroplasts

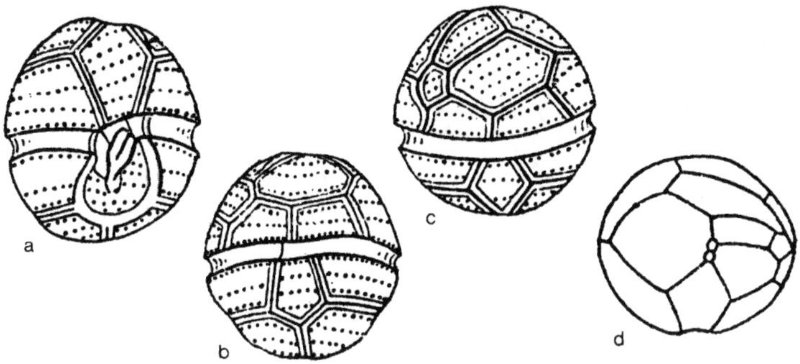

Fig. 201. *Peridiniopsis balticum* (Levander) Bourrelly (after Wołoszyńska).

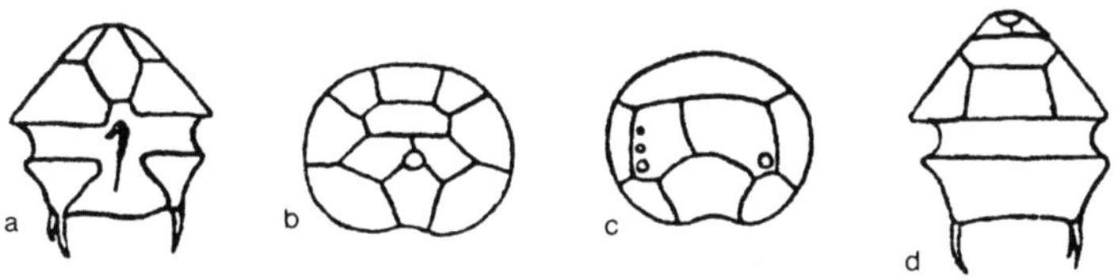

Fig. 202. *Peridiniopsis lindemanni* (Lefèvre) Bourrelly (after Lefèvre).

are numerous, disc or club shaped and parietal. A stigma is present. Vegetative cells are often embeded in a mucilage.
Dimensions: 23–36 μm × 21–36 μm.
Occurrence: ponds, pools; Europe, USA.

Section **elpatiewskyi** (fig. 15)

1a. Cells are pentangular . . . . **5. P. elpatiewskyi**
1b. Cells are spherical or broadly oval . . . . **4. P. charkowiensis**

4. Peridiniopsis charkowiensis (Matvienko) Bourrelly 1968 (fig. 204)
Peridinium charkowiensis Matvienko 1938
Cells are spherical or broadly oval, and slightly flattened dorsiventrally. The epicone is larger than the hypocone. The sulcus narrows and extends into half of the hypocone. The plates are ornamented with irregularly dispersed papillae which are more numerous on the hypocone. Chloroplasts are numerous, discoidal and parietal. A stigma is absent in vegetative stages.
Dimensions: 18–33 μm × 15–30 μm.
Occurrence: peat-bogs; USSR.

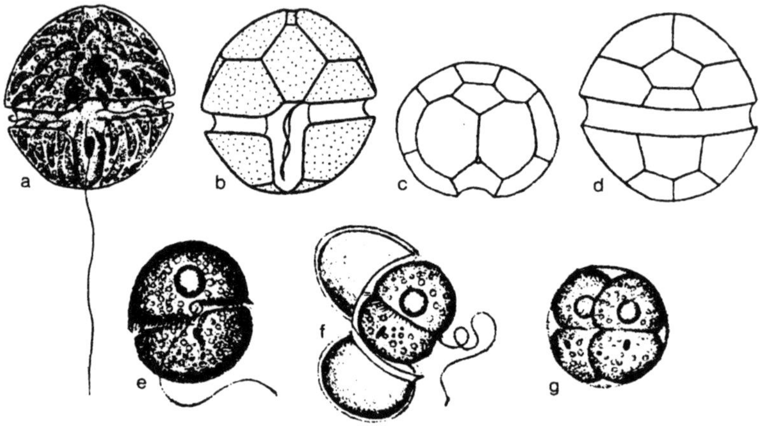

Fig. 203. *Peridiniopsis oculatum* (Stein) Bourrelly (*a, b* after Thompson, *c, d* after Wołoszyńska, *e–g* after Stein).

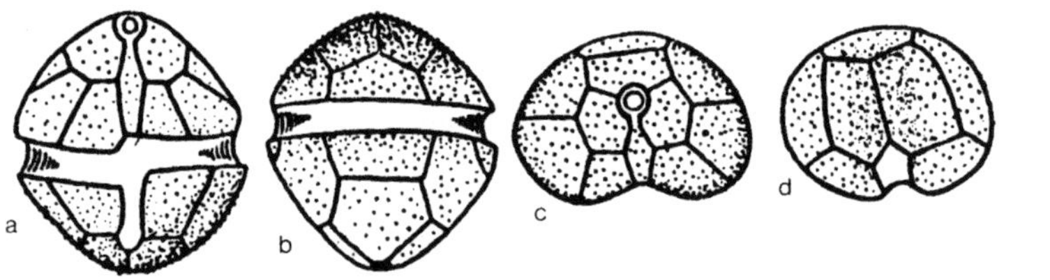

Fig. 204. *Peridiniopsis charkowiensis* (Matvienko) Bourrelly (after Matvienko).

5. Peridiniopsis elpatiewskyi (Ostenfeld) Bourrelly 1968 (fig. 205)
Peridinium umbonatum var. *elpatiewskyi* Ostenfeld 1907, *P. elpatiewskyi* (Ostenfeld) Lemmermann 1910, *Glenodinium elpatiewskyii* (Ostenfeld) Schiller 1937, *Peridinium marchicum* var. *simplex* Wołoszyńska 1916, *P. elpatiewskyi* var. *pseudopenardii* Lindemann 1918, *Peridinium pygmaeum* Lindemann 1918, *P. pygmaeum* f. *brigantinum* Lindemann 1923, *Glenodinium pygmaeum* (Lindemann) Schiller 1937
Cells are pentangular or ovoid in shape and slightly flattened dorsiventrally. The epitheca is conical and larger than the hypotheca, which is rounded and excavated on the antapex. The cingulum circumscribes the cell; the sulcus widens along the hypotheca and reaches the antapex. The plates are concave and ornamented by rows of fine punctae. Several spines or teeth-like projections are present on the hypotheca. The chloroplasts are numerous and discoidal. The hypnospores are somewhat oval with a thick cell wall.
Dimensions: 22–45 μm × 22–35 μm, hypnospores 36 μm × 28 μm.
Occurrence: plankton of lakes and peat bogs; Central Europe, Asia.

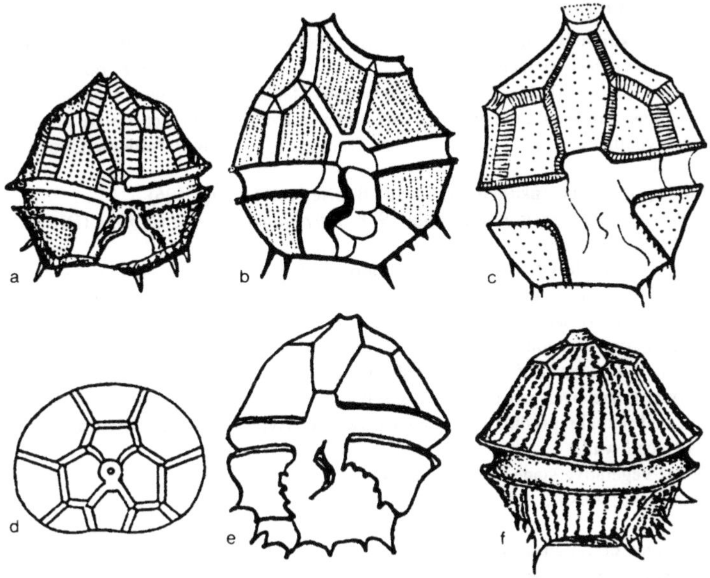

Fig. 205. *Peridiniopsis elpatiewskyi* (Ostenfeld) Bourrelly (*a, d–f* after Lefèvre, *b, c* after Wołoszyńska).

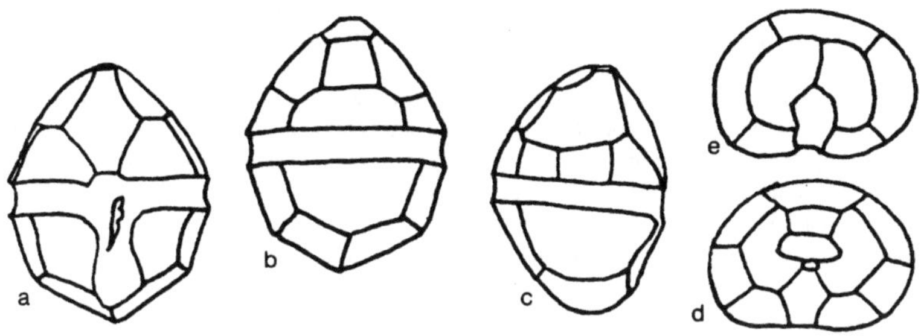

Fig. 206. *Peridiniopsis lubieniensiforme* (Diwald) Bourrelly (after Diwald).

Note: *Peridiniopsis elpatiewskyi* and *Peridinium africanum* differ in that in the first, the two intercalary plates (1a and 2a) are missing and in the shape of their zygotes. Based on zygote morphology *Peridiniopsis elpatiewskyi* and *Peridinium africanum* do not appear to be closely related.

Section **polonicum** (fig. 13)

1a. Cells are extremely flattened dorsiventrally; plates are reticulated with a tooth-like projection on the hypotheca . . . . **7. P. polonicum**

1b. Cells are slightly flattened dorsiventrally with plates bearing rows of papillae or short spines present on the epitheca. No spines or teeth-like projections are present on the hypotheca . . . . **6. P. lubieniensiforme**

6. Peridiniopsis lubieniensiforme (Diwald) Bourrelly 1968 (fig. 206)

Glenodinium lubieniensiforme Diwald 1938, *Peridinium lubieniensiforme* (Diwald) Huber-Pestalozzi 1950

Cells are oval and slightly flattened dorsiventrally. The epitheca, which is helmet-shaped and somewhat pointed, is larger than the hypotheca which is pentangular in shape. The cingulum is relatively wide and spirals to the left; the sulcus is somewhat narrow and extends slightly onto the epitheca and does not reach the antapex on the hypotheca. Plates are ornamented with rows of papillae or short spines.

The chloroplasts are oval and may be large, but few in number, numerous and small. Vegetative and sexual reproduction has been observed. Hypnospores are elliptical with a thick, warty cell wall.

Dimensions: vegetative stage 50–76 μm × 47.5–72.2 μm.

Occurrence: lake; Austria

7. Peridinopsis polonicum (Wołoszyńska) Bourrelly 1968 (fig. 207)

Glenodinium gymnodinium Pénard 1891, *Peridinium polonicum* Wołoszyńska 1916, *P. trilineatum* Lindemann 1919, *Glenodinium gymnodinium* var. *biscutelliforme* Thompson 1950

Cells are oval and extremely flattened dorsiventrally. The epitheca is helmet-shaped and pointed, and is approximately the same size as the broadly rounded hypotheca. The cingulum spirals to the left, the sulcus extends slightly into the epitheca and reaches the antapex. It widens slightly in the hypotheca and its left margin on the antapex bears one flattened tooth-like projection. The plates are almost concave and are reticulated or marked with irregularly-placed areolae. The chloroplasts are small and discoid or ribbon-shaped. A stigma may be present. Hypnospores are spherical, with their thick cell wall covered by dense hair-like projections (spines?). A number of plate pattern variations are known, including variations in dorsal the intercalary plates.

Dimensions: 33–54 μm × 30–51 μm.

Occurrence: plankton of lakes and pools; Europe, North America.

Appendix

1. *Peridiniopsis viguieri* (Lefèvre) Bourrelly 1968 (fig. 208)

Peridinium viguieri Lefèvre 1932, *Glenodinium viguieri* (Lefèvre) Schiller 1937

Cells are somewhat oval or pentangular, and slightly flattened dorsiventrally. The epitheca is large and conical in shape with a relatively large apical pore. The hypotheca is small and excavated on the antapex. The wide cingulum spirals to the left; the sulcus is extremely short and wide, and reaches the antapex. The plates are concave with rows of fine papillae. Sutures on the hypotheca are ornamented with several spines.

Dimensions: 18–25 μm × 15–28 μm.

Occurrence: Madagascar.

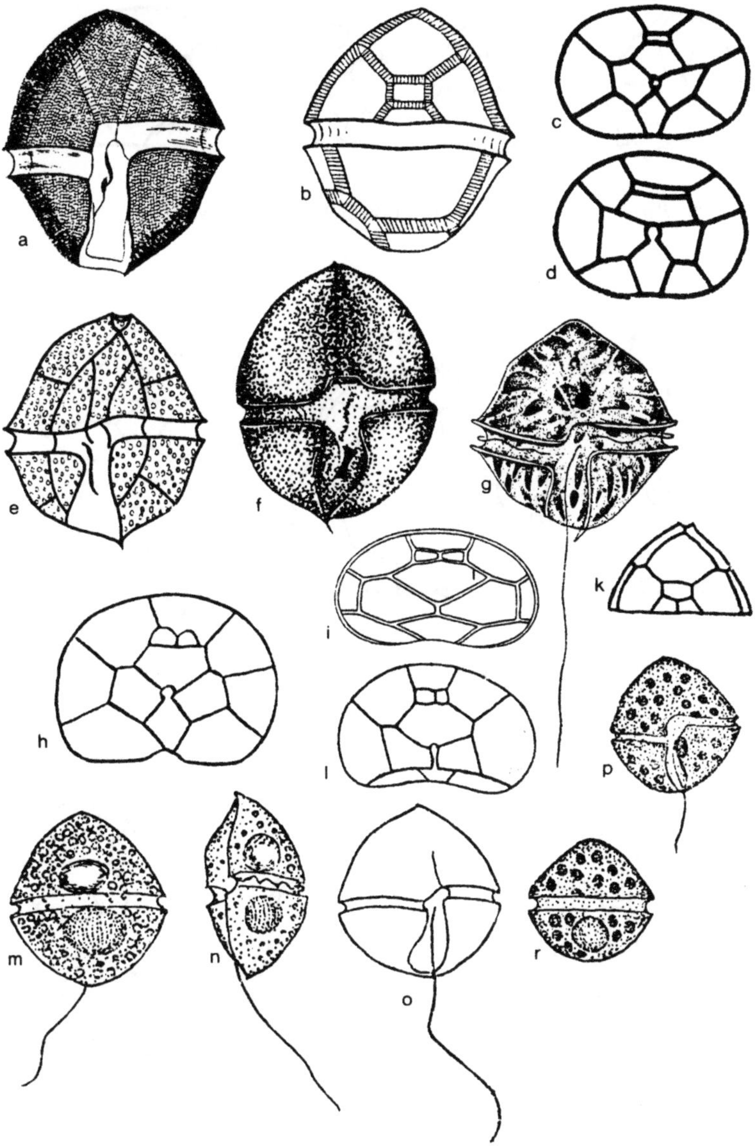

Fig. 207. *Peridiniopsis polonicum* (Wołoszyńska) Bourrelly (*a, b* after Wołoszyńska, *c–f* after Popovský, *g, i* after Thompson, *h* after Bourrelly, *k* after Lindemann, *l* after Lefèvre, *m–o* after Penard, *p, r* after Bachmann).

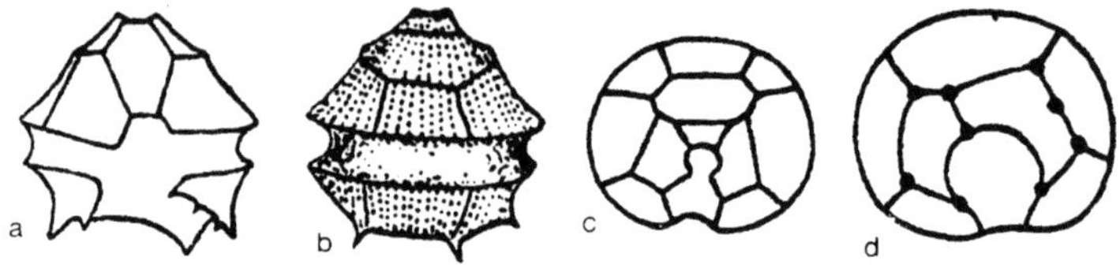

Fig. 208. *Peridiniopsis viguieri* (Lefèvre) Bourrelly (after Lefèvre).

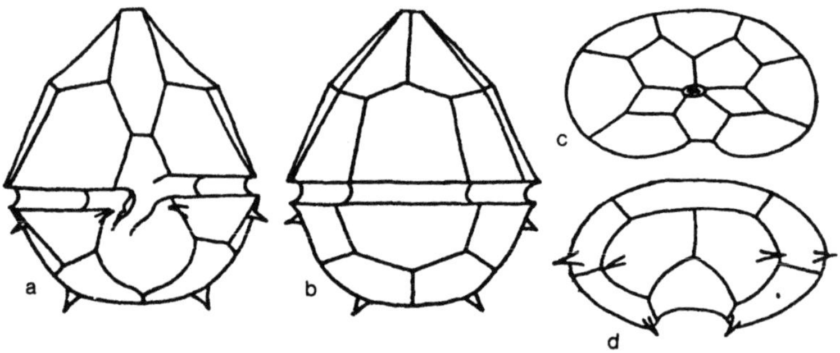

Fig. 209. *Peridiniopsis thompsonii* (Thompson) Bourrelly (after Thompson).

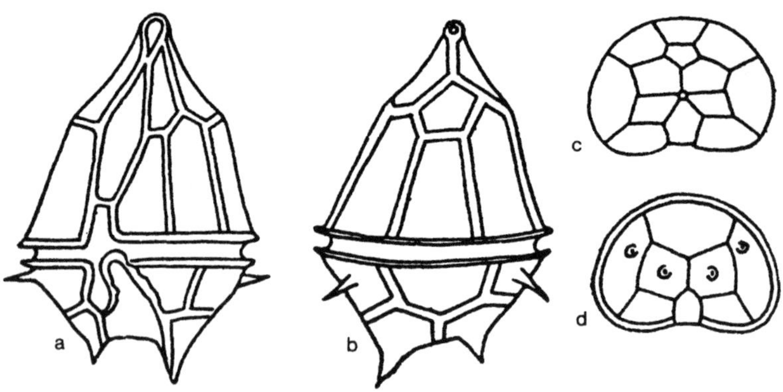

Fig. 210. *Peridiniopsis quadridens* (Stein) Bourrelly (*a*, *d* after Schilling, *b*, *c* after Stein).

Note: *Peridiniopsis elpatiewskyi* appears to be closely related as its 3′ plate is often doubled. *Peridinium africanum* also appears to be related to *Peridiniopsis viguieri* as individuals with the 1 a simplex plate pattern variation have been observed.

Section **thompsonii** (fig. 21)

8. Peridiniopsis thompsonii (Thompson) Bourrelly 1968 (fig. 209)

Glenodinium quadridens (Stein) Schiller 1937 sensu Thompson 1947

Cells are ovoid, and slightly flattened dorsiventrally. The conical epitheca is slightly apiculate and larger than the broadly rounded hypotheca. The cingulum spirals to the left, the sulcus is very broad and does not reach the antapex. Plates 2‴, 4‴, 1⁗, and 2⁗ on the hypotheca bear single spines. Some individuals also bear spines on plates 1‴ and 5‴ which are curved inward toward the sulcus. Chloroplasts are present, but there is no stigma.

Dimensions: 31–38 µm × 18.5 µm.

Occurrence: dam reservoir; USA.

Section **quadridens** (fig. 19)

9. Peridiniopsis quadridens (Stein) Bourrelly 1968 (fig. 210)

Peridinium quadridens Stein 1883, *Glenodinium quadridens* (Stein) Schiller 1937

Cells are double conical. The epitheca is apiculate and larger than the hypotheca. The hypotheca usually bears four sharp spines which are laterally and posteriorily placed (rarely three or five, and sometimes absent entirely). The cingulum circumscribes the cell or slightly spirals to the left; the sulcus extends into the epitheca, it widens in the hypotheca and reaches the antapex. Plates are finely reticulate. Chloroplasts are present. The thick-walled hypnospores are round.

Dimensions: 30–39 µm × 20–33 µm.

Occurrence: mainly in winter plankton of lakes; Europe, Asia, America.

Note: Some authors determine *P. quadridens* as *Peridiniopsis cunningtonii* 3′ 5″ simplex.

Section **borgei** (fig. 20)

1a. Cells rounded, oval, or pear-shaped **2**
1b. Cells pentangular . . . . **13. P. penardiforme**
2a. Cells rounded or oval **3**
2b. Cells apiculated at the apex . . . . **10. P. borgei**
3a. Chloroplasts present . . . . **12. P. kulczynskii**
3b. Chloroplasts absent . . . . **11. P. edax**

10. Peridiniopsis borgei Lemmermann 1904 (figs. 47, 211)

Glenodinium borgei (Lemmermann) Schiller 1937, *Peridinium borgei* (Lemmerman) Lemmermann 1910

Cells oval or pear-shaped, sometimes rounded, and slightly flattened dorsiventrally. The conical epitheca is apiculate at the apex, the hypotheca is broadly rounded. The cingulum circumscribes the cell or spirals to the left. The sulcus slightly extends into the epitheca, widening in the hypotheca and does not reach the antapex. Sulcal margins are sometimes flanged. Plates are reticulated or rugose. Chloroplasts are numerous and ribbon-shaped.

Dimensions: 40–55 µm × 35–41 µm.

Occurrence: lakes and ponds; Europe, Asia, America.

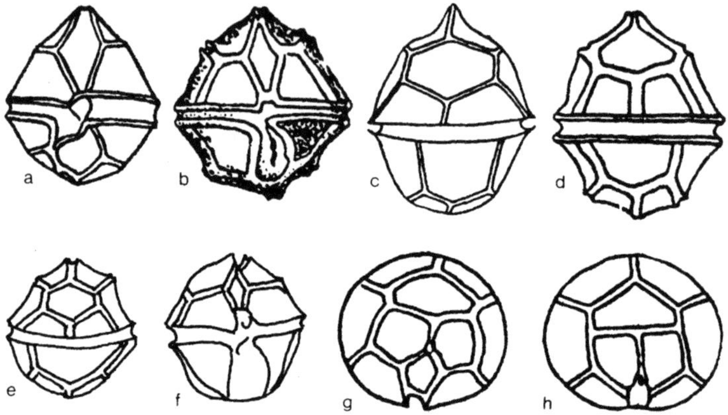

Fig. 211. *Peridiniopsis borgei* Lemmermann (*a, h* after Lemmermann, *b–d, g* after Lindemann, *e–f* after Prescott).

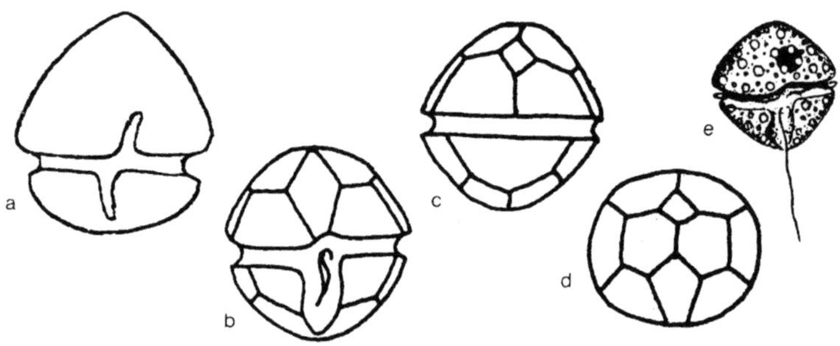

Fig. 212. *Peridiniopsis edax* (Schilling) Bourrelly (*a* after Harris, *b–e* after Thompson).

11. Peridiniopsis edax (Schilling) Bourrelly 1968 (fig. 212)
Glenodinium edax Schilling 1891

Cells are spherical or somewhat oval and slightly flattened dorsiventrally. The helmet-shaped or conical epitheca is larger than the hypotheca, which is broadly rounded. The cingulum spirals to the left, the sulcus is found only in the hypotheca. Plates are without known ornamentation. Chloroplasts and stigma are lacking. The cytoplasm is coarsely granular and contains many refractive bodies at the periphery. Nutrition is holophytic.
Dimensions: 15–34 µm × 12–33 µm.
Occurrence: lakes, ponds, pools; Europe, America.

12. Peridiniopsis kulczynskii (Wołoszyńska) Bourrelly 1968 (fig. 213)
Peridinium kulczynskii Wołoszyńska 1916, *Glenodinium kulczynskii* (Wołoszyńska) Schiller 1937

Cells are almost spherical and slightly flattened dorsiventrally. Both the epi- and hypothecae are rounded and equal in size: the cingulum spirals to the left; the sulcus slightly extends into the epitheca and reaches the antapex on the hypotheca. Plates are without known ornamentation. Chloroplasts are club-shaped and radially arranged. A stigma is sometimes present.
Dimensions: 28–48 µm × 23–46 µm.
Occurrence: lakes and ponds; Europe, America.

13. Peridiniopsis penardiforme (Lindemann) Bourrelly 1968 (fig. 214)
Peridinium penardiforme Lindemann 1918, *Glenodinium penardiforme* (Lindemann) Schiller 1937

Cells are pentangular, and extremely flattened dorsiventrally. The epitheca is conical and apiculate; the hypotheca is excavated at the antapex, both are equal in size. The wide, deep cingulum circumscribes the cell, the sulcus is broad, deep and reaches the antapex. It produces a very marked concavity in the antapical profile. The plates may be smooth, or striately punctate, scrobiculate, or reticulate. The first apical plate is sexangular and not contiguous with the apical pore. Plates 1'''' and 2'''' are squared. Chloroplasts are present. The protoplasm may have a dark granular appearance, such that it is grayish in color. Numerous highly refractive granules may or may not be present in the cell periphery.
Dimensions: 16–35 µm × 9–30 µm.
Occurrence: lakes and ponds; Europe, America.

Appendix

1. *Peridiniopsis cristatum* (Balech) Bourrelly 1968 (fig. 215)
Glenodinium cristatum Balech 1961

Cells are asymmetrically ovoid. The epi- and hypotheca are equal in size. The epitheca is conical and apiculate with a small wing-like process around the apical pore. The hypotheca is rounded. The wide cingulum, circumscribes the cell, the sulcus extends into the hypotheca and does not reach the antapex. The sulcal margins are flanged. A double winged projection is present at the antapex. Plates have irregularly placed papillae. Chloroplasts and pyrenoid are present.
Dimensions: 50 µm × 40 µm.
Occurrence: saline ponds; France.

2. *Peridiniopsis rotunda* Lebour 1922 (fig. 216)
Glenodinium rotundum (Lebour) Schiller 1937, *Peridinium limnophilum* Lindemann 1924, *Peridiniopsis limnophilum* (Lindemann) Bourrelly 1968

Cells are rounded and slightly flattened dorsiventrally. The epi- and hypothecae are equal in size; the apical pore is located on the small sinus. The cingulum with two flanges circumscribes the cell. The sulcus is extremely short and narrow, and does not reach the antapex; a sickle-shaped wing is present on the left sulcal margin. Plates are ornamented with irregularly placed papillae which may be absent. The cytoplasm is colorless or reddish to brownish in color. Chloroplasts are lacking.
Dimensions: 22 µm–28 µm in diameter.
Occurrence: euryhaline, also in inland waters; Ukraine, Poland, Finnland.

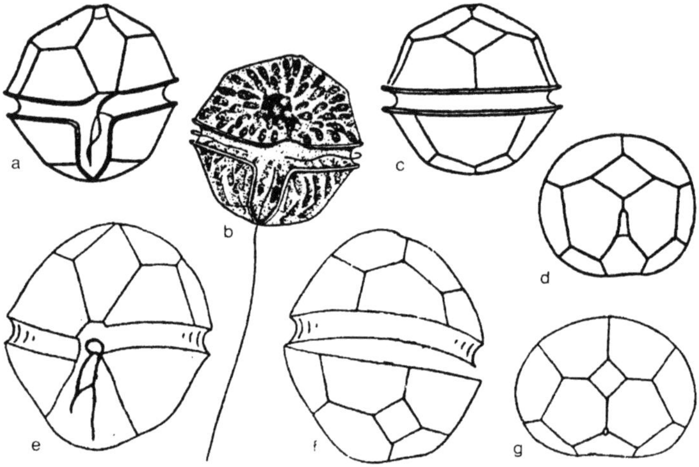

Fig. 213. *Peridiniopsis kulczynskii* (Wołoszyńska) Bourrelly (*a–d* after Thompson, *e–g* after Wołoszyńska).

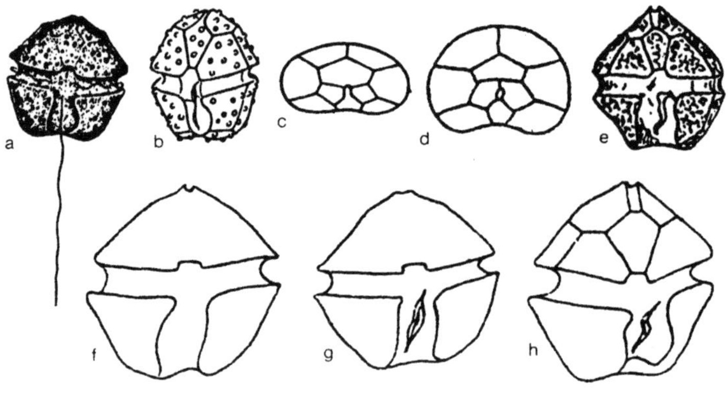

Fig. 214. *Peridiniopsis penardiforme* (Lindemann) Bourrelly (*a–c* after Thompson, *d*, *e* after Lindemann, *f–h* after Lefèvre).

Section **penardii** (fig. 17)

1a. Plate pattern symmetric **2**
1b. Plate pattern asymmetric . . . . **14. P. berolinense**
2a. Cells oval, plate 3′ hexangular . . . . **15. P. penardii**
2b. Cells round, plate 3′ quadrangular . . . . **16. P. sphaeroideum**

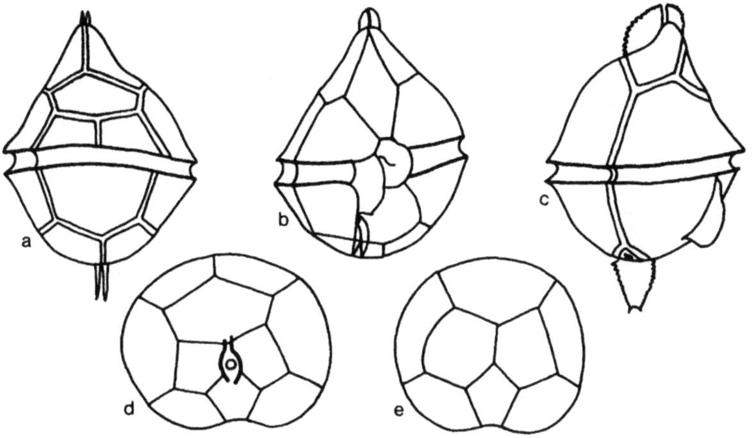

Fig. 215. *Peridiniopsis cristatum* (Balech) Bourrelly (after Bourrelly).

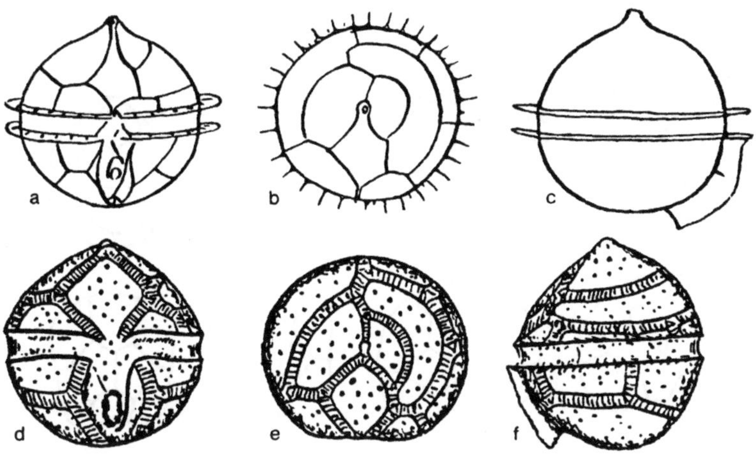

Fig. 216. *Peridiniopsis rotunda* Lebour (*a–c* after Lebour, *d–f* after Lindemann).

14. Peridiniopsis berolinense (Lemmermann) Bourrelly 1968 (fig. 217)
Peridinium berolinense Lemmermann 1900, *Peridinium berolinense* var. *apiculata* Lemmermann 1907, *Glenodinium berolinense* (Lemmermann) Lindemann 1925, *Gymnodinium alatum* Litvinenko 1963, *Peridinium anserinum* Baumeister 1967 nomen nudum

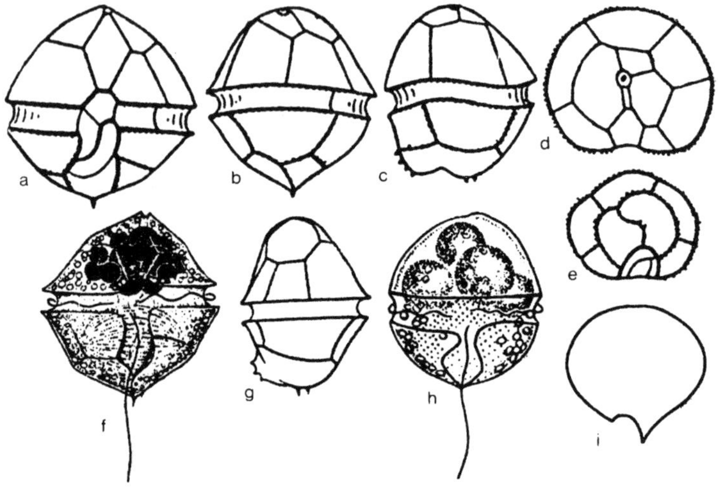

Fig. 217. *Peridiniopsis berolinense* (Lemmermann) Bourrelly (*a–e* after Wołoszyńska, *f, g* after Thompson, *h, i* after Litvinenko).

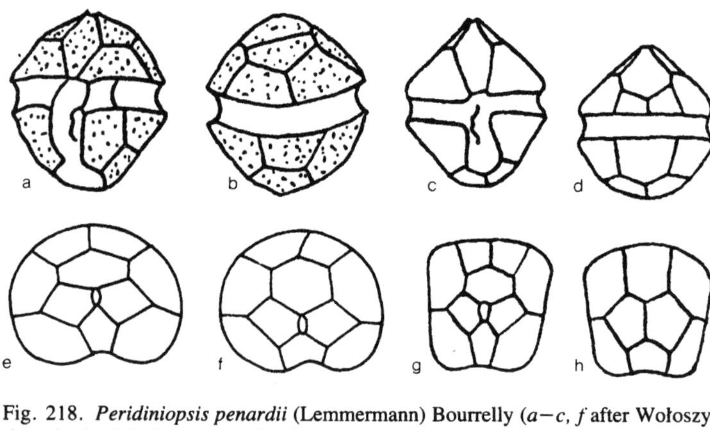

Fig. 218. *Peridiniopsis penardii* (Lemmermann) Bourrelly (*a–c, f* after Wołoszyńska, *d, g, h* after Lindemann, *e* after Lemmermann).

Cells spherical, ellipsoidal or rhomboid in ventral view and slightly flattened dorsiventrally. The epitheca is conical or broadly helmet-shaped; the hypotheca is conical and apiculate bearing one spine. Many spicules are present on the hypothecal plates and plate margins. The cingulum is wide and spirals to the left; the sulcus widens and reaches the antapex. Its left margin is flanged. Chloroplasts may be present.

Numerous, colorless or reddish, highly refractive granules are usually present in the periphery of the cytoplasm.
Dimensions: 21–52 µm × 18–42 µm.
Occurrence: euplankter in lakes, ponds, creeks; Europe, America, Africa.

15. Peridiniopsis penardii (Lemmermann) Bourrelly 1968 (fig. 218)
Peridinium cinctum Penard 1891, *Glenodinium penardii* Lemmermann 1900, *Peridinium penardii* (Lemmermann) Lemmermann 1910, *P. andrzejowskii* Wołoszyńska 1920

Cells are oval and slightly flattened dorsiventrally with a delicate theca. The epitheca is dome-shaped and equal in size with the rounded hypotheca. Cells are slightly flattened laterally such that they are quadrangular in cross-section. The cingulum spirals to the left; the sulcus is broad and nearly reaches the antapex. Plates are ornamented with irregularly placed papillae. The first apical plate is pentangular and contiguous with the apical pore. Plates 1'''' and 2'''' are pentangular. Chloroplasts are numerous, discoid in the shape and parietal, they are sometimes lacking. The thick-walled hypnospores are spherical.
Dimensions: 27–42 µm × 25–35 µm.
Occurrence: euplankter in lakes and ponds; Europe.

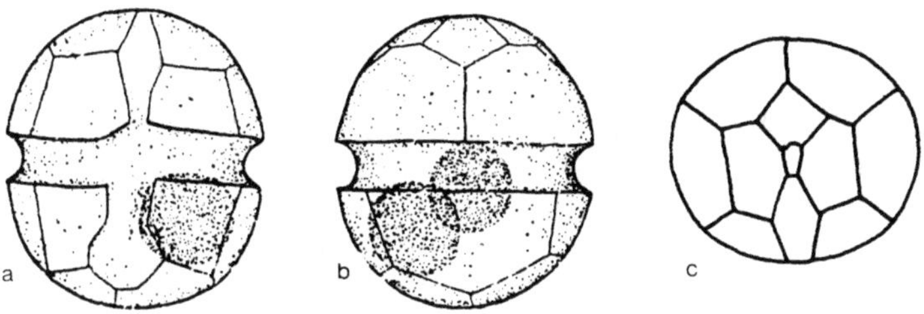

Fig. 219. *Peridiniopsis sphaeroideum* (Christen) Bourrelly (after Christen).

16. Peridiniopsis sphaeroideum (Christen) Bourrelly 1968 (fig. 219)
Peridinium sphaeroideum Christen 1958

Cells are spherical and slightly flattened dorsiventrally. The epi- and hypothecae are broadly rounded. The cingulum spirals to the left; the sulcus extends slightly into the epitheca, widening along the hypotheca where it does not reach the antapex. The plates are very delicate without visible makings. The cytoplasm is colorless with numerous refractive bodies. Chloroplasts are lacking. Nutrition holozoic.
Dimensions: 20–22 µm × 18–20 µm.
Occurrence: lake; Switzerland.

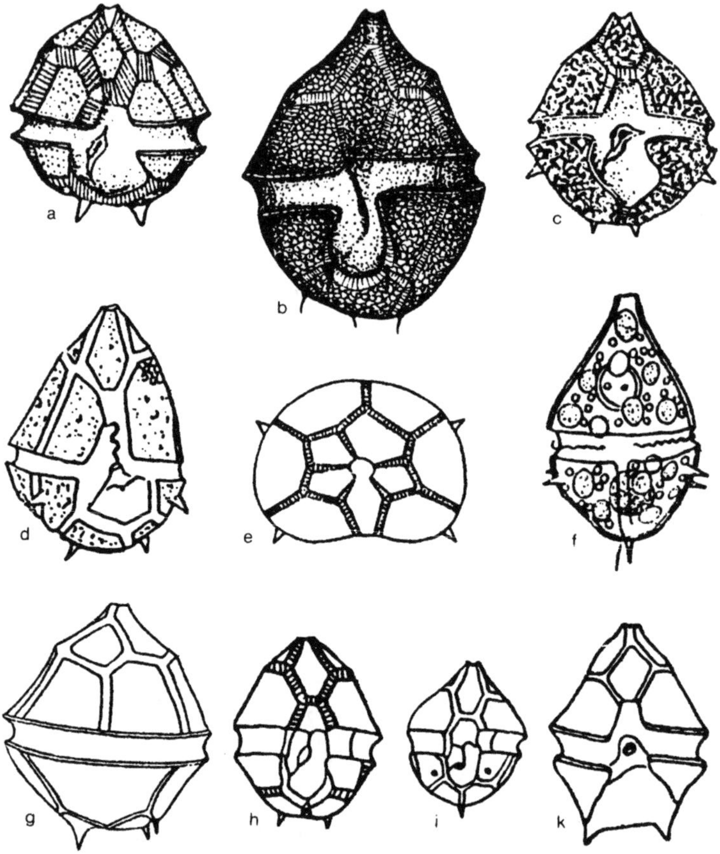

Fig. 220. *Peridiniopsis cunningtonii* Lemmermann (*a, c, e, g, k* after Lefèvre, *b* after Popovský, *d* after Virieux, *f* after Massart, *h, i* after Wołoszyńska).

Section **cunningtonii** (fig. 16)

17. Peridiniopsis cunningtonii Lemmermann 1907 (fig. 220)

Peridinium cunningtonii (Lemmermann) Lemmermann 1910, *P. suttoni* B. M. Griffiths 1922, *Glenodinium sedens* Lindemann 1928, *Peridinium treubii* Wołoszyńska 1912, *P. cunningtonii* var. *pseudoquadridens* Lindemann 1919, *P. wildemani* Wołoszyńska 1923, *P. cunningtonii* var. *quinquecuspidata* (Massart) Nygaard 1926, *Heterocapsa quinquecuspidata* Massart 1920

Cells oval and extremely flattened dorsiventrally. The epitheca is conical; the hypotheca rounded with 2 to 6 spines. The cingulum spirals slightly to the left. The sulcus barely

Fig. 221. *Peridiniopsis dinobryonis* (Wołoszyńska) Bourrelly (*a–h* after Lindemann, *i–p* after Wołoszyńska, *r* after Bourrelly).

extends into the epitheca; it widens along the hypotheca where it does not reach the antapex. The plates are delicate reticulated. Hypothecal plates bear spicules. Chloroplasts are present. Hypnospores are oval with a thick cell wall. Plate pattern variability is well known.

Dimensions: 20–40 µm × 20–30 µm.

Occurrence: in lakes and ponds; cosmopolitan.

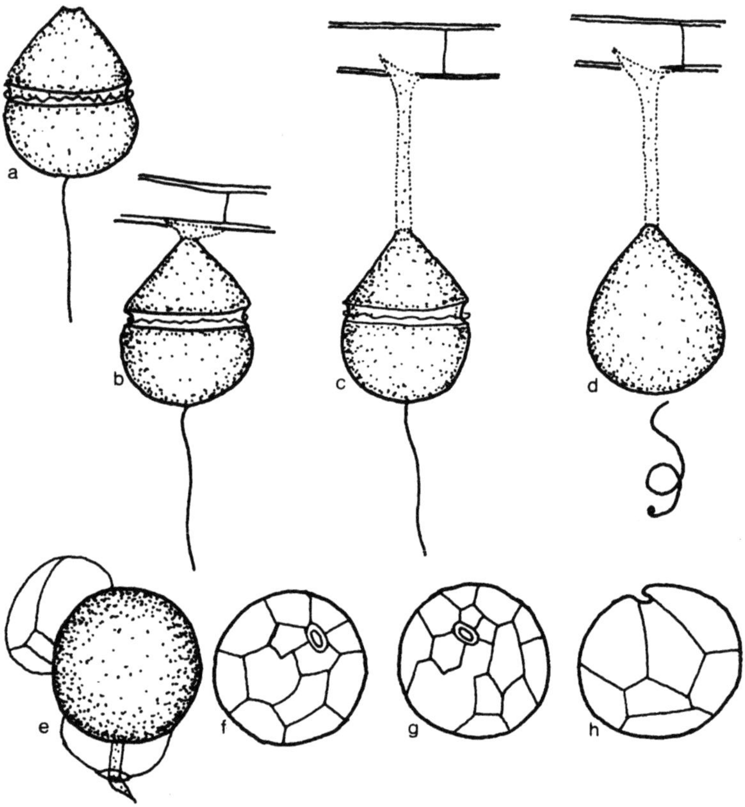

Fig. 222. *Peridiniopsis dinobryonis* (Wołoszyńska) Bourrelly, *a* vegetative cell, *b* early stage of stipe production on a filamentous alga, *c* later stage of stipe formation from the apical pore, *d* flagella and furrows disappear, *e* ecdysis, *f–h* empty thecae (after Baumeister).

Note: some authors determine *P. cunningtonii* as *Peridiniopsis quadridens* 2'4'' complexum.

Section **dinobryonis** (fig. 25)

18. Peridiniopsis dinobryonis (Wołoszyńska) Bourrelly 1968 (figs. 221, 222)
Staszicella dinobryonis Wołoszyńska 1916, *Glenodinium dinobryonis* (Wołoszyńska) Schiller 1937, *Stylodinium tarnum* Baumeister 1943

Spherical cells in which the epitheca is smaller than the large rounded hypotheca. The cingulum slightly spirals to the left. The sulcus extends into the epitheca; it widens along the hypotheca where it almost reaches the antapex. Plates are without ornamentation. Fine papillae are sometimes apparent on the plate sutures. Chloroplasts are present. The sessile stage (*Stylodinium*-like) bears a stalk from the apical pore.
Dimensions: 22–35 μm × 19–38 μm.
Occurrence: euplankter or settling on Dinobryon colonies or filamentous algae; lakes, ponds, Europe.
The following invalidly described species by Schiller (1955): *Glenodinium bieblii, Peridinium* (?) *peisonis* are probably *Peridiniopsis penardii* (Lemmermann) Bourrelly 1968. *Glenodinium peisonis* also invalidly described is probably *Peridiniopsis kulczynskii* (Wołoszyńska) Bourrelly 1968. *Glenodinium fungiforme* is probably *Peridiniopsis polonicum* (Wołoszyńska) Bourrelly 1968.
The species *Peridinium hiemale* Schiller 1955 (no latin diagnosis) was invalidly erected, and should be named *Peridiniopsis hiemale* from the section elpatiewskyi (for details see p. 20, fig. 15, and p. 239).

6. Family **Ceratiaceae** (Schütt 1896) Lindemann 1928

Cells posses 2–4 horns. Chloroplasts are present.
Plate formula: 4′, 0a, 5″; 5‴, 2⁗ (or 4′, 0a, 6″; 5‴, 1p, 1⁗).

1. **Ceratium** Schrank 1793

Bursaria O. F. Müller 1773, *Cercaria* Michaelis 1830, *Tripos* and *Hirundinella* Bory 1824, *Peridinium* Ehrenberg 1838, *Ceratophorus* Diesing 1850, 1866, *Glenodinium* Diesing 1866, *Poroceratium, Biceratium, Amphiceratium* Vanhöffen 1899

Cells are asymmetrical and strongly flattened dorsiventrally. The epivalve ends in one large horn which is shifted slightly to the right. Two or 3 horns are present on the hypovalve depending on temperature variations. The cingulum is equatorial and spirals. The sulcus does not extend into the epivalve. The theca is composed of 16 to 17 reticulate plates with 5 cingular plates and 6 sulcal plates. Chloroplasts are numerous and oval or slightly ribbon-shaped. Formation of rhizopodia and phagotrophy are known. Nutrition may be autotrophic, holophytic or holozoic. Vegetative reproduction is by division of motile cells into two daughter cells. Sexual reproduction is anisogamous. Parthenospores and hypnozygotes are quadrangular or 3-horned in shape.
Important references: Lederbauer 1904, Bachmann 1911, Entz 1927, 1931, Schiller 1937, Huber-Pestalozzi 1950, Skuja 1964, Morling 1979, v. Stosch 1964.

Key to the species

1a. Apical horn long and tapering, as long as or longer than the cell body and the posterior horns **3**
1b. Apical horn shorten than the remainder of the cell **2**
2a. Apical horn sharply curved and tapering to a blunt point . . . . **1. C. carolinianum**
2b. Apical horn straight but directed at an angle from the longitudinal axis; squarely truncate at the apex . . . . **2. C. cornutum**

3a. Epivalve helmet-shaped, all 4 apical plates reache the apex . . **4. C. hirundinella**
3b. Epivalve conical, the plate 4′ does not reache the apex . . . . **3. C. furcoides**

1. Ceratium carolinianum (Bailey) Jørgensen 1911 (fig. 223)
Peridinium carolinianum Bailey 1850, *Dimastigoaulax carolina* Diesing 1866, *Ceratium curvirostre* Huitfeld-Kaas 1900

Cells broadly fusiform in outline; the epivalve is broad just above the cingulum narrowing abruptly forming a stout, curved apical horn, with a shoulder on each side at the base. The cingulum is relatively narrow, spirals to the left and has flanges. The hypovalve is broad; it has a short divergent horn on the left, and a longer central horn which is somewhat obliquely directed. The theca and plates are robust with distinct reticulations. The chloroplasts are oval and parietal. A stigma is absent. Hypnozygotes are irregularly tetrangular and enveloped by a thick, multilayered wall.
Dimensions: 125–213 μm × 73–105 μm.
Occurrence: rarely in dystrophic lakes, common in oligotrophic lakes; Europe.

2. Ceratium cornutum (Ehrenberg) Claparède and Lachmann 1858 (fig. 224)
Peridinium cornutum Ehrenberg 1831, *Dimastigoaulax cornuta* Diesing 1866, *Peridinium carolinianum* var. *elongatum* Skuja 1964

Stout cells broadly fusiform in outline, horns more or less short, and strongly flattened dorsiventrally. The epivalve is broad above the cingulum; the sides rapidly converging and narrowed forming a short, stout anterior horn which extends obliquely and is truncate at the apex. The cingulum is relatively wide and spirals to the right. The hypovalve is broad below the cingulum. It extends into 2 horns; 1 short lateral horn and 1 which is longer and median. The theca and plates are robust with distinct reticulations. The chloroplasts are numerous, oval shaped and parietal. A stigma is lacking. The hypnozygotes are tri- or tetrangular with rounded ends and are enveloped by a thick multilayered wall. Nutrition is autotrophic, holozoic and holophytic.
Dimensions: 97–150 μm × 48–75 μm.
Occurrence: plankton of ponds, shallow lakes; Europe.

3. Ceratium furcoides (Levander) Langhans 1925 (fig. 225)
Ceratium hirundinella var. *furcoides* Levander 1900, *C. reticulatum* Imhoff 1884–87 partim, *C. furcoides* fo. *gracile* Entz 1927; non: *C. hirundinella* fo. *furcoides* Schroeder 1920

Cells are narrowly fusiform in outline and strongly flattened dorsiventrally. The epivalve is narrowly conical; a long horn forms from just above the cingulum. The cingulum is slightly narrow. The body of the hypovalve is broad and short. It is divided into a varying number of posterior horns, usually 2 or 3, rarely 1. The central or median horn is the longest. It is formed by the antapical plates. On epivalve here four apical plates are present of which the fourth does not reach the apex. The plates are coarsely reticulate with fine spicules. Chloroplasts are numerous and oval-shaped. Nutrition is autotrophic. Vegetative and sexual reproduction are known. Hypnozygotes are triangular with one horn on each end.
Dimensions: 123–322 μm × 28–42 (56) μm.
Occurrence: oligotrophic and eutrophic lakes and dam reservoirs; Europe.

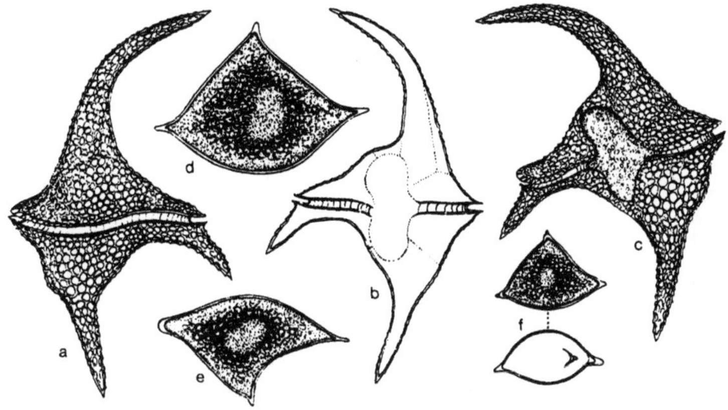

Fig. 223. *Ceratium carolinianum* (Bailey) Jørgensen, *e, f* hypnozygotes (after Skuja).

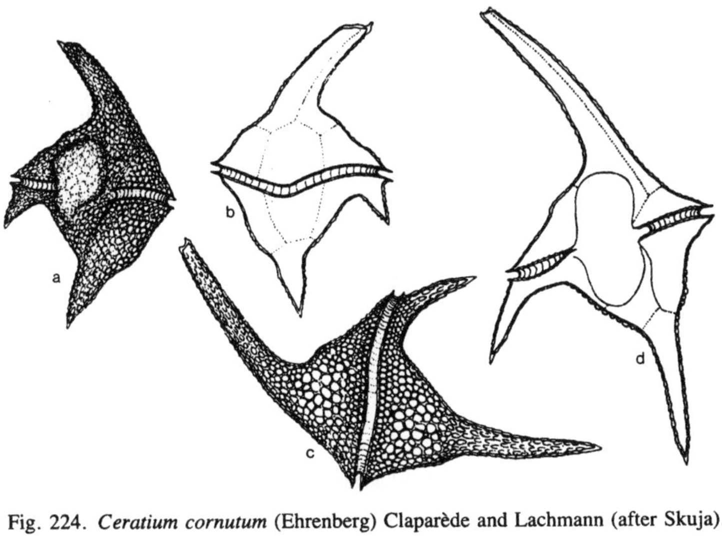

Fig. 224. *Ceratium cornutum* (Ehrenberg) Claparède and Lachmann (after Skuja).

4. Ceratium hirundinella (O. F. Müller) Dujardin 1841 (figs. 1n, 29, 38a–c, 40, 42a, 44, 226)

Bursaria hirundinella O. F. Müller 1773, *C. tetraceras* Schrank 1793, *C. macroceras* Schrank 1802, *C. reticulatum* Imhof 1884/87 partim, *C. longicorne* Perty 1849, *C. cumaonense* Carter 1871, *C. brevicorne* Zacharias 1905, *C. leptoceras* Zacharias 1904, *C. pumilum* Zacharias 1905, *C. handelii* Skuja 1937

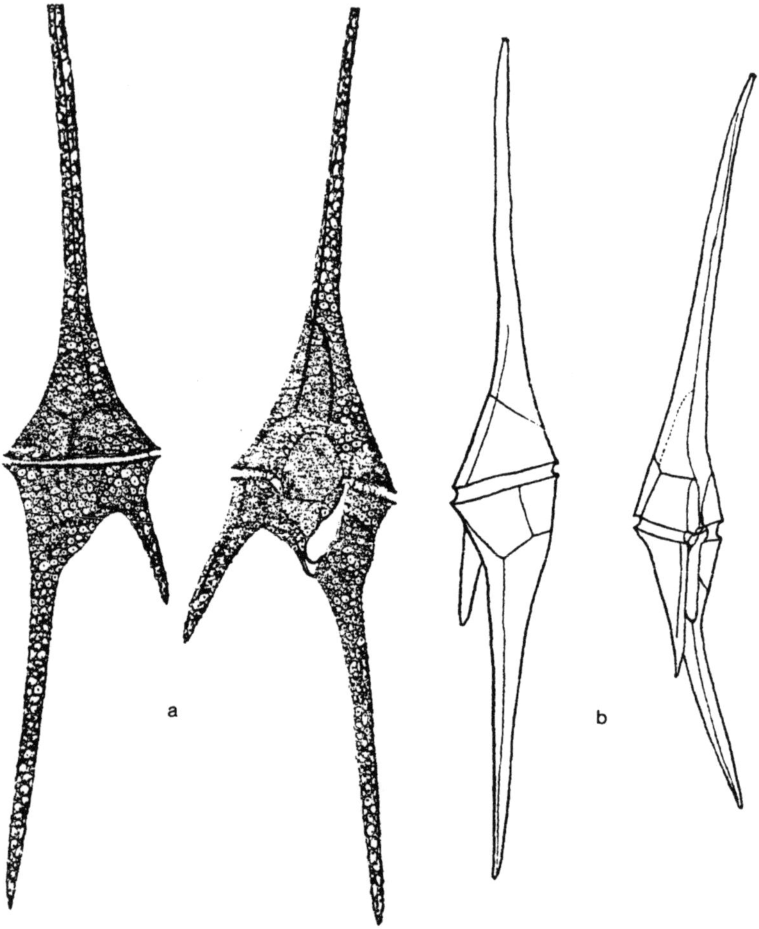

Fig. 225. *Ceratium furcoides* (Levander) Langhans (*a, b* after Entz).

Cells are broadly or narrowly fusiforme in outline, depending upon the degree of divergence of the horns. They are strongly flattened dorsiventrally. The helmet shaped epivalve narrows from just above the cingulum gradually forming a long horn. The cingulum is slightly narrow. The body of the hypovalve is broad and short. It is divided into a varying number of posterior horns, usually 3, sometimes only 1. The central or median horn formed by the antapical plates is the longest. Plates are coarsely reticulate with fine spicules. Chloroplasts are numerous and oval-shaped. Small red corpuscular

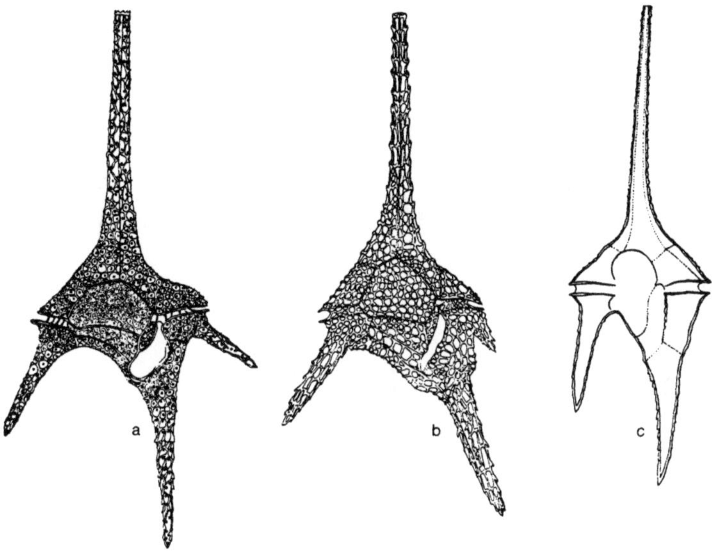

Fig. 226. *Ceratium hirundinella* (O. F. M.) Dujardin (*a, b* after Entz, *c* after Skuja).

bodies are often present. Nutrition is autotrophic and holozoic by means of pseudopodia. Vegetative and sexual reproduction is known. Hypnozygotes are triangular with one horn on each end.

Dimensions: 40–450 μm × 28–55 μm

Several "forms" or "types" are differentiated below (fig. 227):

A. Cells narrow, long to very long with 3 horns, up to 45, rarely up to 55 μm wide. Epicone narrowly helmet-shaped.
   a. Antapical horn in the same angle as the apical horn . type *furcoides* (fig. 227 a) (*C. hirundinella* fo. *furcoides* Schroeder 1920; non: *C. hirundinella* var. *furcoides* Levander 1894)
   b. Antapical horn is shifted to the outside . . . . . type *yuennanense* (fig. 227 c) (*C. handelii* Skuja 1937, *C. hirundinella* fo. *yueannanense* (Skuja) Huber-Pestalozzi 1950)
   c. Antapical horn is shifted inward; is relatively short, and lies parallel with the apical horn . . . . type *brachyceroides* (fig. 227 d) (*C. hirundinella* fo. *brachyceroides* Schroeder 1920; non: *C. brachyceros* Daday 1907, *C. hirundinella* var. *brachyceros* (Daday) Ostenfeld 1909)

B. Cells short and wide with three horns. Cells more than 45 μm wide. Both antapical horns are slightly divergent . . . . type *carinthiacum* (fig. 227 e) (*C. hirundinella* fo. *carinthiacum* (Zederbauer) Bachmann 1911)

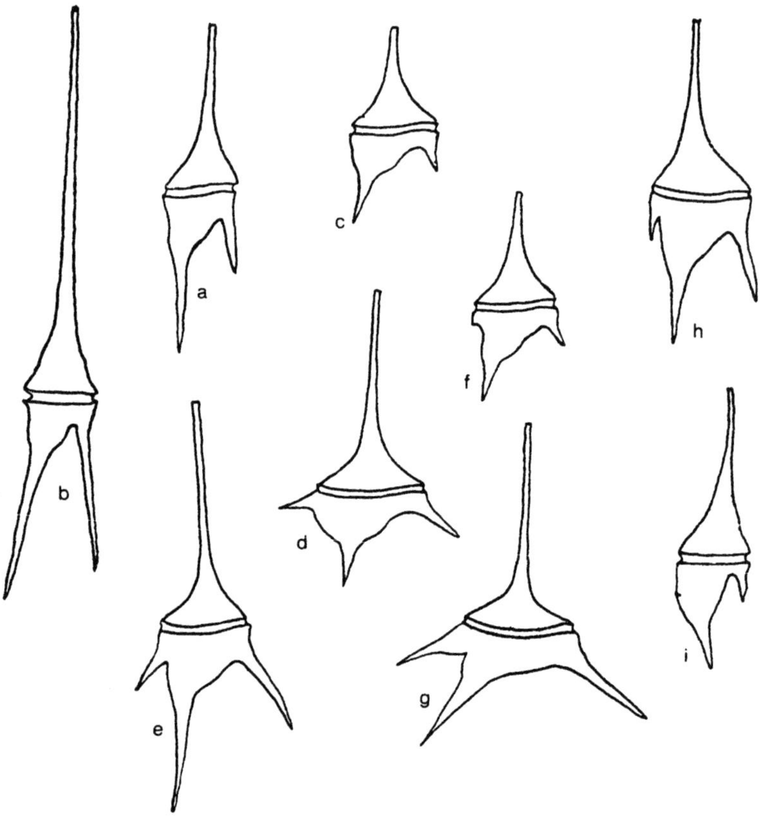

Fig. 227. *Ceratium hirundinella* (O. F. M.) Dujardin, *a* type furcoides, *b* yuennanense, *c* carinthiacum, *d* scotticum, *e* robustum, *f* austriacum, *f* piburgense, *h* gracile, *i* brachyceroides (after Schröder, Huber-Pestalozzi).

C. Cells short and long with 4 horns under optimal conditions. Cells more than 45 µm wide. The helmet-shaped epivalve is shorter than in the other types described in parts A and B.

a. Antapical horns are not or only slightly divergent. Cells are long and not narrow

1. The outer antapical horns are parallel with the apical horn. The right antapical horn is rarely slightly divergent. The dorsal side of the epivalve is not convex; plates and reticulations strongly delineated
. . . . type *gracile* (fig. 227 g)
(*C. hirundinella* fo. *gracile* Bachmann 1911)

2. The outer antapical horns are divergent; the dorsal side of the epivalve convex. Plates are strong or robust; ornamentation distinct and often irregular
 . . . . type *robustum* (fig. 227h)
 (*C. hirundinella* fo. *robustum* (Amberg) Bachmann 1911)

b. Outer antapical horns are moderately divergent. Cells are usually shorter and narrower than the other types. The antapical horn is shifted toward the outside
 . . . . type *austriacum* (fig. 227f)
 (*C. hirundinella* fo. *austriacum* (Zederbauer) Bachmann 1911)

c. Outer antapical horns are strongly divergent
 1. Antapical horn is short and parallel with the apical horn. The hypovalve is very short . . . . type *scotticum* (fig. 227i)
 (*C. hirundinella* fo. *scotticum* (Zederbauer) Bachmann 1911)
 2. The antapical horn is longer and shifted toward the outside. The hypovalve is somewhat longer . . . . . . type *piburgense* (fig. 227k)
 (*C. hirundinella* fo. *piburgense* (Zederbauer) Bachmann 1911)

Occurrence: common in ponds and lakes in the temperate zone that are oligo- to eutrophic.

Appendix

1. *Ceratium brachyceros* Daday 1907 (fig. 228)
 C. hirundinella var. *brachyceros* (Daday) Ostenfeld 1909

Cells robust, distorted, slightly dorsiventrally flattened, and kidney-shaped in cross section. Conical epitheca extending into the apical horn with its concave right margin. The hypotheca is rounded, the antapical, short, apiculate horn is placed on its right side. On the right is placed a right posterior horn, which is somewhat conical and apiculate. Both hypothecal horns are approximately parallel or curved inward slightly. The left posterior horn is lacking. Plates are robust and reticulae with fine spicules present where the net of areole cross.
Dimensions: 90–195 µm × 30–70 µm.
Occurrence: euplankter, tropical zone; Africa, Asia.

7. Family **Gonyaulacaceae** Lindemann 1928

Cells are oval, or spherical and apiculate. The hypothecal plate pattern is the main diagnostic character. Plate formula: (2–4)′, (0a–3a), (5–6)″, (5–6)‴, 1p, (1–2)⁗.
Important reference: Lindemann 1928.

1. **Gonyaulax** Diesing 1866

Peridinium Claparède and Lachmann 1858/61, *Blepharocysta* Ehrenberg 1873, *Roulea* Gourret 1883, *Protoperidinium* Pouchet 1883, *Steiniella* Schütt 1895, *Goniodoma* Murray and Whitting 1899, *Ceratocorys* Murray and Whitting 1899, *Heterodinium* Kofoid 1906, *Amylax* Meunier 1910, 1919

The cells are ellipsoid or oval in shape; the epi- and hypothecae are equal in size and flattened dorsiventrally. The apical pore is extended forming a cone. The cingulum is

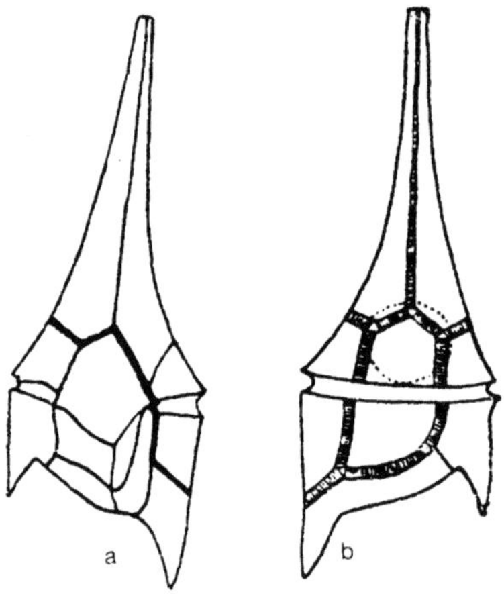

Fig. 228. *Ceratium brachyceros* Daday (after Entz).

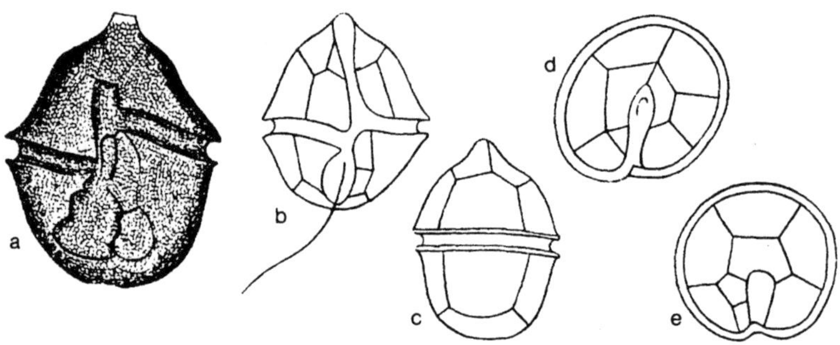

Fig. 229. *Gonyaulax apiculata* (Pénard) Entz (*a* after Wołoszyńska, *b–e* after Entz).

equatorial, deep, and spirals to the left. Cingular and sulcal margins are well developed. The sulcus extends from the apex to the antapex narrowing from the poles. The theca is composed of 14–18 reticulate plates with 6 cingular and 6 sulcal plates. The hypotheca is composed of one antapical and one accessory plate. Chloroplasts may be present or absent. When present, they are numerous, small, disc-shaped and parietal. No stigma is present. Plate formula: 2′, 0a, 5″; 5‴, 1p, 1⁗ or 4′, 0a, 6″; 6‴, 1p, 1⁗.

Important references: Entz 1904, Wołoszyńska 1916, Lindemann 1918, Schiller 1937, Starmach 1974.

1. Gonyaulax apiculata (Penard) Entz 1904 (figs. 26, 229)
Peridinium apiculatum Penard 1891, *Gonyaulax clevei* Ostenfeld 1901, *G. apiculata* var. *clevei* Ostenfeld 1901, *G. polonica* Wołoszyńska 1916, *G. limnetica* Lindemann 1918, *G. austriaca* Schiller 1926, *Kolkwitziella salebrosa* Lindemann 1918, *K. salebrosa* var. *gibbera* Lindemann 1924, *K. gibbera* (Lindemann) Lindemann 1928

Spines are found along the margins of both furrows. A wing or flange is also present on the sulcus in the hypotheca. Small teeth-like projections often occur on the hypotheca.
Dimensions: 30–62 µm × 29–57.5 µm.
Occurrence: euplanktic in lakes; Europe.

Appendix

1. *Thompsodinium* Bourrelly 1970

Cells are spherical or oval, somewhat apiculate and slightly flattened dorsiventrally. The cingulum circumscribes the cell and spirals slightly to the left. The sulcus is wide; it does not reach the antapex. The plate epithecal pattern is symmetrical, plates which are ornamented with rows of fine papillae are quadrangular or pentangular in shape. Irregularly ovate plates are present on the hypotheca. Chloroplasts and stigma ore present.
Plate formula: 3′, 4a, 6″, 5‴, 1p, 2⁗ or 4′, 3a, 6″, 5‴, 2p, 1⁗.
Important references: Thompson 1950, Bourrelly 1970.

1. *Thompsodinium intermedium* (Thompson) Bourrelly 1970 (figs. 27, 230)
Peridinium intermedium Thompson 1950

Dimensions: 32–39 µm × 33–40 µm.
Occurrence: USA, Cuba.

8. Family **Dinosphaeraceae** Lindemann 1928

An apical pore and chloroplasts may be present.
Plate formula: (3–4)′, (1a–2a), (5–7)″, 5‴, 1⁗.
Important reference: Lindemann 1928.

Key to the genera

1a. Apical pore present . . . . **1. Diplopsalis** (p. 213)
1b. Apical pore not present . . . .**2. Dinosphaera** (p. 215)

1. **Diplopsalis** Bergh 1881

non: *Diplopsalis* Sclater 1866

Cells are symmetrical, spherical, oval, or doubly conical. The cingulum is wide; the sulcus extends down the hypotheca and has wing-like flanges mainly on its left margin.
Plate formula: (3–4)′, (0a–2a), (5–7)″, 5‴, 1⁗.
Important references: Entz 1904, Bourrelly 1970, Starmach 1974.

1. Diplopsalis acuta (Apstein) Entz 1904 (figs. 23, 231)
Glenodinium acutum Apstein 1896, *Peridinium latum* Paulsen 1908, *Entzia acuta* (Apstein) Lebour 1922, 1925

Cell shape may resemble that of two cones with their flat surfaces opposing each other (double conical) or the epitheca may be cone-shaped; the hypotheca may be broadly

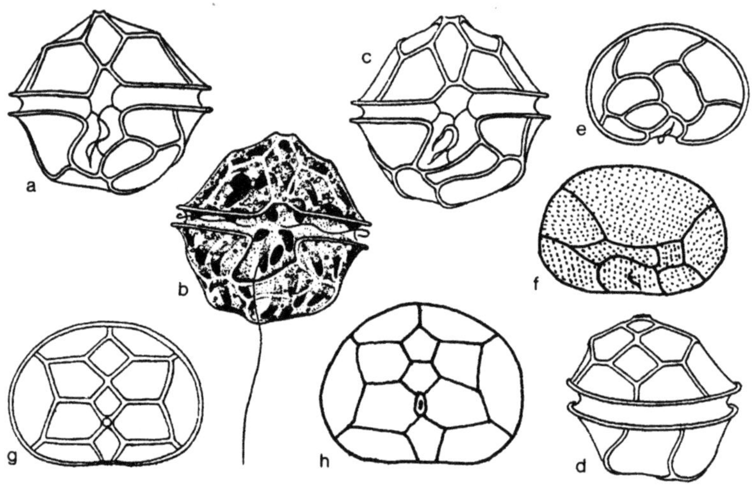

Fig. 230. *Thompsodinium intermedium* (Thompson) Bourrelly (*a–e, g* after Thompson, *f, h* after Popovský).

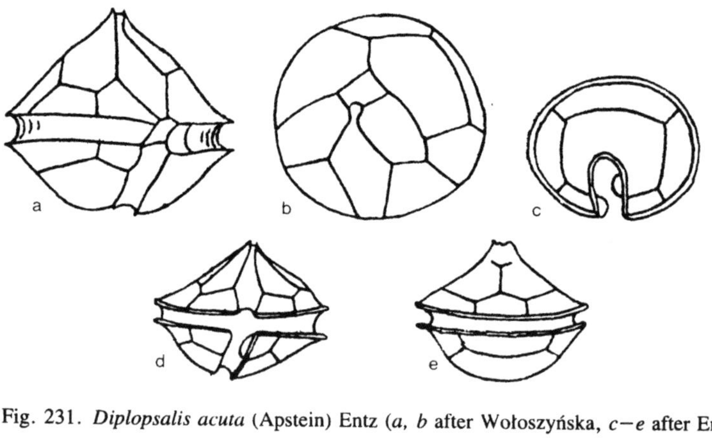

Fig. 231. *Diplopsalis acuta* (Apstein) Entz (*a, b* after Wołoszyńska, *c–e* after Entz).

rounded, or slightly flattened dorsiventrally. The cingulum is equatorial; the sulcus is flanged by a wing-like projection on its left margin. It reaches to the antapex. Plates are not sculptured nor do they have any reticulations. Chloroplasts and stigma are lacking, though the protoplasm appears brown. Plate formula: 4′, (1a–2a), 7″, 5‴, 1⁗.

Dimensions: 30–38 µm × 26–27 µm; 40–48 µm × 43–57 µm. The relationship between lenght and breadth varies widely.
Occurrence: lakes, pools, and brackish water; Europe.

Appendix

1. *Diplopsalis ambiguum* (Thompson) Bourrelly 1970 (fig. 232)
   Glenodinium ambiguum Thompson 1950

Cells ovate to spherical and slightly flattened dorsiventrally. The epitheca is large and rounded; the hypotheca is large, conical, and obliquely cut. The wide cingulum spirals to the left. The sulcus is wide and extends only a short distance on the hypotheca. The theca is smooth, punctuate, or irregularly striated. The plates are irregularly arranged on the epitheca. The numerous, radially arranged chloroplasts are shaped like elongated ovals. A stigma and several red droplets are present. Plate formula: 4′, 0a, 5″, 5‴, 1⁗.
Dimensions: 32–40 µm× 32–40 µm.
Occurrence: creeks, pool, lakes; USA.
Note: Carty and Cox (1986) proposed the new generic name *Kansodinium* for *Diplopsalis ambiguum* (Thompson) Bourrelly 1968.

2. *Diplopsalis ovum* (Matvienko) Bourrelly 1970 (fig. 233)
   Peridinium ovum Matvienko 1938

Cells oval and not flattened dorsiventrally. The helmet-shaped epitheca is slightly larger than the rounded hypotheca. The wide cingulum spirals to the left. The sulcus widens along the hypotheca where it reaches the antapex. Chloroplasts are numerous, discoid and sometimes radially arranged. A stigma is present as are several refractile bodies. Reproduction is by means of two daughter cells following ecdysis.
Plate formula: 4′, 0a, 5″, 5‴, 1⁗.
Dimensions: 16.5–22.5 µm × 12–19.5 µm.
Occurrence: peat bogs; USSR.

2. **Dinosphaera** Kofoid and Michener 1912

Cells are spherical; the epi- and hypotheca are equal in size. The theca is composed of 16 plates with 6 cingular and 4–5 sulcal plates. No apical pore is present. The epithecal plates spiral to the left in apical view along the longitudinal plane. There is only one antapical plate. The cingulum is equatorial and slightly spiralled. The sulcus extends into the hypotheca for only a short distance. Chloroplasts are small, numerous, and parietal. Several refractile droplets are present.
Plate formula: 3′, 1a, 6″, 5‴, 1⁗.
Important references: Kofoid and Michener 1912, Bourrelly 1970.

1. **Dinosphaera palustris** (Lemmermann) Kofoid and Michener 1912 (figs. 24, 234)
   Gonyaulax palustris Lemmermann 1907, *Glenodinium palustre* (Lemmermann) Schiller 1937

Dimensions: 25 µm – 34 µm in diameter.
Occurrence: planktonic in lakes and acid bogs; Germany, USA, France.

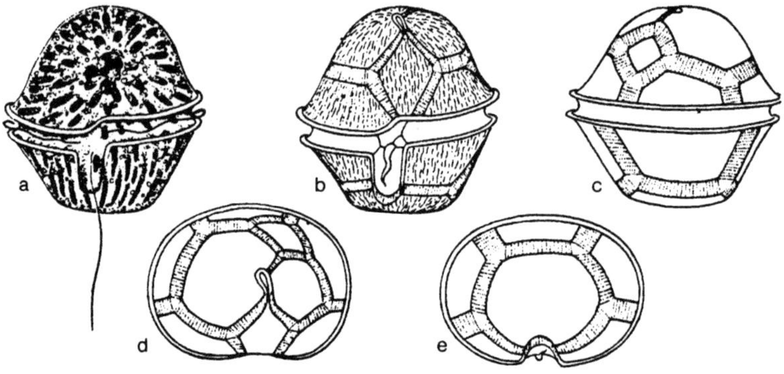

Fig. 232. *Diplopsalis ambiguum* (Thompson) Bourrelly (after Thompson).

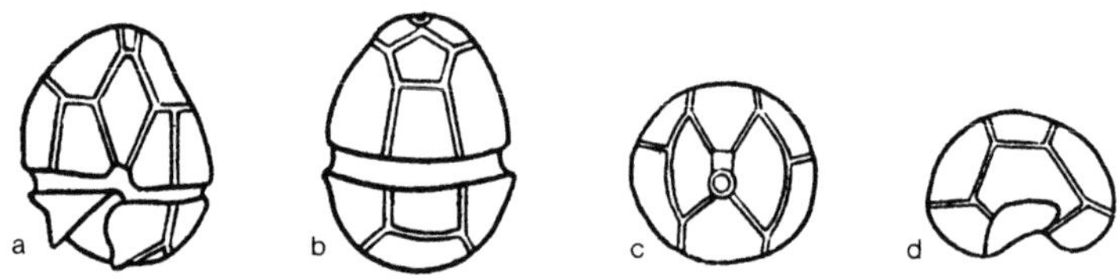

Fig. 233. *Diplopsalis ovum* (Matvienko) Bourrelly (after Matvienko).

Appendix

1. *Peridinium baicalense* Kiselev and Cvetkov 1935 (fig. 235)
Cells are pear-shaped and extremely flattened dorsiventrally. The helme-shaped epitheca is twice as long as the hypotheca, and extends into a flat horn with an apical pore. A short, pyramidal tooth-like projection occurs on the 4a plate. The hypotheca is trapezoidal in shape with four to six pyramidal teeth-like projections on the bottom. The cingulum circumscribes the cell; the sulcus widens greatly along the hypotheca and reaches the antapex. Chloroplasts are probably present.
Plate pattern: 4′, 4a, 7″, 5‴, 2⁗.
Dimensions: 55–72 μm × 42–50 μm.
Occurrence: endemic stenotherm, deep water (25 to 200 m) of Baikal Lake and its drainage area; USSR.
Important references: Kiselev and Cvetkov 1934, Kiselev 1954.

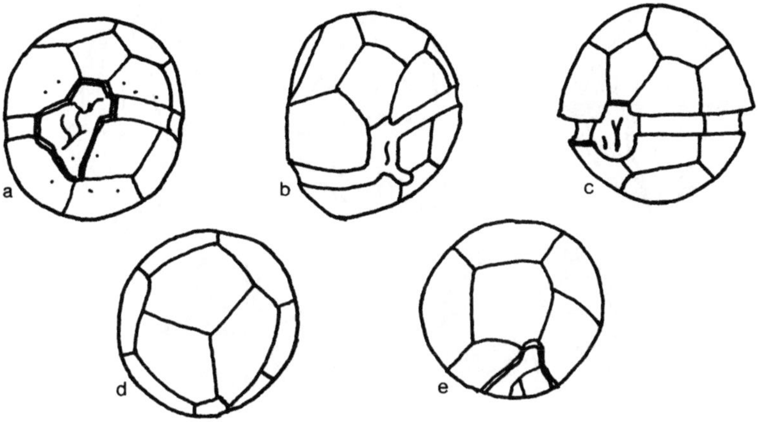

Fig. 234. *Dinosphaera palustris* (Lemmermann) Kofoid and Swezy (*a, d, e* after Kofoid and Michener, *b* after Prescott, *c* after Bourrelly).

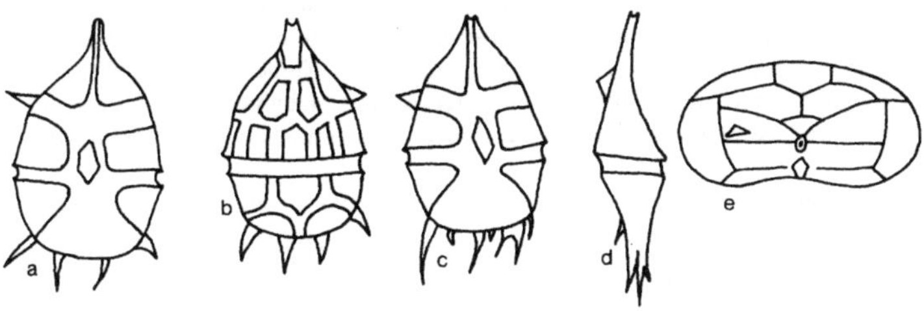

Fig. 235. *Peridinium baicalense* Kiselev and Cvetkov (*a–d* after Kiselev and Cvetkov, *e* constructed projection of epitheca, orig.).

1. **Lophodiniaceae** Lemmermann 1910

1. *Lophodinium* Lemmermann 1910

Glenodinium partim

Cells are biconical in ventral view and slightly flattened dorsiventrally. The apex is bifurcate; the antapex shortly apiculate. The cingulum spirals to the left and is displaced one cingulum width. Its margins are flanged. The sulcus extends slightly into the epitheca and widens along the hypotheca where it reaches the antapex. An apical ridge and slit bisect the epitheca from midway between the cingulum and apex on the ventral and dorsal sides. Parallel, double non-braching longitudinal ridges extend from margins

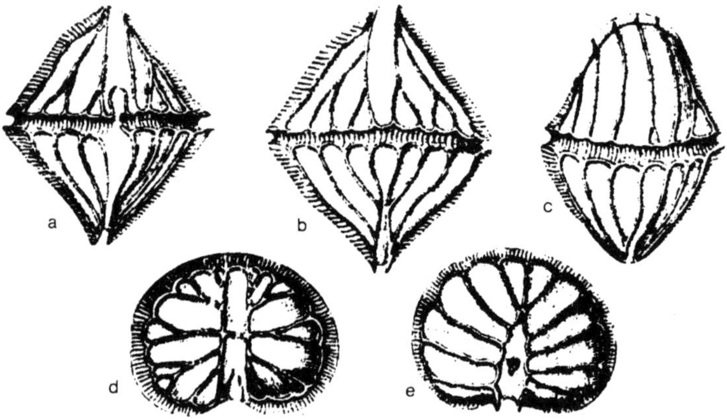

Fig. 236. *Lophodinium polylophum* (Daday) Lemmermann (after Daday).

of the apical ridge to the cingulum. The double longitudinal ridges of the hypotheca occasionally diverge forming smaller divisions. The ends of the double ridge diverge at the cingulum in both the epi- and hypothecae, giving a scalloped appearance to the theca. The theca is composed of hexagonal plates which are continuous over the cell. Hyaline wings are also present on the sides of the theca. Numerous oval chloroplasts and a stigma are present. Reproduction: after ecdysis daughter cells (autospores? zoospores?) produce new theca. Cell enlargement is accomodate by flattening of the ridges.

Important references: Lemmermann 1910b, Schiller 1937, Carty and Cox 1985.

1. *Lophodinium polylophum* (Daday) Lemmermann 1910 (fig. 236)

Glenodinium polylophum Daday 1888, *Lophodinium dadayi* Osorio-Tafall 1942.
Dimensions: 42–85 μm × 31–67 μm.
Occurrence: standing waters in Paraguay and Texas (USA), and a river in Mexico.
Note: the original description and drawings of structures were consequently inverted. It may be related to the genera *Woloszynskia* and *Glenodiniopsis*.

2. Order **Dinococcales** Pascher 1927, 1931

1. Family **Dinococcaceae** Fott 1960

Phytodiniaceae Klebs 1912 sensu Bourrelly 1970

Coccoid-like unicells which occur in the plankton or are attached to a substrate; cell walls flexible. Cells are motile except when in a resting stage. Reproduction can take place by means of amoebae, sporocysts, autospores, or zoospores and gametes. The latter have a cingulum and sulcus and two different types of flagella. The dinokaryon is not visible in the amoeboid stage. Colorless specimens are known. Cells may be phagocytic by haustoria of varying lengths. The life cycle is presumed to be complica-

ted. Stages with axopods and filopods have been described. The presence of armored reproductive stages has been incompletely described and should not be included in the family.
Nutrition: holophytic, autotrophic.
Important references: Klebs 1912, Pascher 1927, 1931, Fott 1960, Bourrelly 1970.

Key to the genera
1a. Cells attached **2**
1b. Cells planktonic **5**
2a. Without a visible stalk . . . . **5. Cystodinedria** (p. 232)
2b. With a visible stalk **3**
3a. Cells with spines or horns **4**
3b. Cells without spines or horns . . . . **4. Stylodinium** (p. 230)
4a. With one or two horns . . . . **1. Dinococcus** (p. 219)
4b. With more than two spines or horns . . . . **2. Tetradinium** (p. 222)
5a. Cells rounded or ellipsoidal . . . . **3. Cystodinium** (p. 222)
5b. Cells are triangular or tetrahedral . . . . **2. Tetradinium** (p. 222)
Note: The following genera; *Hypnodinium, Phytodinium,* and *Dinastridium* are probably zygotes of other species. The genus *Gloeodinium* was included in the genera *Hemidinium* and *Glenodiniopsis.*

1. Dinococcus (Wołoszyńska) Fott 1960

Raciborskia Wołoszyńska 1919, *Raciborskia* (Wołoszyńska) Pascher em. Geitler 1943

Cells are attached and inversely triangular-elipsoidal in shape. In side view, they are bean- or crescent-shaped and ellipsoid or oval when viewed from above. Each pole ends in a single stout spine. Cells are attached to the substrate (filaments of algae and mosses) by a stout stipe. The upper margin appears straight or concave in side view, the 2 lateral margins are straight or slightly convex converging to the thick stalk. Chloroplasts are radially or parietally arranged and oval or club-shaped. A pyrenoid and stigma may be present. Cells form 2–4 zoospores or autospores.
Important references: Wołoszyńska 1919, Geitler 1943, Fott 1960, Pfiester and Lynch 1980.

Key to the species
1a. Pyrenoid present **2**
1b. Pyrenoid absent . . . . **2. D. incus**
2a. Chloroplasts radially arranged . . . . **3. D. oedogonii**
2b. Chloroplasts not radially arranged . . . . **1. D. bicornis**
Note: There is probably only one species of *Dinococcus* which exhibits a great deal of morphological, ecological, and nutritional variability.

1. **Dinococcus bicornis** (Wołoszyńska) Fott 1960 (fig. 237)
Raciborskia bicornis Wołoszyńska 1919, *Cystodinium bicorne* (Wołoszyńska) Huber-Pestalozzi 1950

Cells appear oval in vertical view and crescent-shaped in end view. The poles end in a single, slightly bent spine. Cells are circular in cross section. They are attached to the

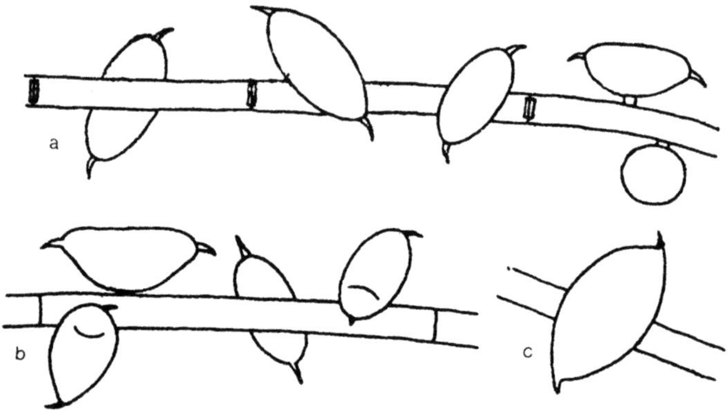

Fig. 237. *Dinococcus bicornis* (Wołoszyńska) Fott (after Wołoszyńska).

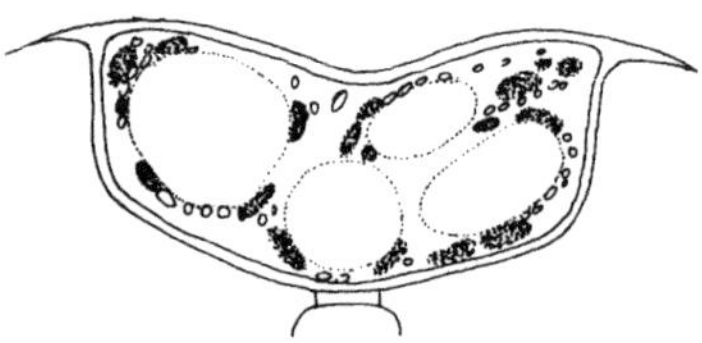

Fig. 238. *Dinococcus incus* (Pascher) Fott (after Pascher).

substrate by means of a very short stipe. A pyrenoid may be present. Nutrition is unknown. Reproduction is by means of two gymnodinium-like zoospores which lack a stigma.
Dimension: 25–35 μm × 9–12 μm.
Occurrence: epiphytic (?) on filamentous algae or aquatic mosses; Europe.

2. **Dinococcus incus** (Pascher) Fott 1960 (fig. 238)
   Raciborskia incus Pascher 1932

The dorsal side of the cell is strongly concave; the ventral side parallels the dorsal. Each pole ends in a single stout spine. Cells are attached to the substrate by a short, thick stipe. Numerous small, discoidal chloroplasts and one to several large vacuoles are present. No pyrenoids are present. Reproduction is by zoospore formation. The zoospore has a wide sulcus and a stigma.
Dimensions: 22–28 μm × 10–14 μm.
Occurrence: epiphytic (?); Czechoslovakia.

3. **Dinococcus oedogonii** (P. Richter) Fott 1960 (figs. 53, 239)
   Rhizophydium oedogonii P. Richter 1897, *Raciborskia oedogonii* (P. Richter) Pascher em. Geitler 1943, *Cystodinium brevipes* Geitler 1928 partim

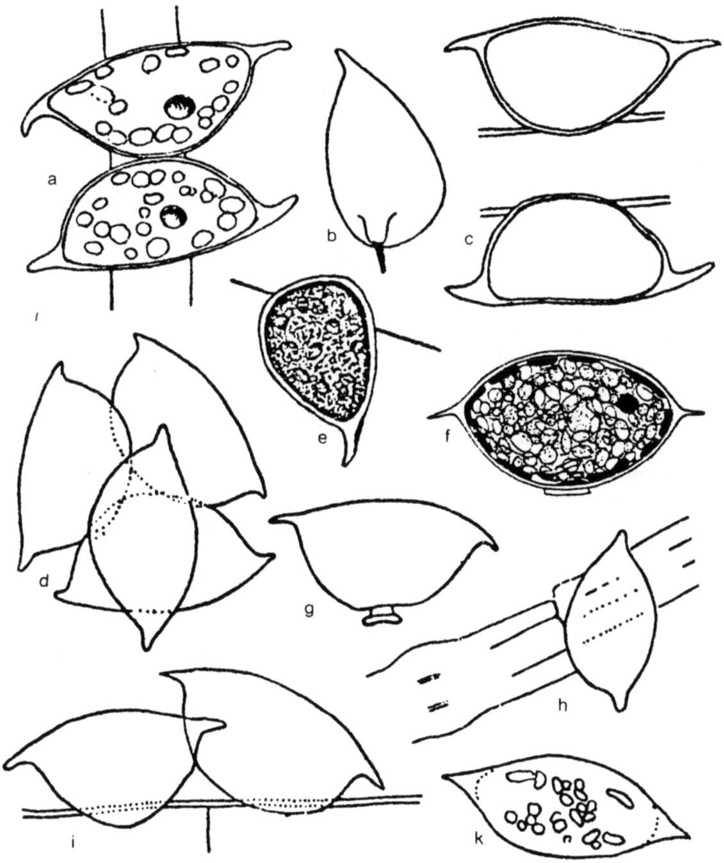

Fig. 239. *Dinococcus oedogonii* (P. Richter) Fott (after Baumeister).

Cells are bean-shaped with rounded, apiculate poles ending in a large spine. They are broadly rounded in side view and relatively straight in dorsal view. Cells are often convex, rarely slightly concave. The stalk is thick and maximally 7 µm high. Chloroplasts are club-shaped and radially arranged with a central pyrenoid and numerous starch grains. A stigma is usually present in vegetative cells. Reproduction is by means of two zoospores or autospores with distinct furrows. Nutrition partially unknown.

Dimensions: 20–35 µm × 13–20 µm.

Occurrence: epiphytic (?) on filamentous algae; Europe.

Note: *D. oedogonii* is related to *Cystodinium bataviense*.

2. **Tetradinium** Klebs 1912

Cells are planktonic or attached by a short stalk, triangular or tetrahedral in shape; the angles are tipped with 1 or 2 short, stout spines. Chloroplasts are numerous and resemble small, ovate discs. Reproduction is by the formation of two swarmers per cell.
Important reference: Klebs 1912.

Key to the species

1a. Cell sides are concave . . . . **1. T. intermedium**
1b. Cell sides are convex . . . . **2. T. javanicum**

1. Tetradinium intermedium Geitler 1928 (fig. 240)
Cells are tetrahedral in shape with slightly concave sides, and two stout spines which are sometimes curved at the angeles. One side of the cell is slightly convex and bears a short stipe by means of which the cell is attached to the algal filament. Chloroplasts are numerous and resemble small ovate discs. One or several vacuoles are often present.
Nutrition: autotrophic (holozoic?, holophytic?).
Dimensions including spines: 33–48 μm in diameter.
Occurrence: oligotrophic lakes; Austria, France, Czechoslovakia.

2. Tetradinium javanicum Klebs 1912 (fig. 241)
T. minus Pascher 1927
Cells are tetrahedral (similar in shape to *Tetraedron*) with slightly convex sides; the angles of which end in 1 or 2 spines. Cells may be attached to filamentous algae by a short stipe. One or several vacuoles are present. Reproduction is by the formation of two gymnodinioid zoospores, which resemble an elongated oval. The epicone is helmet-shaped; the hypocone large and rounded. The cingulum is deeply incised: the sulcus extends slightly into the epicone and the hypocone does not reach the antapex. Numerous parietal, oval chloroplasts and a stigma are present.
Dimensions: vegetative cells 30–60 μm in diameter, zoospores (approx.) 30–40 μm × 18–22 μm.
Occurrence: pools, lakes, peat bogs; Java, Poland, USA.

Appendix

1. *Tetradinium simplex* Prescott 1949 (fig. 242)
Cells are triangular in lateral view with a convex dorsal side. Both horns are curved downward. Two chloroplasts are present.
Dimensions: 12–15 μm × 12–20 μm.
Occurrence: attached to filamentous algae by means of a short stipe; USA.

3. **Cystodinium** Klebs 1912

Glenodinium Schilling 1891, *Gymnodinium* Baumeister 1938 partim, *Gymnocystodinium* Baumeister 1957, *Dinorbiculus* Baumeister 1968

Vegetative cells are free floating, non-motile and elongate, elliptical or lunate in shape. The poles may be rounded or apiculate with large spines. The cell wall is thick; no furrows are visible in the protoplast. There is no stigma and chloroplasts may be absent. Trichocysts have been reported. Reproduction is by the formation of 2–4 zoospores, 2 autospores or 8 to 16 amoebae within the parent cell wall. The amoebae have many axopods and are brown to almost colorless with 2 to 3 discoid chloroplasts. The

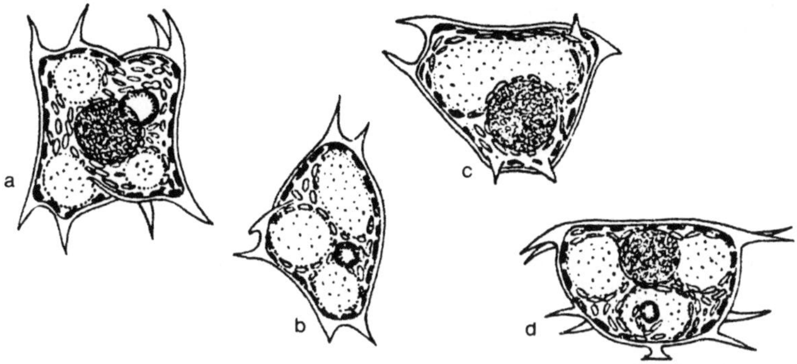

Fig. 240. *Tetradinium intermedium* Geitler (after Geitler).

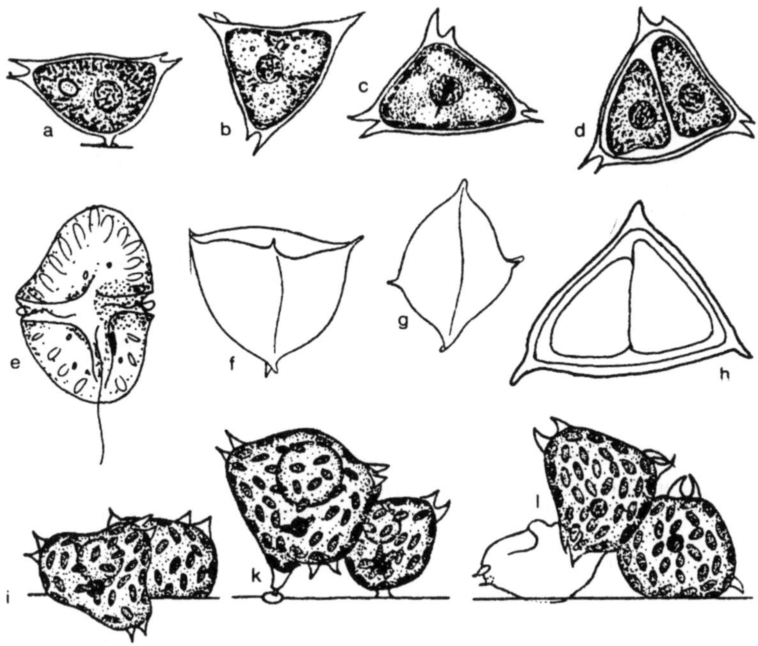

Fig. 241. *Tetradinium javanicum* Klebs (*a–d* after Klebs, *e–h* after Pascher, *i–l* after Thompson).

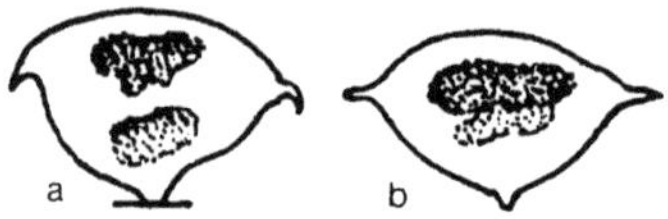

Fig. 242. *Tetradinium simplex* Prescott (after Prescott).

production of 8 gametes has been reported as has the fusion of isogametes. The zoospores and gametes are *Gymnodinium*-like, with or without a distinct stigma. Their chloroplasts are numerous. Zoospores may be motile from several to 30 minutes. A thick-walled resting stage has been observed.

Important references: Klebs 1912, Baumeister 1938, 1957, 1968, Pfiester and Lynch 1980, Popovský and Pfiester 1982, Timpano and Pfiester 1984, 1985b.

Key to the species

1a. Cells with two (sometimes 1) sharply pointed horn(s) . . . . **2. C. cornifax**
1b. Cells without sharply pointed horns **2**
2a. Cells lunate or naviculate with blunt horns . . . . **1. C. bataviense**
2b. Cells somewhat rounded, elliptical or asymmetrical from lateral view **3. C. phaseolus**

1. **Cystodinium bataviense** Klebs 1912 (fig. 53, 243)

C. lunare Pascher 1927

Cells may be lunate or shaped like an elongated oval. Poles are bluntly prolonged and without spines. Numerous small chloroplasts are rod-shaped and parietal. Trichocysts are present in zoospores and vegetative cells. Accumulation bodies have also been observed. Reproduction may be by the formation of: (a) 2–3 autospores; (b) 2–3 zoospores with stigma; (c) of 8–16 amoebae or; (d) zygotes resulting from the fusion of isogametes. The amoebae have numerous axopods and are brown to colorless with 2–3 discoid chloroplasts. Zygote development and germination are presently unknown.

Dimensions: vegetative cells 60–200 μm × 30–60 μm, zoospores 44 bis 48 μm × 38–42 μm, amoebae 10–15 μm in diameter.

Occurrence: peat bogs, clay pools; Central Europe, Java, USA.

Note: *C. bataviense* is related to *Dinococcus oedogonii*.

2. **Cystodinium cornifax** (Schilling) Klebs 1912 (figs. 244, 245)

Cysts in Stein 1883 (of *Peridinium cinctum* and of *P. tabulatum*), *Glenodinium cornifax* Schilling 1891, *Cystodinium unicorne* Klebs 1912, *C. steinii* Klebs 1912, *C. steinii* var. *tenuirostris* Wołoszýnska 1919, *C. steinii* var. *dimidio-minor* Baumeister 1957, *C. iners* Geitler 1928, *Gymnocystodinium iners* (Geitler) Baumeister 1957, *Cystodinium schilleri* Baumeister 1957

Cells are lunate, elongate, or shortly ovate; free floating or attached. One pole, sometimes both, are apiculate with large spines. A cluster of very short bristles (haustoria?) are usually found on one of the spines. Cells often have several large vacuoles, refractile bodies and numerous disc or rod-shaped chloroplasts. Reproduction is by the formation of 2 or 4 zoospores with stigma which are very similar to *Gymnodinium fuscum*.

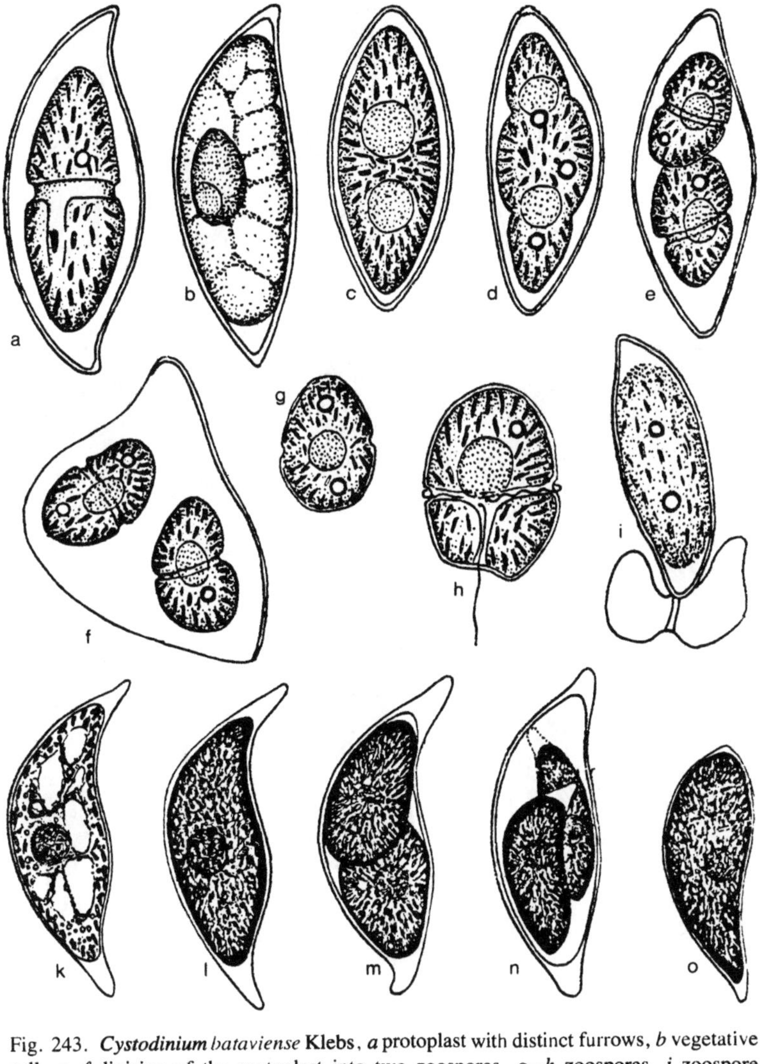

Fig. 243. ***Cystodinium** bataviense* **Klebs**, *a* protoplast with distinct furrows, *b* vegetative cell, *c–f* division of the protoplast into two zoospores, *g–h* zoospores, *i* zoospore developing into a vegetative cell, *k, l, o* vegetative cells, *m, n* division of the protoplast into two autospores (*a–i* after Klebs, *k–o* after Pascher).

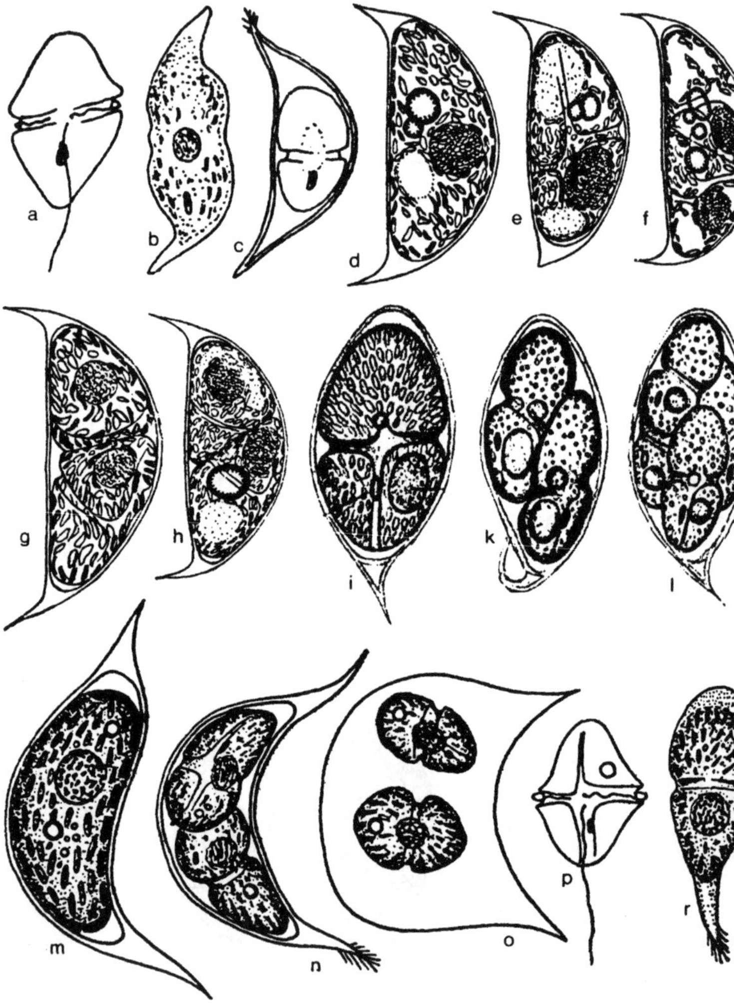

Fig. 244. *Cystodinium cornifax* (Schilling) Klebs, *a* zoospore, *b* zoospore developing into a vegetative cell, *c* zoospore within the parent cell wall, *d* vegetative cell, *e–h* protoplast dividing into two zoospores, *i–l* protoplast dividing into four zoospores, *m* vegetative cell, *n, o* protoplast dividing into two zoospores, *p* zoospore, *r* zoospore developing into a vegetative cell (*a–c* after Schilling, *d–h* after Geitler, *i–l* after Stein, *m–r* after Klebs).

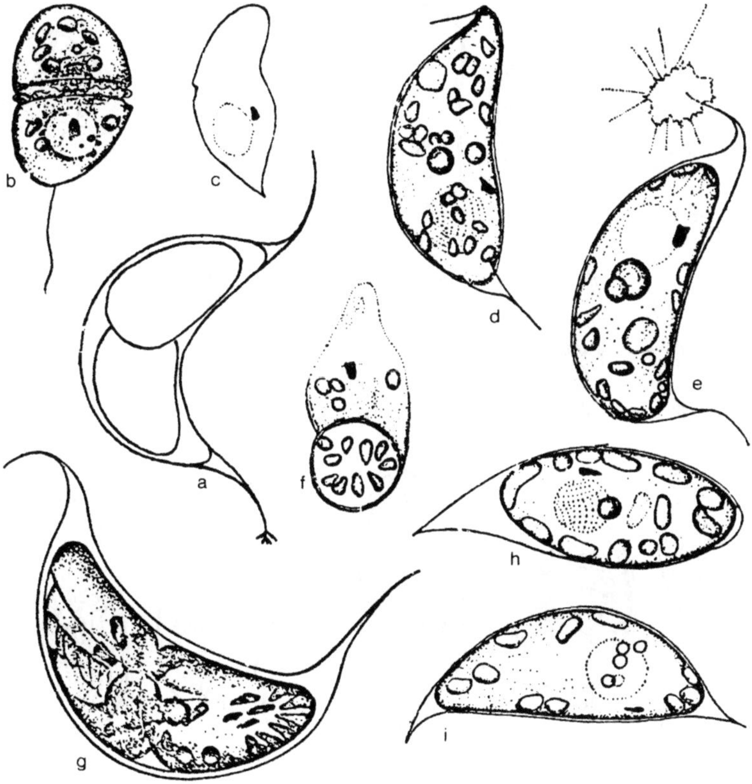

Fig. 245. *Cystodinium cornifax* (Schilling) Klebs, *a* protoplast dividing into two zoospores, *b* zoospore, *c* zoospore developing into a vegetative cell, *d, e, g, i* vegetative cells (*a* after Wołoszyńska, *b–i* after Baumeister).

Dimensions: vegetative cells 33–130 μm × 17–60 μm, zoospores 25 bis 60 μm × 20–48 μm.
Occurrence: clay bogs, ponds, pools, peat bogs; Central Europe.
Note: *Gymnodinium fuscum* is very probably related to *C. cornifax*.

3. **Cystodinium phaseolus** Pascher 1928 (fig. 246)

C. dominii Fott 1933, *C. hyalinum* Pascher 1944, *C. minimum* Pascher 1944, *C. conchaeforme* Baumeister 1962/63, *C. dimorphe* (Baumeister) Huber-Pestalozzi 1950, *Gymnodinium dimorphe* Baumeister 1938, *Gymnocystodinium dimorphe* (Baumeister) Baumeister 1957, *Dinorbiculus dimorphus* Baumeister) Baumeister 1968

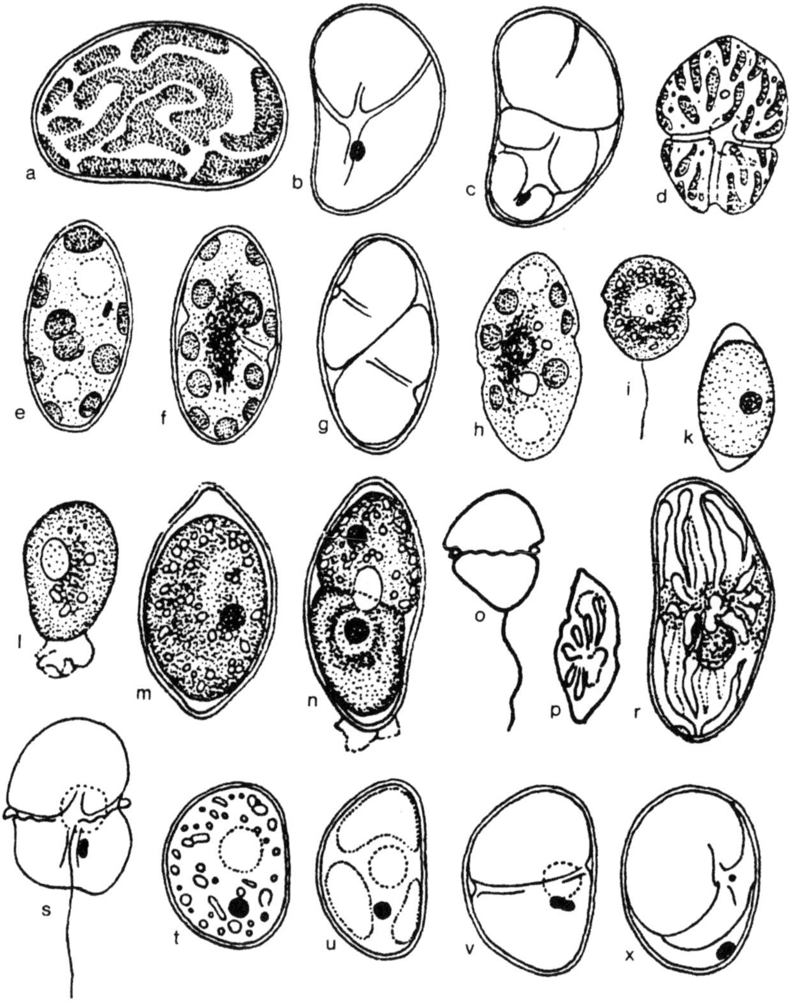

Fig. 246. *Cystodinium phaseolus* Pascher, *a, e, m, r, t, u* vegetative cells, *b, c, f, g, n, v, x* zoospore formation, *d, i, o, s* zoospores, *k, l, p* zoospores developing into vegetative cells (*a–d, s–x* after Pascher, *e–h* after Fott, *i–r* after Baumeister).

Cells are rounded, elliptical, or asymmetrical from lateral view. Chloroplasts are ribbon-shaped or discoidal, anastomosing, if they are present. Several large vacuoles and numerous refractile bodies are visible. Reproduction is by formation of *Gymnodinium*-like zoospores, autospores and gametes (?). A stigma may be present. Sexual fusion has not been observed. One stage of the life cycle is known to be sessile. The resting stage is enveloped by a very thick cell wall.
Dimensions: vegetative cells 13–53 µm × 8–30 µm.
Occurrence: plankter in swampy lakes and pools; Ohrid Lake, Europe.
Note: *C. phaseolus* is related to *Cystodinedria inermis*.

Insufficiently and invalidly described genera and species:

Cystodinium partim

Cystodinium bisetosum (Lindemann) Huber-Pestalozzi 1950 *(Gymnodinium bisetosum* Lindemann 1928)
- *C. dirhamphos* Baumeister 1957
- *C. papilionaceum* Baumeister 1957
- *C. guttatum* Baumeister 1957
- *C. heterorhamphos* Baumeister 1957
- *C. novaculosum* (Baumeister) Huber-Pestalozzi 1950 *(Gymnodinium novaculosum* Baumeister 1939)
- *C. pallidulum* Baumeister 1957
- *C. grabensei* Baumeister 1957
- *C. closterium* Pascher 1927
- var. *crassa* Bourrelly 1961
- *C. interspersum* Baumeister 1957
- *C. achroum* Baumeister 1957
- *C. reniforme* Baumeister 1957
- *C. familiaris* Baumeister 1967
- *C. uliginosum* Baumeister 1967
- *C. frustumiforme* Baumeister 1967
- *C. woloszynskae* Baumeister 1967
- *C. helveticum* Baumeister 1967
- *C. circulum* Baumeister 1967
- *C. tholicum* Baumeister 1967
- *C. fallerum* Baumeister 1967

Renodinium Baumeister 1968
- *R. closterium* (Pascher) Baumeister 1968
- *R. phaseolus* Baumeister 1968
- *R. reniforme* Baumeister 1968
- *R. familiaris* Baumeister 1968
- *R. frustumiforme* Baumeister 1968

Ovodinium Baumeister 1968
- *O. achroum* Baumeister 1968
- *O. minimum* Baumeister 1968
- *O. hyalinum* Baumeister 1968

Dinorbiculus Baumeister 1968

D. dimorphus Baumeister 1968

Important references: Huber-Pestalozzi 1950, Baumeister 1957b, 1967a, 1968.

4. **Stylodinium** Klebs 1912

inclus. *Dinopodiella* Pascher 1944

The *Stylodinium*-phase cell is spherical to kidney-shaped and attached to the substrate by a short or long stipe (haustorium). Stipe shape varies greatly during the life cycle. The cellulose wall is thick, flexible, and structureless. The chloroplasts are ribbon-shaped and parietal. A pyrenoid and stigma may be present. Red bodies, storage products and one food vacuole are present.

The stipe (haustorium) may be thick or thin with a visible channel in its central region. It widens at the top and is disc-shaped at the bottom. Thick and short stipes may serve for locomotion. The stalk may also be very minute or lacking.

The *Stylodinium* phase divides producing three known motile stages. Division may result in: (a) two amoeboid *(Vampyrella*-like) cell which settle on a substrate and develop into the *Stylodinium*-phase cell. The amoeboid cell is orange with numerous filopodia and two flagellar like filopodia; (b) two helizoid *(Actinophrys*-like) cells which also settle on the substrate and develop into the *Stylodinium* phase cell. The helizoid cell is spherical in shape with numerous axopodia and two flagellar-like axopodia; (c) four anisogametes may be formed, two of which are *Gymnodinium*-like and the other two *Gyrodinium*-like. These gametes have a dinokaryon, two chromatophores, a stigma, and one or two reddish spheres. The cingulum spirals to the left; the sulcus nearly reaches the antapex. The *Gymnodinium*-shaped gametes are larger than the *Gyrodinium*-like ones; (d) two-four *Gymnodinium*-like zoospores may be formed. They have a stigma, a narrow cingulum and an indistinct sulcus. The cells are spherical or ellipsoidal and may be slightly flattened. The zoospore has been reported to be capable of changing into the amoeboid stage. In the first two motile stages (amoeboid and heliozoid) the typical dinokaryon and chloroplasts are not visible at the light microscopic level but several contractile vacuoles are present.

The resting stage is *Stylodinium*-like. It is immobile and at first its protoplasm can be seen to contain a large number of small refractile droplets, one large refractile droplet and the dinokaryon. Red refractile bodies are often present. The protoplast eventually balls up and becomes surrounded by a new thick wall.

Important references: Klebs 1912, Wołoszyńska 1916, Baumeister 1943, Pascher 1944, Pfiester and Popovský 1979, Popovský and Pfiester 1982, Cachon et al. 1965, Horiguchi and Chihara 1983.

1. **Stylodinium globosum** Klebs 1912 (fig. 247)

St. truncatum Klebs 1912, *St. cerasiforme* Pascher 1927, *St. sphaera* Pascher 1944, *St. longipes* Thompson 1949, *Dinopodiella phaseolus* Pascher 1944, *Dinopodiella gracilis* Pascher 1944, *St. phaseolus* (Pascher) Starmach 1974, *St. gracilis* (Pascher) Starmach 1974; non: *Stylodinium globosum* Lindemann 1927.

Nutrition: holophytic, autotrophic.

Fig. 247. *Stylodinium globosum* Klebs, 1, 8, 9, 10, 11, 14, 15, 18 vegetative cells, 2, 12, 21 zoids release, 3, 4, 7 *"Vampyrella"* stage, 13, 17 *"Actinophrys"* stage, 5, 6 formation of four zoospores, 18, 20 formation of four anisogametes, 9, 11, 15 phagocytosis of a filamentous alga, 16 resting stage (after Popovský and Pfiester).

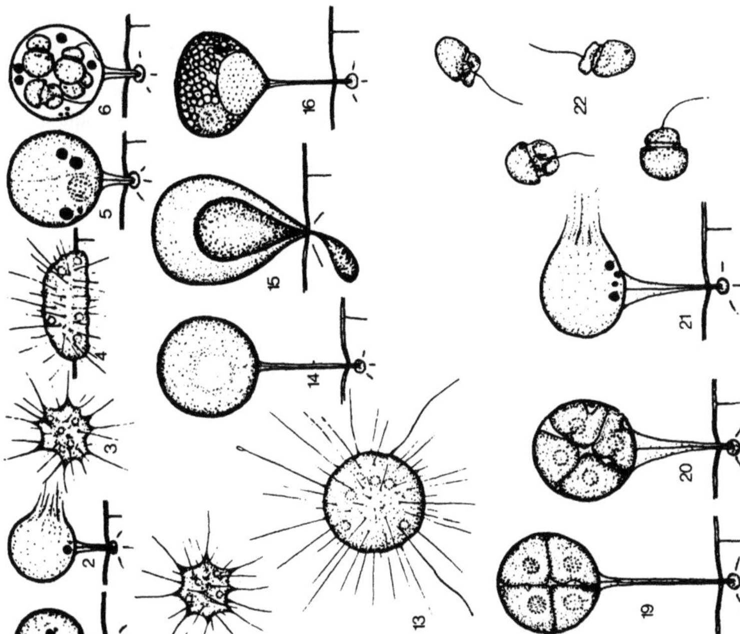

Fig. 247

Dimensions: *Stylodinium*-like 20–60 µm × 17–60 µm, *Vampyrella*-like 25 bis 30 µm in diameter, *Actinophrys*-like 50–60 µm in diameter, gametes 25–30 µm × 17–25 µm, zoospores 12–35 µm × 12–30 µm.
Occurrence: The *Stylodinium* phase is found attached mainly to green filamentous algae although there are reports of it attached to *Closterium* and *Utricularia*. The motile stages occur in and among filamentous algae.
Note: *Stylodinium tarnum* is *Peridiniopsis dinobryonis* (fig. 222). *St. polymorphum* Baumeister 1957 is probably a species of *Peridinium* or *Peridiniopsis*.
Invalidly described genera and species belonging to the genus *Stylodinium*:
Bacadinium Baumeister 1968

B. sphaerum (Pascher) Baumeister 1968. *(Stylodinium sphaera* Pascher 1944)

Coridinium Baumeister 1968

C. truncatum (Klebs) Baumeister 1968 *(Stylodinium truncatum* Klebs 1912)

Guttadinium Baumeister 1968

G. cerasiforme (Pascher) Baumeister 1968 *(Stylodinium cerasiforme* Pascher 1927)

G. lindemannii (Lindemann) Baumeister 1968 *(Stylodinium globosum* sensu Lindemann 1927, *St. lindemannii* (Lindemann) Baumeister 1943)

G. bavariense Baumeister 1968

G. longipes (Thompson) Baumeister 1968 *(Stylodinium longipes* Thompson 1949)

Glenobacadinium Baumeister 1968

G. tarnum (Baumeister) Baumeister 1968 *(Stylodinium tarnum* Baumeister 1943)

G. polymorphum (Baumeister) Baumeister 1968 *(Stylodinium polymorphum* Baumeister 1957)

Important reference: Baumeister 1968.

5. **Cystodinedria** Pascher 1944

Cystodinium partim, *Raciborskia* partim, *Dinococcus* partim; inclus. *Phytodinedria* Pascher 1944

Cystodinedria phase: Cells are oval or ellipsoidal in shape and attached to a substrate. The cell wall is flexible, composed of cellulose, smooth of with numerous spines of short finger-like projections. The chloroplasts are discoid, ribbon-like of may be absent. A stigma and pyrenoid may be present. A food vacuole is present.
Division: The *Cystodinedria* phase divides producing four motile and one non-motile stages. Division may result in (a) the production of two to four zoospores in which a stigma is present, and the cells are ellipsoidal with a rounded of helmet-shaped epicone and a conical hypocone. The epicone is usually larger than the hypocone. The cingulum is narrow; the sulcus nearly reaches to the antapex. Chloroplasts are present. (b) the production of four probable isogametes identical in structure though not in size to the zoospores. (c) two *Vampyrella*-like (non-heliozoid) amoebae. The amoeboid cell is orange with numerous filopodia and two flagellar-like filopodia. (d) two to four *Actinophrys*-like cells. These cells are spherical with numerous axopodia and two flagellar-like axopodia. (e) numerous, non-motile microspores are also produced at times. They are spherical or ellipsoidal. In the latter three stages, chloroplasts and the dinokaryon are not visible.
The *Cystodinedria*-like cell may directly develop into the following forms: (a) an *Amoeba verrucosa*-like stage in which the cell is naked and relatively non-transparent. It

is morula-shaped with lobopodia; (b) a *Vampyrella*-like (helizoid form). Cells are rounded with a well-developed, but flexible cell wall covered with numerous short, solid, spines. This wall may be cast off at anytime; chloroplasts and a dinokaryon are not visible in the *Amoeba verrucosa* or *Vampyrella*-like stages; (c) a free-living, unattached *Cystodinium*-phase.
Important references: Pascher 1944, Fott 1960, Popovský 1961, Bourrelly 1970, Pfiester and Popovský 1979, Popovský and Pfiester 1982.

1. **Cystodinedria inermis** (Geitler) Pascher 1944 (fig. 248)
   Cystodinium brevipes Geitler 1928 partim, *Raciborskia inermis* Geitler 1943, *Phytodinedria setosa* Pascher 1944, *Ph. aeruginea* Pascher 1944, *Ph. adpressa* Pascher 1944, *Ph. hemisphaera* Pascher 1944, *Ph. procubans* Pascher 1944, *Ph. hyalina* Pascher 1944, *Ph. reniformis* Baumeister 1957, *Ph. cocconiforme* Baumeister 1957, *Cystodinedria brunnea* Pascher 1944, *C. obtusata* Pascher 1944, *C. maxima* Popovský 1961, *C. setosa* (Pascher) Bourrelly 1970, *C. aeruginea* (Pascher) Bourrelly 1970, *C. hyalina* (Pascher) Bourrelly 1970, *C. reniformis* (Baumeister) Starmach 1974, *C. cocconiforme* (Baumeister) Starmach 1974, *Cystodinedria adpressa* (Pascher) Bourrelly 1970, *C. hemisphaera* (Pascher) Bourrelly 1970, *C. procubans* (Pascher) Bourrelly 1970, *Dinococcus inermis* (Geitler) Fott 1960.

Dimensions: *Cystodinedria*-phase 8–88 µm × 6–62 µm, zoospores 10 bis 56 µm × 9–40 µm, gametes 15–17 µm × 10–12 µm, *Vampyrella*-like non-heliozoidal 8–30 µm in diameter, *Actinophrys*-like 25–60 µm in diameter, microspores 6–8 µm in diameter, *Amoeba verrucosa*-like approx. 60 µm in diameter, *Vampyrella*-like, heliozoidal approx. 60 µm in diameter,
Nutrition: autotrophic and holophytic.
Occurrence: observed on filamentous cyanophytes and chlorophytes, as well as *Sphagnum, Lemna, Utricularia* and detritus.
Invalidly described genera and species belonging to the genus *Cystodinedria: Cystodinedria oculata* Baumeister 1967, *Phytodinedria echinata* Baumeister 1967, *Renodinium* Baumeister 1968, *R. cocconiforme* (Baumeister) Baumeister 1968.
Important references: Baumeister 1967 a, 1968.

The following taxa have been described but are probably resting stages (zygotes?) of other described dinoflagellate species. They are recorded here for historical accuracy.

1. *Phytodinium* Klebs 1912

Cells are spherical or oval and surrounded by a thick cell wall. Chloroplasts are oval and parietal. Both oil and starch bodies are present. Reproduction is by means of two autospores.
1. *Phytodinium simplex* Klebs 1912 (fig. 249)
Phytodinium globosum Pascher 1927, *Ph. aureum* Starmach 1963
Dimensions: 14–50 µm × 8–45 µm.
Occurrence: pools; Central Europe.

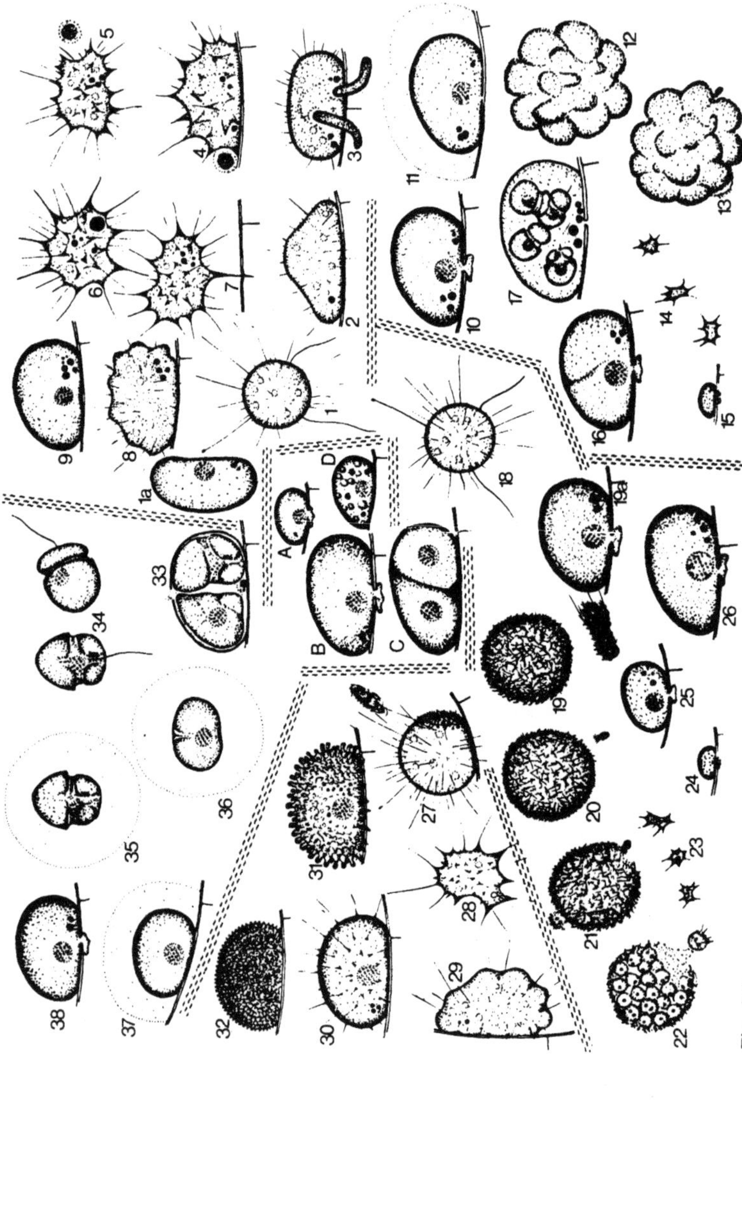

Fig. 248

2. *Hypnodinium* Klebs 1912

Spherical, planktonic cells with numerous, small chloroplasts and a single parietal stigma. Several large vacuoles surround the central nucleus. Reproduction is by formation of two *Gymnodinium*-like cells which have both furrows and a stigma but lack flagella.
1. *Hypnodinium sphaericum* Klebs 1912 (fig. 250)
Dimensions: 72– 122 µm.
Occurrence: Central Europe, USA.

3. *Dinastridium* Pascher 1927

Originally described as a dinococcal genus which reproduces by autospore or zoospore formation. Four species were described: *D. sexangulare* Pascher 1927 (hypnospore of *Gymnodinium chiastosporum)* (fig. 79); *D. verrucosum* Baumeister 1957 (the hypnospore of *Woloszynskia pascheri)* (fig. 136) ; *D. armata* (Baumeister) Bourrelly 1970 (a hypnospore of?). The last species was described as *Bourrellyella armata* Baumeister 1957. Its cells are spindle-shaped, with blund ends bearing three large spines.
Dimensions: 28 × 13 µm including spines.
Occurrence: Germany, Denmark, Czechoslovakia, Great Britain.
Important references: Klebs 1912, Pascher 1927, Cridland 1958, Starmach 1963.

Appendix.

1. *Rufusiella insignis* (Hassall) Loeblich Jr. and Loeblich III 1966 (fig. 251)
   Urococcus insignis (Hassall 1845) Lindley 1847

Cells are embedded in multilayered mucilage and located at the ends of layered mucilagenous tubes. Reproduction is by means of *Gymnodinikum*-like zoospores. Morphologically similar to the chlorococcalean alga *Hormotila* Borzi 1883, and to some of the non-motile stages of *Hemidinium nasutum* var. *tatricum*Wołoszyńska and *Glenodiniopsis steinii* (Lemmermann) Wołoszyńska.
Occurrence: Europe, N. America.
Important references: Loeblich Jr. and Loeblich III 1966, Richards 1962.

The genus *Dinamoebidium* Pascher 1917 (syn. *Dinamoeba* Pascher 1916) with its freshwater taxon *D. coloradense* Bursa 1970 (fig. 252) definitely belongs in the order *Dinococcales*. Vegetative cells are amoeboid in shape, free-swimming or attached to a

---

Fig. 248. *Cystodinedria inermis* (Geitler) Pascher, A–D, 1a, 9–11, 15, 19a, 24–26, 30, 37, 38 vegetative cells, C, 13, 22 zoid formation, 1, 18, 27 *"Actinophrys"* stage, 12, 13 *"Amoeba verrucosa"* stage, 1, 2, 3 change from *"Actinophrys"* to *"Vampyrella"* stage, 27, 28 change from *"Amoeba verrucosa"* over *"Actinophrys"* into *"Vampyrella"* stage, 4–8, 14, 23, 28, 29 *"Vampyrella"* stage, 19, 22 *"Vampyrella"* stage in non-heliozoid form, 22 a sporozoid, formation of microspores, 16, 17 formation of four isogametes, 33 formation of two zoospores, 34–37 zoospores and settlement on filamentous alga, 31–32 resting stage, 2, 3, 7, 8, 10, 15, 24–26, 29, 30, 38 phagocytizing protoplasts of a filamentous alga, 12, 13, 20, 21 phagocytizing *Oedogonium* zoospores (after Popovský and Pfiester).

249

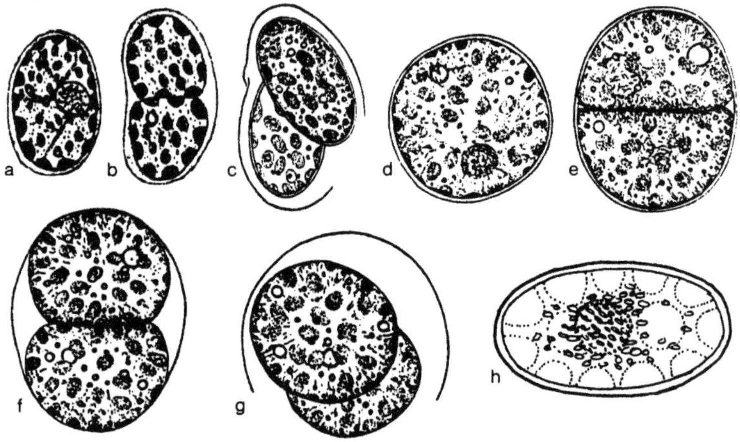

250

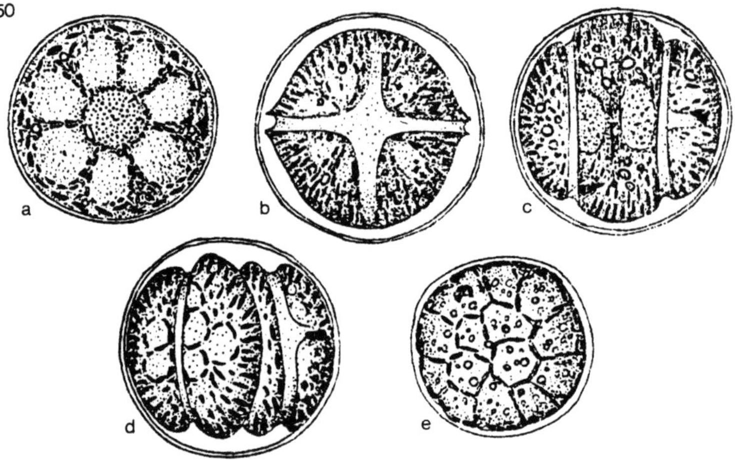

Fig. 249. *Phytodinium simplex* Klebs (*a, b* after Klebs, *c–g* after Pascher, *h* after Starmach).

Fig. 250. *Hypnodinium sphaericum* Klebs (after Klebs).

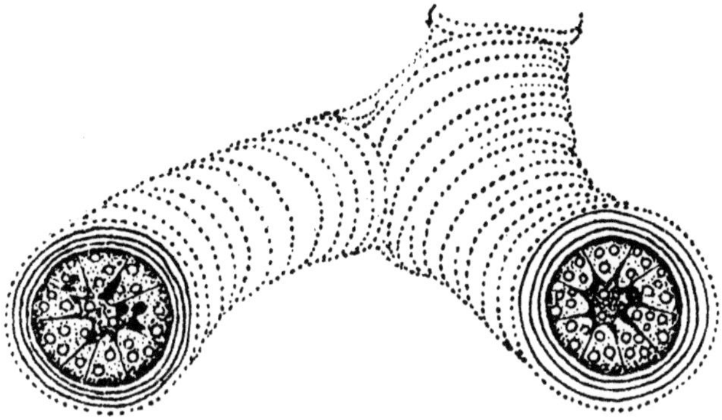

Fig. 251. *Rufusiella insignis* (Hassall) Loeblich Jr. and Loeblich III (after Thompson).

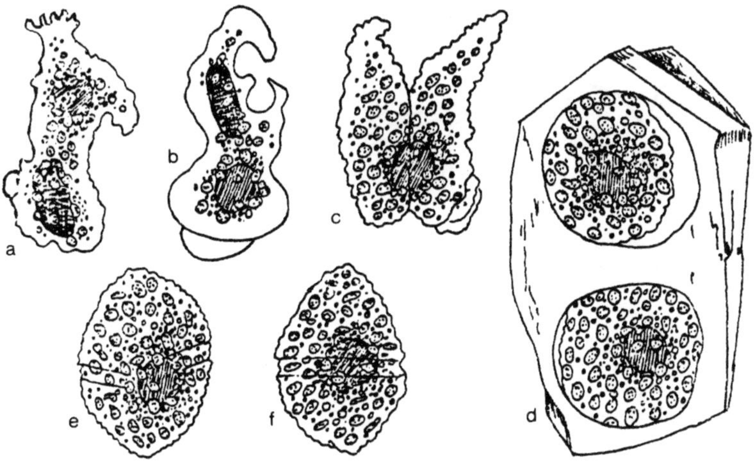

Fig. 252. *Dinamoebidium coloradense* Bursa, *d* resting stages attached to pebble (after Bursa).

substrate. Cells have numerous colorless bodies, red bodies and chloroplasts. Reproduction is by the division of the vegetative cell into two parts or by the formation of *Gymnodinium*-like zoospores, which are oval and partly metabolic. The cingulum circumscribes, the cell. A sulcus is not visible. Flagella and stigma have not been observed. Resting stages are discoid in shaped and attached.

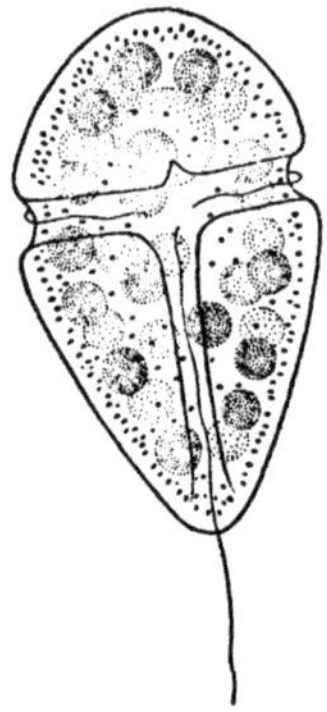

Fig. 253. *Gymnodinium eucyaneum* Hu (after Hu).

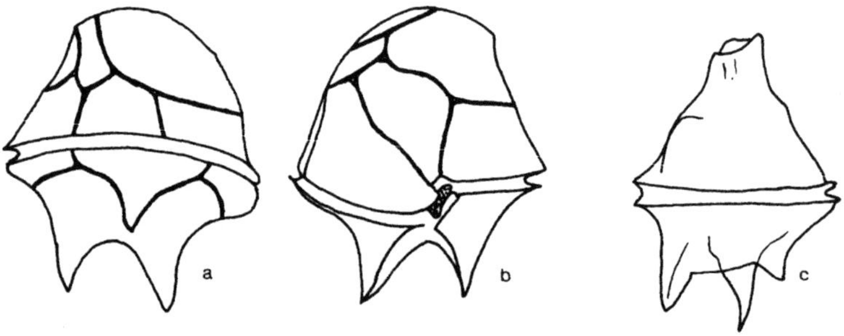

Fig. 254. *Peridinium lingii* Thomasson, *a* dorsal view, *b* ventral view, *c* lateral view (*a, b* after Ling and Tyler, *c* after Thomasson).

Dimensions: vegetative cells 28 bis 46.5 µm in diameter, zoospores 19.8–38.5 µm × 13–23 µm.
Occurrence: brook; USA.
Important references: Pascher 1917, Bursa 1970.

Addenda

Gymnodinium eucyaneum Hu 1983 (fig. 253)
Gymnodinium cyaneum Hu, Yu, and Zhang 1980
The cells are ovoid and dorsiventrally flattened. The epicone is somewhat spherical and smaller than the conical hypocone which has a rounded antapex. The cingulum spirals slightly to the left. The narrow sulcus extends slightly on the epicone; it does not reach the antapex. Numerous discoid, blue green chloroplasts (cyanells?) are present.
Dimensions: 29–45 µm × 16–19 µm.
Occurrence: in summer plankton of a carp-pond, Wuchan; China.
Note: *Gymnodinium eucyaneum* is very similar to descriptions of *Gymnodinium aeruginosum* by Stein 1883.

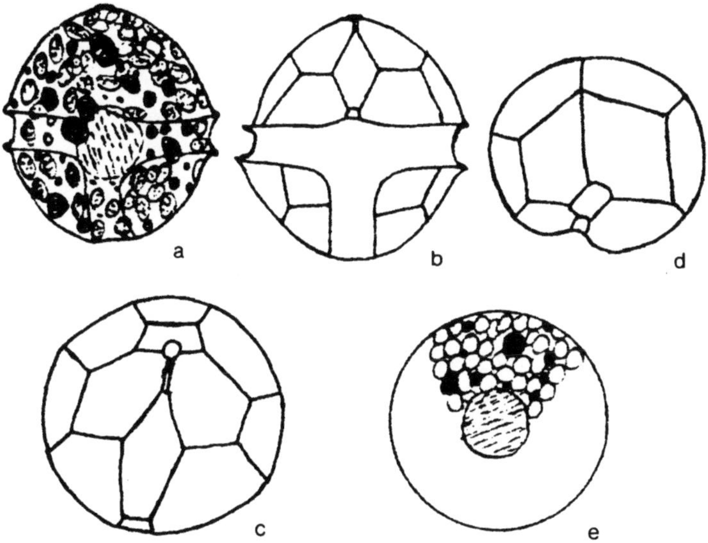

Fig. 255. *Peridiniopsis hiemale* Schiller, *a, b* ventral, *c, e* apical, *d* antapical views (after Schiller).

Peridinium lingii Thomasson 1986 (fig. 254)
Peridinium sp. Ling and Tyler 1980, 1983
Cells are ovoid with three posterior horns. The epitheca is rounded and larger than the hypotheca with its three horns. The cingulum spirals to the left; the sulcus extends slightly onto the epitheca. The sulcus widens in the hypotheca where it reaches the antapex. The plates are ornamented with ridges. Chloroplasts?
Dimensions: 116–118 μm × 96–99 μm.
Occurrence: in plankton of ox-bow ponds; Northern Australia.

Peridiniopsis hiemale Schiller 1955 (fig. 255)
Peridinium hiemale Schiller 1955
Cells are spherical or oval in shape and not flattened dorsiventrally. The slightly larger epitheca and hypotheca are hemispherical in shape. The shallow cingulum circumscribes the cell. The sides of the sulcus are parallel; it does not extend on the epitheca but almost reaches to the antapex. Plates are not visibly ornamented. The yellow-green chloroplasts are small and numerous. Refractive bodies are present in the cytoplasm. Several specimens are recorded which have a lower number of postcingular plates.
Dimensions: 40–48 μm × 38–48 μm.
Occurrence: Ruster Canal of Lake Neusiedlersee; Austria.

Ceratium rhomvoides Hickel 1987 (fig. 256)
The dorsiventrally flattened cells are rhomboidal in outline when seen in ventral view. The depressed cingulum divides the cell into a long cuneiform epitheca and a shorter

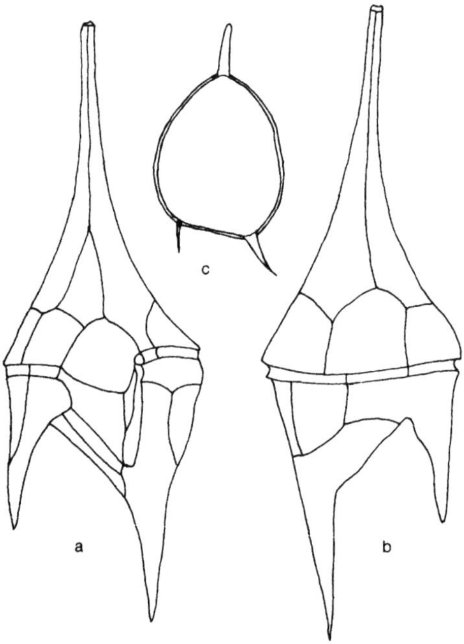

Fig. 256. *Ceratium rhomvoides* Hickel, *a* ventral, *b* dorsal views, *c* hypnozygote (after Hickel).

hypotheca with an antapical and a right postcingular horn; rarely a short accessory horn is present. The plates are distinctly reticulated. The apical plates surround an apical pore except for plate 4′ which is short and does not reach the apex. The cingulum is composed of 6 plates unequal in lenght. Trichocyst pores are surrounded by polygonal-like ridges. Chloroplasts are present. Rounded, triangular shaped cysts (hypnozygotes?) with relatively short spines are known. Asexual and sexual reproduction is known.
Dimensions: vegetative cells 123–203 µm × 32–56 µm, cysts 72–101 µm × 47–65 µm (without spines).
Occurrence: in plankton of lake Plusssee; Federal Republic of Germany.

References

Adachi, R.: Studies on a dinoflagellate *Peridinium polonicum* Wołoszyńska. I. The structure of skeleton. J. Fac. Fish. Univ. Mie **6:** 317–326 (1965).

Alexeieff, A.: Matériaux pour servir à l' étude des Protistes coprozoïtes. Arch. Zool. Exp. Gén. **68:** 609–698 (1929).

Allman, G. J.: Observations on *Aphanizomenon flos-aquae* and a species of Peridinea. Quart. J. Microscop. Sci. **3:** 21–25 (1855).

Anderson, D. M.: Effects of temperature conditioning on development and germination of *Gonyaulax tamarensis* (Dinophyceae) hypnozygotes. J. Phycol. **16:** 166–172 (1980).

– Kulis, D. M., Binder, B. J.: Sexuality and cyst formation in the dinoflagellate *Gonyaulax tamarensis:* cyst yield in batch cultures. J. Phycol. **20:** 418–425 (1984).

– Wall, D.: Potential importance of benthic cysts of *Gonyaulax tamarensis* and *G. excavata* in initiating toxic dinoflagellate blooms. J. Phycol. **14:** 224–234 (1978).

Anisimova, N. V.: Novye Peridineae, najdennye v solennych vodoemach Staroj Russy (Novgorod. Gub.). Russk. Gidrobiol. Ž. **5:** 188–193 (1926).

Apstein, C.: Das Süßwasserplankton. Methode und Resultate der quantitativen Untersuchung, p. 200. Lipsius & Tischer, Kiel, Leipzig 1896.

Bachmann, H.: Das Phytoplankton des Süsswassers mit besonderer Berücksichtigung des Vierwaldstättersees. p. 213. G. Fischer, Jena 1911.

– Eröffnungsrede der 105. Jahresversammlung: Das Phytoplankton der Schweizerseen (Vergleichende Planktonstudien XVI.). Verh. Schweiz. Naturf. Ges. **105.** Jahresversammlung vom 1. bis 4. Oktober 1924 in Luzern; II.: 15–30 (1924).

Bailey, J. W.: Microscopical observations made in South Carolina, Georgia and Florida. Smithsonian Inst. Contr. Knowl. **2:** 1–48 (1850).

Balech, E.: *Glenodinium cristatum* sp. nov. Neotropica **7** (23): 47–51 (1961).

Baumeister, W.: Überwinterung und Hüllenstruktur bei *Hemidinium nasutum* Stein. Arch. Protistenk. **91:** 456–461 (1938).

– *Gymnodinium dimorphe* spec. nov. Arch. Protistenk. **91:** 462–464 (1938).

– Über ein Vorkommen von *Stylodinium globosum* Klebs im Hochmoorgebiet von Seeon. Int. Rev. ges. Hydrobiol. **39:** 391–412 (1939).

– Die Dinoflagellaten der Kreise Pfarrkirchen und Eggenfelden (Gau Bayreuth) 2. Das Sumpfgebiet im Walde südlich Altersham. Arch. Protistenk. **97:** 344–364 (1943).

– Neue Dinococcalen aus dem niederbayerischen Hügelland zwischen Isar und Inn I. Arch. Prostistenk. **102:** 21–43 (1957).

– Zur Kenntnis der Dinophyceen des Seeoner Gebietes I. Arch. Protistenk. **102:** 1–20 (1957).
– Neue Dinococcalen aus dem niederbayerischen Hügelland zwischen Isar und Inn II. Arch. Protistenk, **102:** 241–257 (1957).
– Zur Kenntnis der Dinophyceen des Seeoner Gebietes II. Arch. Protistenk. **102:** 258–264 (1957).
– Dinophyceen aus perennierenden Gewässern des Schwingrasenmoores bei Burgberg, sowie aus Alpsee, Freibergsee und dem Moorweiher in Oberstdorf (Allgäu). Arch. Protistenk. **106:** 535–552 (1963).
– Phykologische und zoologische Notizen über Gewässer der niederbayerischen Landgemeinden Taubenbach, Julbach, Rent und Gumpersdorf (Kreis Pfarrkirchen). Arbeitsstätte zur Erforschung des Lebens in Kleingewässern. Mitt. **4:** 1–24 (1967) [Mimeographed].
– Zur Kenntnis der Dinophyceen des Seeoner Gebietes (III.). Arbeitsstätte zur Erforschung des Lebens in Kleingewässern. Mit. **5:** 1–6 (1967) [Mimeographed].
– Zum System der Dinophyceen – Ordnung Dinococcales. 1. Die Familien Glutinodiniaceae und Dinorbiculaceae. Arbeitsstätte zur Erforschung des Lebens in Kleingewässern. Mit. **6:** 1–13 (1968). [Mimeographed].
– Zum System der Dinophyceen – Ordnung Dinococcales. 2. Die Familie Stylodiniaceae. Arbeitsstätte zur Erforschung des Lebens in Kleingewässern. Mitt. **7:** 1–11 (1968) [Mimeographed].
Beam, C. A., Himes, M.: Evidence for sexual fusion and recombination in the dinoflagellate *Crypthecodinium/Gyrodinium cohnii/*. Nature **250:** 435–436 (1974).
– – Sexual isolation and genetic diversification among some strains of *Crypthecodinium cohnii*-like dinoflagellates. Evidence of speciation. J. Protozool. **24:** 532–539 (1977).
– – Sexuality and meiosis in dinoflagellates. In: Lewandowsky, M., Hutner, S. H. (eds.): Biochemistry and Physiology of Protozoa II. Academic Press, London, New York: 171–206 (1980).
Bergh, R. S.: Bidrag til Cilioflagellaternes naturhistoria. Nat. Viden. Medd. Copenhagen 1881: 60–76 (1881).
– Der Organismus der Cilioflagellaten. Eine phytogenetische Studie. Morph. Jb. **7** (2): 178–288 (1881).
– Über die systematische Stellung der Gattung *Amphidinium* Claparède et Lachmann. Zool. Anz. **5:** 693–695 (1882).
Bibby, B. T., Dodge, J. D.: The encystment of a freshwater dinoflagellate: a light and electron-microscopical study. Brit. Phycol. Bull. **7:** 85–100 (1972).
Bicudo, C. E. M., Skvortzov, B. V.: Contribution to the knowledge of Brazilian Dinophyceae – free-living unarmoured genera. Rickia **5:** 5–21 (1970).
Biecheler, B.: Recherches sur les Péridiniens. Bull. Biol. France Belgique Suppl. **36:** 1–149 (1952).
Bold, H. C., Wynne, M. J.: Introduction to the Algae. Structure and Reproduction, p. 706. Prentice-Hall, New Jersey 1978.
Bory de St. Vincent, J.: Encyclopédie méthodique. Histoire naturelle des Zoophytes ou animaux rayonnés. II (2). Agasse, Paris 1824.
Bourrelly, P.: Notes sur les Péridiniens d'eau douce. Protistologica **4.:** 5–14 (1968).
– Note sur le *Peridiniopsis borgei* Lemmermann. Phykos **7:** 1–2 (1968).

– Les algues d'eau douce III. Les algues bleues et rouges, les Eugléniens, Péridiniens et Cryptomonadines, p. 512. N. Boubée, Paris 1970.

Brugerolle, G., Mignot, J. P.: Observations sur le cycle, l'ultrastructure et la position systématique de *Spiromonas perforans* ‹Bodo perforans Hollande 1938›, flagellé parasite de *Chilomonas paramaecium,* ses relations avec les Dinoflagellés et les Sporozoaires. Protistologica **15:** 183–196 (1979).

Bursa, A: The freshwater Dinoflagellate *Woloszynskia limnetica* n. sp. Membrane and Protoplasmatic Structures. J. Protozool. **5:** 299–304 (1958).

– *Actiniscus canadensis* n. sp., *A. pentasterias* Ehrenberg v. *arcticus* n. var., *Pseudactiniscus apentasterias* n. gen., n. sp., marine relicts in Canadian Arctic Lakes. J. Protozool. **16:** 411–418 (1969).

– *Dinamoebidium coloradense* spec. nov. and *Katodinium auratum* spec. nov. in Como Creek, Boulder Country, Colorado, Arctic and Alpine Research **2:** 145–151 (1970).

Bütschli, O.: Protozoa. Abteilung II. Mastigophora. 617–1097. In: Bronn, H. G. (ed.): Bronns Klassen und Ordnungen des Thier-Reichs. **1.** Akad. Verlagsges., Leipzig 1883–1887. [906–1029, 1885.]

Cachon, J., Cachon, M., Bouquaheux, F.: *Stylodinium gastrophilum* Cachon, péridinien dinococcoïde parasite de Siphonophores. Bull. Inst. Océanogr. Monaco **65** (1359): 1–8 (1965).

Cao Vien, M.: Sur l'existence de phénomènes sexuels chez un péridinien libre, l'*Amphidinium carteri.* C. R. Acad. Sci. (D) **264:** 1006–1008 (1967).

– Sur la germination du zygote et sur un mode particulier de multiplication végétative chez le péridinien libre, l'*Amphidinium carteri.* C. R. Acad. Sci. (D) **267:** 701–703 (1968).

Carefoot, J. R.: Culture and heterotrophy of the freshwater dinoflagellate *Peridinium cinctum* fo. *ovoplanum* Lindemann. J. Phycol. **4:** 129–131 (1968).

Carty, S., Cox, E. R.: Observations on *Lophodinium polylophum* (Dinophyceae). J. Phycol. **21:** 396–401 (1985).

– – *Kansodinium* gen. nov. and *Durinskia* gen. nov.: two genera of freshwater dinoflagellates (Pyrrhophyta). Phycologia **25** (2): 197–204 (1986).

Chadefaud, M.: Les végétaux non vasculaires (Cryptogamie). In: Chadefaud, M., Emberger, L. (eds.): Traité de Botanique, systématique I, p. 1016. Masson, Paris 1960.

Chatton, E.: Diagnoses préliminaires de Péridiniens parasites nouveaux. Bull. Soc. Zool. France **37:** 85 (1912).

– Les Péridiniens parasites; morphologie, reproduction, éthologie. Arch. Zool. Exp. Gén. **54:** 1–157 (1920).

– Classe des Dinoflagellés ou Péridiniens, 309–406. In: Grassé, P. P. (ed.): Traité de zoologie I. Phylogénie, Protozoaires: Généralités. Flagellés. p. 1071. Masson, Paris 1952.

Chodat, R.: Matériaux pour l'histoire des algues de la Suisse. Bull. Soc. Bot. Genève **29:** 66–114 (1921).

– Zender, J.: Algues de la région du Grand-St-Bernard. Bull. Soc. Bot. Genève (2) **15:** 33–48 (1923).

Christen, H. R.: *Gymnodinium Nygaardii* sp. nov. Ber. Schweiz. Bot. Ges. **68:** 44–49 (1958).

– Neue und wenig bekannte Flagellaten aus dem Hausersee (Schweiz). Hydrobiologia **10:** 66–84 (1958).

– Flagellaten aus dem Schützenweiher bei Veltheim. Mitt. Naturw. Ges. Winterthur **29:** 167–189 (1959).

– Über die Gattung *Katodinium* Fott (= *Massartia* Conrad). Schweiz. Z. Hydrol. **23:** 309–342 (1961).

Christensen, T.: Alger. In: Böcher, T. W., Lange, M., Sørensen, T. (eds.): Systematisk Botanik II/2, p. 178. Muksgaard, København 1962.

Claparède, E., Lachmann, J.: Études sur les infusoires et les rhizopodes. Mém. Inst. Nat. Génèvois **5:** 1–260 (1858); **6:** 261–482 (1860);**7:** 1–291 (1861).

Coats, W., Tyler, M., Anderson, D.: Sexual process in the life cycle of *Gyrodinium uncatenum* (Dinophyceae): a morphogenetic overview. J. Phycol. **20:** 351–361 (1984).

Conrad, W.: Recherches sur les Flagellates de nos eaux saumâtres. 1ère partie: Dinoflagellates. Arch. Protistenk. **55:** 63–98 (1926).

– Notes Protistologiques IX. Sur trois Dinoflagellates de l'eau saumâtre. Bull. Mus. Royal Hist. Nat. Belgique **15** (12): 1–10 (1939).

– Notes Protistologiques XXII. *Amphidinium mucicolum.* Bull. Mus. Royal Hist. Nat. Belgique **17** (40): 1–2 (1941).

– Notes Protistologiques XXX. Sur quelques organismes sphagnicoles. Bull. Mus. Royal Hist. Nat. Belgique **19:** 1–4 (1946).

Cridland, A. A.: A note on two freshwater dinoflagellates. New Phytol. **57** (3): 285–287 (1958).

Daday, E.: Systematische Übersicht der Dinoflagellaten des Golfes von Neapel. Természetrajzi Füz. **11:** 1–98 (1888).

– Plankton-Tiere aus dem Victoria-Nyanza. Sammelausbeute von A. Borgert 1904–1905. Zool. Jb. Abt. Syst. **25:** 245–262 (1907).

Dale, B.: Dinoflagellate resting cysts: "benthic plankton". In: Fryxell, G. A. (ed.): Survival Strategies of the Algae, p. 144. Cambridge University Press, London 1982.

Dangeard, P. A.: Mémoire sur la famille des Péridiniens. Botaniste **29:** 3–182 (1938).

Desortová, B.: Productivity of individual algal species in natural phytoplankton assemblage determined by means of autoradiography. Algol. Studies **17** (Arch. Hydrobiol. Suppl. **49**): 415–449 (1976).

Diesing, K. M.: Systema Helminthium, I, II p. XVI, 680; VI, 591. Braumüller, Vindobonae 1850.

– Revision der Prothelminthen. Abt. Mastigophoren. Sitzber. Akad. Wiss. Wien **52:** 287–402 (1866).

Diwald, K.: Die ungeschlechtliche und geschlechtliche Fortpflanzung von *Glenodinium lubieniensiforme.* Flora **32:** 174–192 (1938).

Dodge, J. D.: The Dinophyceae. In: Godward, M. B. E. (ed.): The Chromosomes of the Algae, 95–115. Arnold, London 1966.

– A dinoflagellate with both a mesocaryotic and eucaryotic nucleus I. Fine structure of the nuclei. Protoplasma **73:** 145–157 (1971).

– Fine structure of the Pyrrophyta. Both. Rev. **37:** 481–508 (1971).

– Crawford, R. M.: Observations on the fine structure of the eyespot and associated organelles in the dinoflagellate *Glenodinium foliaceum.* J. Cell. Sci. **5:** 479–493 (1969).

– – The fine structure of *Gymnodinium fuscum* (Dinophyceae). New Phytologist **68:** 613–618 (1969).

– – A review of the theca fine structure in the Dinophyceae. J. Linn. Soc. Bot. **63:** 53–67 (1970).

– – A fine-structural survey of dinoflagellate pyrenoids and food-reserves. J. Linn. Soc. Bot. **64** (2): 105–115 (1971).

Doflein, F., Reichenow, E.: Lehrbuch der Protozoenkunde I, II. 6. Aufl. p. 408; p. 1213. G. Fischer/VEB G. Fischer, Jena 1949, 1952.

Dujardin, F.: Histoire naturelle des Zoophytes, Infusoires, comprenant la physiologie et la clasification de ces animaux et la manière de les étudier à l'aide du microscope. In: Nouvelles Suites à Buffon, p. 684. Roret, Paris 1841.

Dürr, V. G.: Elektronenmikroskopische Untersuchungen am Panzer von Dinoflagellaten. I. *Gonyaulax polyedra.* Arch Protistenk. **122:** 55–87 (1979).

Eddy, S.: The fresh-water armored or thecate dinoflagellates. Trans. Amer. Microscop. Soc. **49** (4): 277–321 (1930).

Ehrenberg, C. G.: Beiträge zur Kenntniss der Organisation der Infusorien und ihrer geographischen Verbreitung, besonders in Sibirien. Phys. Abh. K. Akad. Wiss. Berlin, aus dem Jahre 1830: 1–88 (1830).

– Über die Entwicklung und Lebensdauer der Infusionsthiere. Abh. Akad. Wiss. Berlin, aus dem Jahre 1831: 1–154 (1832).

– Dritter Beitrag zur Erkenntniss grosser Organisation in der Richtung des kleinsten Raumes. Abh. Akad. Wiss. Berlin, aus dem Jahre 1833: 145–336 (1834).

– Die Infusionsthierchen als vollkommene Organismen, p. 547. Voß, Berlin, Leipzig 1838.

– Über noch jetzt zahlreich lebende Thierarten der Kreidebildung und den Organismus der Polythalamien. Abh. Akad. Wiss. Berlin, aus dem Jahre 1839: 81–174 (1840).

– Über die Verbreitung des jetzt wirkenden kleinsten organischen Lebens in Asien, Australien und Afrika und über die vorherrschende Bildung auch des Oolithkalkes der Juraformation aus kleinen polythalamischen Thieren. Abh. Akad. Wiss. Berlin: 100–106 (1843).

– Die systematische Charakteristik der neuen mikroskopischen Organismen des tiefen Atlantischen Ozeans. Monatsber. Akad. Wiss. Berlin: 236–250 (1854).

– Mikrogeologie – das Erden und Felsen schaffende Wirken des unsichtbar kleinen selbständigen Lebens auf der Erde, p. 374 [Namenregister p. 31, Index to Atlas p. 88], Voß, Berlin, Leipzig 1854.

Entz, G.: Die Fauna der kontinentalen Kochsalzwasser. Math. nat. Ber. Ungarn **19:** 89–124 (1904).

– Beiträge zur Kenntnis der Peridineen I. Zur Morphologie und Biologie von *Peridinium Borgei* Lemmermann. Arch. Protistenk. **56:** 397–416 (1926).

– Beiträge zur Kenntnis der Peridineen II resp. VIII. Studien an Süsswasser-Ceratien (Morphologie, Variation, Biologie). Arch. Protistenk. **58:** 344–440 (1927).

– Über gehemmte Lebens- u. Absterbeerscheinungen einiger Dinoflagellaten. Arb. Ungar. Biol. Forschungsinst. I. Abt. **1:** 206–246 (1930).

– Analyse des Wachstums und der Teilung einer Population sowie eines Individuums der Protisten *Ceratium hirundinella* unter den natürlichen Verhältnissen. Arch. Protistenk. **74:** 311–361 (1931).

Ettl, H.: Beitrag zur Kenntnis der Süßwasseralgen Dänemarks. Bot. Tidsskr. **74:** 179–222 (1980).

– Grundriß der allgemeinen Algologie, p. 549. VEB G. Fischer, Jena 1980.

Evitt, W. R.: Sporopollenin Dinoflagellate Cysts, their Morphology and Interpretation, p. 333. Amer. Ass. Stratigr. Palynol. Found., Texas 1985.

Fott, B.: Die Schwebeflora des Ohrid-Sees. Bull. Inst. Jard. Bot. Univ. Belgrade **2:** 153–175 (1933).

– Eine neue *Gymnodinium-* und *Massartia*-Art. Studia Bot. Čechica **1:** 100–104 (1938).

– Taxonomie drobnohledné flory našich vod. Preslia **29:** 278–319 (1957).

– Taxonomische Übertragungen und Namensänderungen unter den Algen. Preslia **32:** 142–154 (1960).

– Algenkunde, 2. Aufl., p. 581. VEB G. Fischer, Jena 1971.

Garbini, A.: Intorno al plancton dei laghi di Mantova. Mem. Accad. Verona **74:** 258–314 (1898).

Geitler, L.: *Gymnodinium amphidinioides,* eine neue blaugrüne Peridinee. Bot. Arch. **6:** 110–112 (1924).

– Neue Gattungen und Arten von Dinophyceen, Heterokonten und Chrysophyceen. Arch. Protistenk. **63:** 67–83 (1928).

– Neue oder wenig bekannte Mikroorganismen aus der Umgebung von Lunz, I. Zwei neue Dinophyceenarten. Arch. Protistenk. **61:** 1–8 (1928).

– Koloniebildung und Beeinflussung der Unterlage bei zwei Dinococcalen *(Raciborskia oedogonii* und *R. inermis).* Beih. Bot. Cbl. **62**A: 160–174 (1943).

– „Pseudostigmen" bei dem blaugrünen *Gymnodinium amphidinioides.* Ber. Deutsch. Bot. Ges. **82:** 639–641 (1969).

Grell, K. G.: Protozoology, p. 554. Springer Verlag, Berlin, Heidelberg, New York 1973.

Griffiths, B. M.: The heleoplankton of three Berkshire Pools. J. Linn. Soc. Bot. **46:** 1–11 (1922).

Harris, T. M.: A contribution to the knowledge of the British freshwater Dinoflagellata. Proc. Linn. Soc. London **152:** 1–33 (1939/1940).

Hasimoto, Y., Okaichi, T.: Glenodinine, an ichthyotoxic substance produced by a dinoflagellate, *Peridinium polonicum.* Bull. Jap. Soc. Sci. Fish. **34:** 528–534 (1968).

Hassall, A. H.: A History of the British Freshwater Algae, Including Descriptions of the Desmidiaea and Diatomaceae. I, II. p. viii, 462, 24. Highley & Bailliere, London 1845 [also: Longman, Brown, Green, Longmans & Roberts, London 1857].

Hickel, B.: The population structure of *Ceratium* in a small eutrophic lake. Verh. Internat. Verein. Limnol. **22:** 2845–2849 (1985).

– Morphology and life cycle of *Ceratium rhomvoides* nov. sp. (Dinophyceae) from the lake Plusssee (F.R.G.). Hydrobiologia **161:** 49–54 (1988).

– Sexual reproduction and life cycle of *Ceratium furcoides* (Dinophyceae) in situ in the lake Plusssee (F.R.G.). Hydrobiologia **161:** 41–48 (1988).

Himes, M., Beam, C. A.: Further studies of breeding restrictions naming *Crypthecodinium cohnii*-like Dinoflagellates. Evidence of a new interbreeding group. J. Protozool. **25:** 378–380 (1978).

Höll, K.: Ökologie der Peridineen. Pflanzenforschung **11:** 1–105 (1928).

Hofeneder, H.: Über die animalische Ernährung von *Ceratium hirundinella* O. F. Müller und über die Rolle des Kernes bei dieser Zellfunktion. Arch. Protistenk. **71:** 1–32 (1930).

Hofker, J.: Über *Noctiluca scintillans* (Macartney). Arch. Protistenk. **71:** 59–78 (1930).

Hollande, A.: *Bodo perforans* n. sp., flagellé nouveau parasite externe du *Chilomonas paramaecium* Ehrenberg. Arch. Zool. Exp. Gén. **79:** 75–81 (1938).

Holt, J., Pfiester, L. A.: A survey of auxotrophy in five freshwater dinoflagellates (Pyrrhophyta). J. Phycol. **17:** 415–416 (1981).

Horiguchi, T., Chihara, M.: *Stylodinium littorale,* a new marine dinococcalean alga (Pyrrhophyta). Phycologia **22:** 2–8 (1983).

Hu, H.: On *Gymnodinium cyaneum* Hu. Chin. J. Ocean. Limnol. **1** (2): 198–199 (1983).

– Yu, M., Zhang, X.: Discovery of phycobilin in *Gymnodinium cyaneum* Hu sp. nov. and its phylogenetic significance. Kexue Tongbao **25** (10): 882–884 (1980).

Huber, G.: Monographische Studien im Gebiete der Montigglerseen (Südtirol) mit besonderer Berücksichtigung ihrer Biologie III. Die Region des offenen Wassers. Arch. Hydrobiol. **1:** 123–210 (1905).

– Formanomalien bei *Ceratium hirundinella* O. F. M. Int. Rev. Ges. Hydrol. Hydrogr. Biol. Suppl. **6:** 1–40 (1914).

– Nipkow, F.: Experimentelle Untersuchungen über die Entwicklung und Formbildung von *Ceratium hirundinella* O. F. Müller. Z. Bot. **14:** 357–371 (1922).

– – Experimentelle Untersuchungen über die Entwicklung und Formbildung von *Ceratium hirundinella* O. F. Müller. Flora N. F. **116:** 114–215 (1923).

Huber-Pestalozzi, G.: Das Phytoplankton des Süsswassers. Cryptophyceen, Chloromonadineen, Peridineen. Die Binnengewässer **16** (3): p. 310. Schweizerbart, Stuttgart 1950.

Huitfeld-Kaas, H.: Die limnetischen Peridineen in norwegischen Binnenseen. Skr. Vidensk.-Selsk. Christiania, Math.-Naturvidensk. Kl. 1900: 1–7 (1900).

Imhof, O. E.: Resultate meiner Studien über die pelagische Fauna kleinerer und grösserer Süsswasserbecken der Schweiz. Z. Wiss. Zool. **40:** 154–178 (1884).

– Weitere Mitteilung über die pelagische und Tiefseefauna der Süsswasser-Becken. Zool. Anz. **8:** 160 (1885).

– Über die mikroskopische Tierwelt hochalpiner Seen. Zool. Anz. **10:** 13–30 (1887).

Irish, A. E.: *Gymnodinium helveticum* Penard f. *achroum* Skuja, a case of phagotrophy. Brit. Phycol. J. **14:** 11–15 (1979).

Jakobs, D. L.: A new parasitic dinoflagellate from freshwater fish. Trans. Amer. Microsc. Soc. **65:** 1 (1946).

Javornický, P.: Několik nových a málo známých bičíkovců z kmene Pyrrhophyta. Univ. Carol. Pragensis, Biologica **3** (3): 251–268 (1957).

– Two scarcely known genera of the class Dinophyceae: *Bernardinium* Chodat and *Crypthecodinium* Biecheler. Preslia **34:** 98–113 (1962).

– Unarmoured Dinoflagellata from two small Mazurian Lakes. Phycologia **5:** 53–60 (1965).

– Some interesting algal flagellates. Folia Geobot. Pytotaxonom. **2:** 43–67 (1967).

– Two species of *Katodinium* Fott from the Abisko National Park. Svensk Bot. Tidskr. **64:** 297–302 (1970).

Jollos, V.: Dinoflagellatenstudien. Arch. Protistenk. **19:** 178–206 (1910).

Joseph, G.: Über Grotten-Infusorien. Zool. Anz. **2:** 114–118 (1879).

Kent, W.: A manual of the Infusoria; including a description of all known flagellate, ciliate and tentaculiferous Protozoa, British and foreign, and an account of the organisation and affinities of the sponges, p. 193. Bogue, London 1880/81.

Killian, Ch.: Le cycle évolutif du *Gloeodinium montanum* (Klebs). Arch. Protistenk. **50:** 50–66 (1924).

Kiselev, I. A.: Kriptomonadovye i peridinei evropejskogo Severa SSSR (Pyrrophyta). Trudy Bot. Inst. Akad. Nauk SSSR, Ser. 2, Sporov. rasten. **7:** 13–164 (1951).

– Pirofitovye vodorosli. Opredelitel' presnovodnych vodoroslej SSSR **6, p.** 212. Sov. Nauka, Moskva, Leningrad 1954.

– Cvetkov, N. W.: Zur Morphologie und Ökologie von *Peridinium baicalense* n. sp. Beih. Bot. Cbl. **53** (B): 518–524 (1935).

Klebs, G.: Ein kleiner Beitrag zur Kenntnis der Peridineen. Bot. Z. **42:** 721–733, 737–745 (1884).

– Über Flagellaten und algenähnliche Peridineen. Verh. Naturhist. Med. Ver. Heidelberg, N. F. **11:** 369–451 (1912).

Koch, W.: Verzeichnis der Sammlung von Algenkulturen am Pflanzenphysiologischen Institut der Universität Göttingen. Arch. Mikrobiol. **47:** 402–432 (1964).

Kofoid, C. A., Michener, J. R.: On the struture and relationship of *Dinosphaera palustris* (Lemm.). Univ. Calif. Publ. Zool. **11:** 21–28 (1912).

– Swezy, O.: The free-living unarmored Dinoflagellata. Mem. Univ. Calif. **5** (8): p. 562 (1921).

Kristiansen, J.: Rødfarvning af Horn Sø ved Lemvig forårsaget af dinoflagellaten *Oxyrrhis marina*. Flora & Fauna **80:** 65–69 (1974).

Kudo, R. R.: Protozoology, 5th ed., 2nd print, p. 1174. Ch. C. Thomas, Illinois 1971.

Kützing, F.: Species algarum, p. 922. F. A. Brockhaus, Leipzig 1849.

Lackey, J. B.: Some fresh-water Protozoa with blue chromatophores. Biol. Bull. **71:** 492–497 (1936).

Langhans, V. H.: Gemischte Populationen von *Geratium hirundinella* (O. F. M.) Schrank und ihre Deutung. Arch. Protistenk. **52:** 585–602 (1925).

Lantzsch, K.: Studien über das Nannoplankton des Zugersees und seine Beziehung zum Zooplankton. Z. Wiss. Zool. **108:** 631–692 (1914).

Largajolli, V.: La varietà *oculata* del *Glenodinium pulvisculus* (Ehr.) Stein. Nuova Notarisia **18:** 1–5 (1907).

Lauterborn, R.: Über die Winterfauna einiger Gewässer der Oberrheinebene, mit Beschreibung neuer Protozoen. Biol. Zbl. **14:** 390–398 (1894).

– Diagnosen neuer Potozoen aus dem Gebiete des Oberrheins. Zool. Anz. **19:** 14–18 (1896).

– Die Vegetation des Oberrheins. Verh. Naturhist. Med. Ver. Heidelberg N. F. **10:** 450–502 (1910).

Leadbeater, B., Dodge, J. D.: The fine structure of *Woloszynskia micra* sp. n., a new marine dinoflagellate. Brit. Phycol. Bull. **3** (1): 1–17 (1966).

– – An electron microscope study on dinoflagellate flagella. J. Gen. Microbiol. **46:** 305–314 (1967).

Lebour, M.: Plymouth Peridinians, I, II, III. J. Mar. Biol. Ass. Plymouth N. S. **12:** 795–818 (1922).

– The dinoflagellates of nothern seas. J. Mar. Biol. Ass. U. K. **15:** 1–172 (1925).

Lefèvre, M.: Contribution à la flora des Péridiniens de France. Rev. Algol. **2:** 327–342 (1925).
– Contribution à la flora des algues d'eau douce du nord de la France. Bull. Soc. Bot. France **62,** sér. 5 (1): 689–699 (1925).
– Sur les variations tabulaires ches les Péridiniens d'eau douce et leur notation. Diagnoses d'espèces et de varietés nouvelles. Bull. Mus. Hist. Nat. Paris **33** (1): 118–122 (1927).
– Notes sur le *Peridinium Cunningtonii* Lemm. et sur quelques formes affinies. Ann. Protistol. **1:** 119–126 (1928).
– Notules systématiques. I. *Peridinium morzinense* nom. nov. Ann. Protistol. **1:** 137 (1928).
– Monographie des espèces d'eau douce du genre *Peridinium.* Arch. Bot. Mem. Caen **2** (5): 1–208 (1932).
Lemmermann, E.: Das Phytoplankton sächsischer Teiche. Forschungsber. Biol. Stat. Plön **7:** 96–135 (1899).
– Neue Schwebalgen aus der Umgegend von Berlin. Ber. Deutsch. Bot. Ges. **18:** 24–32 (1900).
– Beiträge zur Kenntnis der Planktonalgen, VIII. Peridiniales aquae dulcis et submarinae. Hedwigia **39:** 115–121 (1900).
– Algenflora eines Moortümpels bei Plön. Forschungsber. Biol. Stat. Plön **8:** 64–73 (1901).
– Das Plankton schwedischer Gewässer. Ark. Bot. **2** (2): 1–209 (1904).
– Beiträge zur Kenntnis des Planktonalgen XXI. Das Phytoplankton sächsischer Teiche. 2. Beitrag. Forschungsber. Biol. Stat. Plön **12:** 164–168 (1905).
– Über das Vorkommen von Süßwasserformen im Phytoplankton des Meeres. Arch. Hydrobiol. **1:** 409–427 (1906).
– *Gonyaulax palustris,* eine neue Süßwasserperidinee. Beih. Bot. Cbl. **21,** Abt. II: 296–300 (1907).
– Algen aus der Riviera von Lentini, Sizilien. (Algologische Beiträge VI). Arch. Hydrobiol. **4:** 165–192 (1908).
– Beiträge zur Kenntnis der Planktonalgen XXVII. Planktonalgen aus dem Schliersee. Arch. Hydrobiol. **5:** 324–329 (1910).
– Peridiniales. Kryptogamenflora der Mark Brandenburg **3,** 1: 536–712. Borntraeger, Berlin 1910.
Levander, K. M.: Notizen über die Täfelung der Schalenmembran des *Glenodinium cinctum.* Zool. Anz. **15:** 405 (1892).
– Materialien zur Kenntnis der Wasserfauna in der Umgebung von Helsingfors mit besonderer Berücksichtigung der Meeresfauna. I. Protozoa. Acta Soc. Fauna Flora Fennica **12** (2): 1–115 (1894).
– Zur Kenntnis des Lebens in den stehenden Kleingewässern auf den Skäreninseln. Acta Soc. Fauna Flora Fennica **18** (6): 1–107 (1900).
– Zur Kenntnis der Fauna und Flora finnischer Binnenseen. Acta Soc. Fauna Flora Fennica **19** (3): 4–55 (1900).
– Eine neue farblose *Peridinium*-Art, *Peridinium achromaticum* n. sp. Medd. Soc. Fauna Flora Fennica **28:** 49–50 (1902).
Lindemann, E.: Untersuchungen über Süßwasserperidineen und ihre Variationsformen. II. Arch. Naturgesch. **84**A (8): 121–194 (1918).

– Untersuchungen über Süßwasserperidineen und ihre Variationsformen. Arch. Protistenk. **39:** 209–264 (1919).

– Eine Entwicklungshemmung bei *Peridinium borgei* und ihre Folgen. Arch. Protistenk. **40:** 378–382 (1923).

– Über finnische Peridineen. Arch. Hydrobiol. **15:** 1–4 (1924).

– Peridineen aus Seen der Schweiz. Bot. Arch. **10:** 205–208 (1925).

– Peridineen des Oberrhein und seiner Altwässer. Bot. Arch. **11** (5): 474–481 (1925).

– Dinoflagellatae (Peridineae). In: Eyferth, B., Schoenichen, W. (eds.): Einfachste Lebensformen des Tier- und Pflanzenreiches, 5. Aufl., p. 519. Bermühler, Berlin 1925/1926.

– Neue Peridineen. Hedwigia **68:** 291–296 (1928).

– Peridineae (Dinoflagellatae). In: Engler, A., Prantl, K. (eds.): Die natürlichen Pflanzenfamilien, 2. Aufl., Bd. 2, p. 104. Engelmann, Leipzig 1928.

– Experimentelle Studien über die Fortpflanzungserscheinungen der Süßwasserperidineen auf Grund von Reinkulturen. Arch. Protistenk. **68:** 1–104 (1929).

– Die Peridineen der Deutschen Limnologischen Sunda-Expedition nach Sumatra, Java und Bali. Arch. Hydrobiol. Suppl. **8** (4): 1–691 (1931).

Lindström, K., Rodhe, W.: Selenium as a micronutrient for the dinoflagellate *Peridinium cinctum* fa. *westii*. Mitt. Internat. Verein. Limnol. **21:** 168–173 (1978).

Ling, H. U., Tyler, P. A.: Freshwater algae of the Alligator Rivers Region, Northern Territory of Australia. A report to the office of Supervising Scientist: p. 121 (1980) [Mimeographed].

– – A limnological survey of the Alligator River Region. 6. Freshwater algae, exclusive of diatoms, of the Alligator Rivers Region; Northern Territory. Supervising Scientist Alligator Rivers Region, (Australia), Res. Rep. **3:** 1–300 (1983) [Mimeographed].

Litvinenko, R. M.: K izučeniyu bescvetnych peridinej okrestnostej Char'kova. Trudy Naučno-Issl. Inst. Biol. Fak. Univ. Char'kov **141** (37): 76–81 (1963).

– Analiz vydovogo skladu perydynyjevych vodorostej Ukrajiny. Ukrajin. Bot. Ž. **30** (2): 169–174 (1973).

Loeblich, A. R. jr., Loeblich, A. R. III: Index to the genera and sections of the Pyrrhophyta. Stud. Tropic. Oceanogr. **3:** 1–94 (1966).

Loeblich, A. R. III: Dinoflagellatae nomenclature. Taxon **14:** 15–18 (1965).

– Notes on the divisions Chlorophyta, Chrysophyta, Pyrrhophyta and Xanthophyta and the family Paramastigaceae. Taxon **16:** 230–237 (1967).

– A new marine dinoflagellate genus *Cachonina,* in axenic culture from Salton Sea, California, with remarks on the genus *Peridinium*. Proc. Biol. Soc. Washington **81:** 91–96 (1968).

– The amphiesma or dinoflagellate cell covering. Proc. North Americ. Paleont. Convent. G: 867–929 (1969).

– Dinophyceae. In: Parker, S. P. (ed.): Synopsis and Classification of Living Organisms I, p. 101–115. McGraw-Hill Book Co., New York 1982.

Loeblich, A. R., Klotz, C. C., Roberts, T. M., Tutle, R. C., Allen, J. R.: The dinoflagellate nucleus. J. Phycol. **10:** 14–15 (1974).

Malone, T. C.: The 1976 *Ceratium tripos* bloom in the New York Bight: causes and consequences. NOAA Techn. Report NMFS Circ. 410, U.S. Dept. Commerce: 1–4 (1978).

Mapletoft, H., Montgomery, M., Waters, J., Wells, P. A.: A new freshwater dinoflagellate. New Phytol. **65**: 54–58 (1966).

Marchesoni, V.: Sulla posizione sistematica del *Glenodinium* determinante l'arrossamento del Lago di Tovel. Studi Trentini Sci. Natur. A **22**: 11–18 (1941).

Maskell, W. M.: On the fresh-water Infusoria of the Wellington District. Trans. New-Zealand Inst. **20**: 3–19 (1887).

Massart, J.: Sur la motilité des Flagellates. Bull. Acad. Royal Belgique, Cl. Sci. **6**: 116 (1920).

Matvijenko, O. M.: Materialy do vyvčennja vodorostej URSR 1. Vodorosti Kljukvenogo bolota. Učen. Zap. Charkivs'k. Univ. **14** (3): 29–70 (1938).

– Litvinenko, R. M.: Pirofitovi vodorosti – Pyrrophyta. Vyznačnyk prisnovodnych vodorostej Ukrajins'koji RSR **3** (2), p. 386. Naukova Dumka, Kijiv 1977.

Meunier, A.: Microplankton des Mers de Barents et de Kara. Duc d'Orleans Campagne Arctique 1907, p. 355. Bulens, Bruxelles 1910.

– Microplankton de la Mer Flamande III. Les Péridiniens. Mém. Mus. Roy. Hist. Nat. Belgique **8**: 1–116 (1919).

Michaelis, G. A.: Über das Leuchten der Ostsee nach eigenen Beobachtungen p. 52. Perthes & Besser, Hamburg 1830.

Möbius, M.: Über einige brasilianische Algen. Ber. Deutsch. Bot. Ges. **10**: 17–26 (1892).

– Über die Tiere der schleswig-holsteinischen Austernbänke, ihre physikalischen und biologischen Lebensverhältnisse. Sitzungsber. Königl. Preuss. Akad. Wiss. Berlin 1893: 67–92 (1893).

Morey-Gaines, G., Ruse, R. H.: Encystment and reproduction of the predatory dinoflagellate, *Polykrikos kofoidii* Chatton (Gymnodiniales). Phycologia **19**: 230–236 (1980).

Morling, G.: *Ceratium carolinianum* and the related *Ceratium carolinianum* var. *elongatum*, and *Ceratium cornutum*. A study of distribution and morphology. Nova Hedwigia **31**: 937–968 (1979).

Müller, O. F.: Vermium terrestrium et fluviatilium, seu Animalium Infusorium, Helmithorium et Testaceorum, non marinorum, succincta historia. Vol. **1.** Pars prima. p. 34, 135. Faber, Havniae et Lipsiae 1773.

– Animalcula Infusoria fluviatilia et marina. Opus posthumum cura O. Fabricii, p. 8, 56, 367. Nicol. Möller, Havniae 1786.

Nichols, H. W.: Growth media – freshwater. 8–24. In: Stein, J. R. (ed.): Handbook of Phycological Methods, p. 448. Cambridge Univ. Press, London 1973.

Nygaard, G.: Plankton from lakes of the Malayan region. Vidensk. Meddel. Dansk Naturhist. Foren. København **82**: 197–240 (1926).

– Dansk plante plankton, p. 52. Gyldendal, København 1945.

– Hydrobiological studies on some danish ponds and lakes. Kongel. Danske Vidensk. Selsk. Biol. Skr. **7** (1): p. 263 (1949).

Ohno, N.: Beobachtungen an einer Süsswasser-Peridinee. J. Coll. Sci. Imp. Univ. Tokyo **32**: 77–92 (1911).

Osorio-Tafall, B. F.: Estudios sombre el plancton de Mexico I. El genero *Lophodinium* Lemm. (Dinophyceae, Peridiniales). Ciencia, Rev. Hispanoamer. Cienc. P. Aplicadas **3**: 114–119 (1942).

Ostenfeld, C. H.: Phytoplankton fra det Caspiske Hav. Vidensk. Meddel. Dansk Naturhist. Foren. København 1901: 129–215 (1901).

– Beiträge zur Kenntnis der Algenflora des Kossogol-Beckens in der nordwestlichen Mongolei, mit spezieller Berücksichtigung des Phytoplanktons. Hedwigia **46:** 365–420 (1907).

– The phytoplankton of the Aral Sea and iths effluents. Wissenschaftliche Ergebnisse der Aralsee-Expedition. St. Petersburg **8:** 123–225 (1908).

– Notes on the phytoplankton of Victoria Nyanza, East Africa. Bull. Mus. Comp. Zool. Harvard Coll. **52** (10): 171–182 (1909).

Owen, K. C., Norris, D. R.: Cysts and life cycle considerations of the thecate dinoflagellate *Fragilidium*. J. Coastal Res. **1:** 263–266 (1985).

Pascher, A.: Flagellaten und Algen. Ber. Deutsch. Bot. Ges. **32:** 136–160 (1914).

– Studien über die rhizopodiale Entwicklung der Falgellaten. Arch Protistenk. **36:** 81–118 (1916).

– Über eine neue Amöbe (*Dinamoeba varians)* mit dinoflagellatenartigen Schwärmern. Arch. Protistenk. **36:** 118–136 (1916).

– Flagellaten und Rhizopoden in ihren gegenseitigen Beziehungen. Versuch einer Ableitung der Rhizopoden. Arch. Protistenk. **38:** 1–87 (1917).

– Die braune Algenreihe aus der Verwandtschaft der Dinoflagellaten (Dinophyceen). Arch. Protistenk. **58:** 1–54 (1927).

– Von einer neuen Dinococcale (*Cystodinium phaseolus*) mit zwei verschiedenen Schwärmertypen. Arch. Protistenk. **63:** 241–254 (1928).

– Systematische Übersicht über die mit Flagellaten in Zusammenhang stehenden Algenreihen und Versuch einer Einreihung dieser Algenstämme in die Stämme des Pflanzenreiches. Beih. Bot. Zbl. (II, 2) **48:** 317–332 (1931).

– Über drei auffallend konvergente, zu verschiedenen Algenreihen gehörende epiphytische Gattungen. Beih. Bot. Zbl. (I) **49:** 549–564 (1932).

– Über neue, protococcoide, festsitzende Algengattungen aus der Verwandtschaft der Dinoflagellaten. Über zwei farblose, protococcoide Algen aus der Reihe der Dinophyceen und über Schädigungen durch Epiphytismus. Beih. Bot. Zbl. **57** A (3): 376–404 (1944).

Paulsen, O.: Peridiniales. In: Brandt, K., Apstein, C. (eds.): Nordisches Plankton **18:** 1–124. Lipsius & Tischer, Kiel, Leipzig 1908.

Pavillard, J.: Le genre *Diplopsalis* Bergh et les genres voisins. Montpellier 1913.

Pearlmutter, N. L., Timpano, P.: The biology of *Chlamydomyxa montana*. Ultrastructure of the cyst. Protoplasma **122:** 68–74 (1984).

Pechlaner, R.: Finstertaler Seen (Kühtai, Österreich) II. Das Phytoplankton. Arch. Hydrobiol. **63:** 145–193 (1967).

– Factors that control the production rate and biomass of phytoplankton in high-mountain lakes. Mitt. Internat. Verein. Limnol. **19:** 125–145 (1971).

Penard, E.: Les Péridiniacées du Lac Léman. Bull. Trav. Soc. Bot. Genève **6:** 1–63 (1891).

Perty, M.: Über verticale Verbreitung mikroskopischer Lebensformen. Mitt. Naturf. Ges. Bern 1849: 17–45 (1849).

– Zur Kenntniß kleinster Lebensformen, p. 228. Jent & Reinert, Bern 1852.

Peters, H.: Die Bevölkerung des südatlantischen Ozeans mit Ceratien. Wissenschaftliche Ergebnisse der Deutschen Atlantik-Expedition auf dem Forschungs- und Vermessungsschiff „Meteor" 1925–27, **12** (1): 1–68 (1934).

Pfiester, L. A.: Sexual reproduction of *Peridinium cinctum* f. *ovoplanum* (Dinophyceae). J. Phycol. **11:** 259–265 (1975).

– Sexual reproduction of *Peridinium willei* (Dinophyceae). J. Phycol. **12**: 234–238 (1976).

– Sexual reproduction of *Perdinium gatunense* (Dinophyceae). J. Phycol. **13**: 92–95 (1977).

– Sexual Reproduction. In: Spector, D. (ed.): Biology of Dinoflagellates, 181–199. Academic Press, London 1984.

– Lynch, R. A.: Amoeboid stages and sexual reproduction of *Cystodinium bataviense* and its similarity to *Dinococcus* (Dinophyceae). Phycologia **19**: 178–183 (1980).

– – Skvarla, J. J.: Occurrence, growth, and SEM portrait of *Wolozynskia reticulata* Thompson (Dinophyceae). Trans. Amer. Microsc. Soc. **99**: 213–217 (1980).

– Popovský, J.: Parasitic, amoeboid Dinoflagellates. Nature **379**: 421–424 (1979).

– Skvarla, J. J.: Heterothallism and thecal development in the sexual life history of *Peridinium volzii* (Dinophyceae). Phycologia **18**: 13–18 (1979).

– – Comparative ultrastructure of vegetative and sexual thecae of *Peridinium limbatum* and *Peridinium cinctum* (Dinophyceae). Amer. J. Botany **67**: 955–958 (1980).

– Timpano, P., Skvarla, J. J., Holt, J. R.: Sexual reproduction and meiosis in *Peridinium inconspicuum* Lemmermann (Dinophyceae). Amer. J. Botany **71**: 1121–1127 (1984).

Playfair, G. I.: Plankton of the Sydney water-supply. Proc. Linn. Soc. New South Wales **37**: 512–552 (1913).

– Peridineae of New South Wales. Proc. Linn. Soc. New South Wales **44**: 793–818 (1919/20).

Popovský, J.: *Cystodinedria maxima* sp. nov. Preslia **33**: 291–296 (1961).

– Einige Dinoflagellaten des Tatra-Nationalparks. TANAP, Sborn. prác o tatranskom národnom parku **10**: 269–275 (1967).

– A contribution to the knowledge of Dinoflagellates from Bohemia. Preslia **40**: 251–263 (1968).

– Some thecate dinoflagellates from Cuba. Arch. Protistenk. **112**: 252–258 (1970).

– Some interesting freshwater dinoflagellates from Central Europe. Arch. Protistenk. **113**: 277–284 (1971).

– Some remarks to the life cycle of *Gloeodinium montanum* Klebs and *Hemidinium nasutum* Stein (Dinophyceae). Arch. Protistenk. **113**: 131–136 (1971).

– Remarks to the biology of *Chlamydomyxa labyrinthuloides* Archer. Algol. Studies **23** (Arch. Hydrobiol. Suppl. **56**): 147–153 (1979).

– Another case of phagotrophy by *Gymnodinium helveticum* Penard f. *achroum* Skuja. Arch. Protistenk. **125**: 73–78 (1982).

– Problems in the determination of freshwater dinoflagellates (Dinophyceae). Schweiz. Z. Hydrobiol. **45**: 365–372 (1983).

– *Gymnodinium austriacum* Schiller (Dinophyceae), a case of phagotrophy. Algol. Studies **40** (Arch. Hydrobiol. Suppl. **71**): 425–429 (1985).

– Pfiester, L. A.: The life histories of *Stylodinium sphaera* Pascher and *Cystodinedria inermis* (Geitler) Pascher (Dinophyceae), two freshwater facultative predators-autotrophs. Arch. Protistenk. **125**: 115–127 (1982).

Pouchet, G.: Contribution à l'histoire des cilio-flagellés. J. Anat. Physiol. Paris **19**: 399–455 (1883).

– Nouvelle contribution à l'histoire des Péridiniens marins. J. Anat. Physiol. Paris **21**: 28–88 (1883).

– Troisième contribution à l'histoire des Péridiniens. J. Anat. Physiol. Paris **21:** 525–534 (1885).
– Quatrième contribution à l'histoire des Péridiniens. J. Anat. Physiol. Paris **23:** 87–112 (1887).
Prager, J. C.: Fusion of the family Glenodiniaceae into the Peridiniaceae, with notes on *Glenodinium foliaceum* Stein. J. Protozool. **10:** 195–204 (1963).
Prescott, G. W.: Algae of the Western Great Lakes Area, p. 946. Cranbrook Institute of Science **31,** Bloomfield, Michigan 1951.
Pringsheim, E. G.: Pure Cultures of Algae, p. 119. Cambridge Univ. Press, London 1946.
– Die Grundlagen eines taxonomischen Systems der Algen. Z. Pflanzenphysiol. **54:** 99–105 (1966).
– Phycology in the field and in the laboratory. J. Phycol. **3:** 93–95 (1967).
Proškina-Lavrenko, A. I.: Novye rody i vidy vodoroslej iz solenych vodoemov SSSR I (Algae nonnullae novae I). Bot. Mater. Otd. Sporovych Rast. **5:** 142–154 (1945).
Raikov, I. B.: Jadro prosteišich. Morfologija i evolucija, p. 327. Leningrad 1978.
Richards, B. C.: The Morphology, Cytology and Life History of *Urococcus insignis* (Hass.) Kütz. Unpubl. M. S. Thesis Univ. Kansas, 56 pp. 1962.
Roberts, T. M., Tuttle, R. C., Allen, J. R., Loeblich, A. R. III, Klotz, L. C.: New genetic and physicochemical data on structure of dinoflagellate chromosomes. Nature **248:** 446–447 (1974).
Rosowski, J., Parker, B. C. (eds.): Selected Papers in Phycology, p. 876. Univ. Nebraska Press, Lincoln 1971.
Sako, Y., Ishida, Y., Kadota, H., Hata, Y.: Sexual reproduction and cyst formation in the freshwater dinoflagellate *Peridinium cunningtonii*. Bull. Jap. Soc. Sci. Fish. **50:** 743–750 (1984).
Saville-Kent, W.: A manual of the Infusoria **1, 2,** p. 472, p. 473. Bogue, London 1880/82.
Schiller, J.: Der thermische Einfluß und die Wirkung des Eises auf die planktischen Herbstvegetationen in den Altwässern der Donau bei Wien. Arch. Protistenk. **56:** 1–62 (1926).
– Die planktischen Vegetationen des adriatischen Meeres. C. Dinoflagellata. I. II. Adiniferidea, Dinophysidaceae. Systematischer Teil. Arch. Protistenk. **61:** 45–91; **62:** 119–166 (1928).
– Dinoflagellatae (Peridineae). In: Kolkwitz, R. (ed.): Rabenhorst's Kryptogamen-Flora von Deutschland, Österreich und der Schweiz, 2. Aufl., **10** (3) 1–2, p. 617, p. 390. Akadem. Verlagsges., Leipzig 1933/37.
– Über winterliche pflanzliche Bewohner des Wassers, Eises und des darauf liegenden Schneebreies. I. Österr. Bot. Z. **101:** 236–284 (1954).
– Untersuchungen an den planktischen Protophyten des Neusiedlersees 1950–1954. Wissenschaftliche Arbeiten aus dem Burgenland **9:** 1–66 (1955).
Schilling, A. J.: Kleiner Beitrag zur Technik der Flagellatenforschung. Z. Wiss. Mikroskop. **8:** 314 (1891).
– Dinoflagellatae (Peridineae). In: Pascher, A. (ed.): Die Süßwasserflora Deutschlands, Österreichs und der Schweiz, Bd. 3, p. 66. Gustav Fischer Verlag, Jena 1913.
Schlösser, U. G.: Sammlung von Algenkulturen. Pflanzenphysiologisches Institut der Universität Göttingen. Ber. Deutsch. Bot. Ges. **95:** 181–276 (1982).

Schrank, F. von Paula: Mikroskopische Wahrnehmungen. Naturforscher **27**: 26–37 (1793).

– Briefe naturhistorischen, physikalischen und ökonomischen Inhaltes an Herrn B. S. Nau. Erlangen 1802.

Schröder, B.: Die neun wesentlichen Formtypen von *Ceratium hirundinella* O. F. M. Arch. Naturgesch. **84** A (8): 222–230 (1920).

Schussnig, B.: Grundriss der Protophytologie, p. 310. VEB G. Fischer, Jena 1954.

– Handbuch der Protophytenkunde, II, p. 1144. VEB G. Fischer, Jena 1960.

Schütt, F.: Die Peridineen der Plankton-Expedition I. Ergebnisse der Plankton-Expedition der Humboldt-Stiftung, Kiel und Leipzig, **4**: 1–170 (1895).

– Peridiniales. In: Engler, A., Prantl, K. (eds.): Die natürlichen Pflanzenfamilien, **1** (1B), p. 153. Engelmann, Leipzig 1896.

Seligo, A.: Untersuchungen über Flagellaten. Beitr. Biol. Pflanzen **4**: 145–180 (1885).

Shi, Z., Wei, Y., Hu, H.: An observation on the nucleus and nuclear substance of *Gymnodinium eucyaneum* Hu. Oceanologia et Limnologia Sinica **14** (2): 160–167 (1983).

Shyam, R., Sarma, Y. S. R. K.: *Woloszynskia stoschii* and *Gymnodinium indicum*, two new freshwater dinoflagellates from India: morphology, reproduction and cytology. Plant Syst. Evol. **124**: 205–212 (1975).

Skuja, H.: Süsswasseralgen von den westestnischen Inseln Saaremna und Hiiumaa. Acta Horti Bot. Univ. Latv. **4**: 1–76 (1930).

– Algae. In: Handel-Mazetti, H. v. (ed.): Symbolae Sinicae I. Ergebnisse der Expedition nach Südwest-China, 1–106. Wien 1937.

– Beitrag zur Algenflora von Lettland II. Acta Horti Bot. Univ. Latv. **11/12**: 41–169 (1939).

– Taxonomie des Phytoplanktons einiger Seen in Uppland, Schweden. Symb. Bot. Upsal. **9** (3): 1–399 (1948).

– Taxonomische und biologische Studien über das Phytoplankton schwedischer Binnengewässer. Nova Acta Reg. Soc. Sci. Upsal. **16** (3): 1–404 (1956).

– Grundzüge der Algenflora und Algenvegetation der Fjeldgegenden um Abisko in Schwedisch-Lappland. Nova Acta Reg. Soc. Sci. Upsal. **18** (3): 1–645 (1964).

Skvorcov, B. V.: Eine neue Süsswasserart der Gattung *Amphidinium* Claparède et Lachmann aus der Nordmandschurei. Russ. Hydrobiol. Z. Wolgastation Saratow **4**: 146–148 (1925).

– Über einige Peridinaceen aus der Nordmandschurei. Hedwigia **67**: 122–124 (1927).

Spector, D. L., Pfiester, L. A., Triemer, R. E.: Ultrastructure of the dinoflagellate *Peridinium cinctum* f. *ovoplanum* II. Light and electron microscopic observation on fertilization. Amer. J. Bot. **68**: 34–43 (1981).

Spero, H. J., Morée, M. D.: Phagotrophic feeding and its importance to the life cycle of the holozoic dinoflagellate, *Gymnodinium fungiforme*. J. Phycol. **17**: 43–51 (1981).

Starmach, K.: New species of the Order Dinococcales (Pyrrhophyta) from the Twardowski Cave in Kraków. Acta Hydrobiol. Kraków **5**: 337–342 (1963).

– Cryptophyceae, Dinophyceae, Raphidophyceae. Flora Słodkowodna Polski **4**: 1–520, Państwowe Wydawnictwo Naukowe, Warszawa, Kraków 1974.

Starr, R. C.: The culture collection of algae at Indiana University. Amer. J. Bot. **51**: 1013–1044 (1964).

– The culture collection of algae at the University of Texas at Austin. J. Phycol. **14,** suppl.: 47–100 (1978).

– James, D. E.: Sources of algal cultures, 316–351. In: Rosowski, J. R., Parker, B. C.: Selected Papers in Phycology, p. 817. University of Nebraska Press, Lincoln 1971.

Stein, F. v.: Der Organismus der Infusionsthiere. III. Flagellaten I, p. 154. Engelmann, Leipzig 1878.

– Der Organismus der Infusionsthiere. III. Flagellaten II, p. 30. Engelmann, Leipzig 1883.

Stein, J. R. (ed.): Handbook of Phycological Methods. Culture Methods and Growth Measurements, p. 429. Cambridge Univ. Press, Cambridge, London, Melbourne 1973.

Stokes, A. C.: A preliminary contribution toward a history of the freshwater Infusoria of the United States. J. Trenton Nat. Hist. Soc. **1** (3): 71–319 (1888).

Stosch, H. A. v.: Zum Problem der sexuellen Fortpflanzung in der Peridineengattung *Ceratium.* Helgol. Wiss. Meeresunters. **10:** 140–152 (1964).

– Sexualität bei *Ceratium cornutum* (Dinophyta). Naturwissenschaften **52:** 112–113 (1965).

– Dinoflagellaten aus der Nordsee II. *Helgolandinium subglobosum* gen. et spec. nov. Helgol. Wiss. Meeresunters. **19:** 569–577 (1969).

– La signification cytologique de la «cyclose nucléaire» dans le cycle de vie des Dinoflagellés. Bull. Soc. Bot. France Mém. 1972: 201–212 (1972).

– Observations on vegetative reproduction and sexual life cycle of two freshwater dinoflagellates, *Gymnodinium pseudopalustre* Schiller and *Woloszynskia apiculata* sp. nov. Brit. Phycol. J. **8:** 105–134 (1973).

Suchlandt, O.: Dinoflagellaten als Erreger von rotem Schnee. Ber. Deutsch. Bot. Ges. **34:** 242–246 (1916).

Taylor, F. J. R.: On dinoflagellate evolution. BioSystem **13:** 65–108 (1980).

Thomasson, K.: Rotorua phytoplankton reconsidered (North Island of New Zealand). Int. Rev. ges. Hydrobiol. **59** (5): 703–727 (1974).

– Algal vegetation in North Australian Billabongs. Nowa Hedwigia **42** (2–4): 301–378 (1986).

Thompson, R. H.: Freshwater dinoflagellates of Maryland. Chesapeake Biol. Lab. Maryland **67:** 1–28 (1947).

– Immobile Dinophyceae I. New records and a new species. Amer. J. Bot. **36:** 301–308 (1949).

– A new genus and new species of fresh-water Pyrrhophyta in the Desmokontae and Dinophyceae. Lloydia **13:** 277–299 (1950).

Timpano, P., Pfiester, L. A.: Colonization of the epineuston by *Cystodinium bataviense* (Dinophyceae): behavior of the zoospore. J. Phycol. **21:** 56–62 (1985a).

– – Fine structure of the immobile dinococcalean *Cystodinium bataviense* (Dinophyceae). J. Phycol. **21:** 458–466 (1985b).

– – Observation on "*Vampyrella penula – Stylodinium sphaera*" and the ultrastructure of the reproductive cyst. Amer. J. Bot. **73:** 1341–1350 (1986).

Tomas, R. N., Cox, E. R.: Observations on the symbiosis of *Peridinium balticum* and its intracellular alga. I. Ultrastructure. J. Phycol. **9:** 304–323 (1973).

– – Steidinger, K. A.: *Peridinium balticum* (Levander) Lemmermann, an unusual dinoflagellate with mesocaryotic and eucaryotic nucleus. J. Phycol. **9:** 91–98 (1973).

Turpin, D. H., Dobell, P. E. R., Taylor, F. J. R.: Sexuality and cyst formation in pacific strains of the toxic dinoflagellate *Gonyaulax tamarensis*. J. Phycol. **14:** 235–238 (1978).

Tuttle, R. V., Loeblich, A. R. III: Genetic recombination in the dinoflagellate *Crypthecodinium cohnii*. Science **185:** 1061–1062 (1974).

– – Sexual reproduction and segregation analysis in the dinoflagellate *Crypthecodinium cohnii*. J. Phycol. **11:** 1–15 (1975).

Utermöhl, H.: Das Nannoplankton ostholsteinischer Seen. Verh. Internat. Verein. Limnol. **1:** 86–93 (1923).

– Einige Bemerkungen über den Formenkreis von *Gymnodinium mirabile* Penard. Schrift. Süßw. Meeresk. Büsum **1** (1923).

– Limnologische Phytoplanktonstudien. Arch. Hydrobiol. Suppl. **5**, p. 527. Stuttgart 1925.

Vanhöffen, E.: Das Genus *Ceratium*. Zool. Anz. **19** (499): 133–134 (1896).

Wailes, G. H.: Dinoflagellates from British Columbia with descriptions of new species. Study from stations of the Biological Board of Canada. Art, Hist. & Sci. Assoc. Vancouver Museum Notes **3** (1): 20–31 (1928).

– *Peridinium vancouverense* nom. nov. Ann. Protistol. **3:** 110 (1931).

– Protozoa and algae from Mt. Hopeless, B. C., with description of new species of *Peridinium*. Art, Hist. & Sci. Assoc. Vancouver Museum Notes Suppl. 7: 1–4 (1933).

Walker, L. M.: Evidence for a sexual cycle in the Florida red tide dinoflagellate, *Ptychodiscus brevis (=Gymnodinium breve)*. Trans. Amer. Microsc. Soc. **101:** 287–293 (1982).

– Steidinger, K. A.: Sexual reproduction in the toxic dinoflagellate *Gonyaulax monilata*. J. Phycol. **15:** 312–315 (1979).

Wall, D., Dale, B.: Modern dinoflagellate cysts and evolution of the Peridiniales. Micropaleontology **14:** 265–304 (1968).

Watanabe, A., Hattori, A.: Cultures and Collections of Algae. Proc. U. S. Japan. Conf., Hakone, Japan. Japan. Soc. Plant Physiol. 100 pp. (1966).

Wawrik, F.: Sommerliche Planktonaspekte: Seltene und neue Algen aus Teichen des Waldviertels in Niederösterreich. Nova Hedwigia **36:** 775–794 (1982).

Wedemayer, G. J., Wilcox, L. L., Graham, L. E.: *Amphidinium cryophilum* sp. nov. (Dinophyceae) a new frehswater dinoflagellate I. Species description using light and scanning electron microscopy. J. Phycol. **18:** 13–17 (1982).

West, G. S.: A biological investigation of the Peridineae of Sutton Park, Warwickshire. New Phytol. **8:** 181–196 (1909).

– Algae. Myxophyceae, Peridineae, Bacillarieae, Chlorophyceae, together with a brief summary of the occurrence and distribution of freshwater algae. Cambridge Botanical Handbooks, p. 475. Cambridge Univ. Press, London 1916.

Wislouch, S. M.: Przyczynek do biologii solnisk i genezy szłamóv łaczniczych na Krymie. Acta Soc. Bot. Poloniae **2:** 99–129 (1924).

Wołoszyńska, J.: Das Phytoplankton einiger javanischer Seen, mit Berücksichtigung des Sawa-Planktons. Bull. Acad. Sci. Cracovie (B) 1912: 649–709 (1912).

– Polnische Süsswasser-Peridineen. Bull. Acad. Sci. Cracovie (B) 1915: 260–285 (1916).

– Neue Peridineen-Arten, nebst Bemerkungen über den Bau der Hülle bei *Gymno-* und *Glenodinium*. Bull. Acad. Sci. Cracovie (B) 1917: 114–122 (1917).

– Glony stawów i młak tatrańskich I. Bull. Acad. Sci. Cracovie **58** (B): 196–200 (1919).

– Javanische Süsswasserperidineen. Acta Soc. Bot. Poloniae **1:** 263–266 (1923).

– Notatki algologiczne (Algologische Notizen). 1. *Amphidinium wigrense* n. sp. 2. *Peridinium* sp. Spraw. Stac. Hydrobiol. Wigrach **1** (4): 1–9 (1925).

– Przyczynki do znajomości polskich brózdnic słodkowodnych. Acta Soc. Bot. Poloniae **3** (1): 1–16 (1925).

– Dinoflagellatae polskiego Bałtyku i błot nad Piaśnicą. Arch. Hydrobiol. Ichthyol. **3** (1928): 153–278 (1929).

– Beitrag zur Kenntnis des Phytoplanktons tropischer Seen. Arch. Hydrobiol. Ichthyol. **5:** 159–169 (1930).

– Brózdnice torfowiska "Kopytowiec" pod Poturzyca koło Sokala. Acta Soc. Bot. Poloniae **7:** 499–505 (1930).

– Die Algen der Tatraseen und Tümpel II. Über zwei Gymnodinien aus den Seen "Morskie Oko" und "Czarny Staw pod Rysami". Bull. Acad. Pol. Sci. Lett. B 1935: 1–9 (1935).

– Die Algen der Tatraseen und Tümpel III. Peridineen im Winterplankton einiger Tatraseen. Arch. Hydrobiol. Ichthyol. **10:** 188–196 (1936).

– Bruzdnice Tatr i Karpat Wschodnich. Acta Soc. Poloniae **21:** 311–316 (1952).

Yoshimatsu, S.: Sexual reproduction of *Protogonyaulax catenella* in culture. Bull. Plankton Soc. Jap. **28:** 131–139 (1981).

Zacharias, O.: Zur Kenntnis der niederen Flora und Fauna holsteinischer Moorsümpfe. Forschungsber. Biol. Stat. Plön **10:** 223–289 (1903).

Zederbauer, E.: *Ceratium hirundinella* in den österreichischen Alpenseen. Österr. Bot. Z. **4** (5): 1–10 (1904).

Zimmermann, W.: Neue und wenig bekannte Kleinalgen von Neapel I–IV. Z. Bot. **23:** 419–442 (1930).

Zingmark, R. G.: Sexual reproduction in the dinoflagellate *Noctiluca miliaris* Suriray. J. Phycol. **6:** 122–126 (1970).

– Ultrastructural studies on two kinds of mesocaryotic dinoflagellate nuclei. Amer. J. Bot. **57:** 586–592 (1970).

Taxonomical Index

Page numbers in boldface types refer to the main discussion of the taxa; names in italics are synonymes.

MIX
Papier aus verantwortungsvollen Quellen
Paper from responsible sources
FSC® C105338

If you have any concerns about our products,
you can contact us on
ProductSafety@springernature.com

In case Publisher is established outside the EU,
the EU authorized representative is:
Springer Nature Customer Service Center GmbH
Europaplatz 3, 69115 Heidelberg, Germany

Printed by Libri Plureos GmbH
in Hamburg, Germany